Springer Texts in Statistics

Advisors:
Stephen Fienberg   Ingram Olkin

Springer Texts in Statistics

Alfred	Elements of Statistics for the Life and Social Sciences
Chow and Teicher	Probability Theory: Independence, Interchangeability, Martingales Second Edition
Christensen	Plane Answers to Complex Questions: The Theory of Linear Models
du Toit, Steyn, and Strumpf	Graphical Exploratory Data Analysis
Kalbfleisch	Probability and Statistical Inference: Volume I: Probability. Second Edition
Kalbfleisch	Probability and Statistical Inference: Volume 2: Statistical Inference. Second Edition
Keyfitz	Applied Mathematical Demography. Second Edition
Kiefer	Introduction to Statistical Inference
Madansky	Prescriptions for Working Statisticians
Peters	Counting for Something: Statistical Principles and Personalities

Albert Madansky

Prescriptions for Working Statisticians

With 25 Illustrations

Springer-Verlag
New York Berlin Heidelberg
London Paris Tokyo

Albert Madansky
Graduate School of Business
University of Chicago
Chicago, IL 60637
U.S.A.

Mathematics Subject Classification (1980): 62-01

Library of Congress Cataloging-in-Publication Data
Madansky, Albert
Prescriptions for working statisticians.
(Springer texts in statistics)
Includes index.
1. Statistics. I. Title. II. Series.
QA276.M27 1988 519.5 87-28346

Typeset by Asco Trade Typesetting Ltd., Hong Kong.
Printed and bound by R. R. Donnelley & Sons, Harrisonburg, Virginia.
Printed in the United States of America.

9 8 7 6 5 4 3 2 1

ISBN 0-387-96627-7 Springer-Verlag New York Berlin Heidelberg
ISBN 3-540-96627-7 Springer-Verlag Berlin Heidelberg New York

To my daughters—Susan, Cynthia, Noreen, and Michele

רבות בנותי עשו חיל

adapted from Proverbs 31:29

Contents

Preface

The first course in statistics, no matter how "good" or "long" it is, typically covers inferential procedures which are valid only if a number of preconditions are satisfied by the data. For example, students are taught about regression procedures valid only if the true residuals are independent, homoscedastic, and normally distributed. But they do not learn how to check for independence, homoscedasticity, or normality, and certainly do not learn how to adjust their data and/or model so that these assumptions are met.

To help this student out[1] I designed a second course, containing a collection of statistical diagnostics and prescriptions necessary for the applied statistician so that he can deal with the realities of inference from data, and not merely with the kind of classroom problems where all the data satisfy the assumptions associated with the technique to be taught. At the same time I realized that I was writing a book for a wider audience, namely all those away from the classroom whose formal statistics education ended with such a course and who apply statistical techniques to data. Thus the book begins with four chapters of *diagnostics*, answering in turn the following questions:

- How do I find out whether my data came from normal distributions?
- How do I find out whether my data are homoscedastic?
- How do I find out whether my data are independent?
- How do I find out whether there are outliers in my data?

I know of no book that treats the material in Chapters 1, 2, and 3, let alone combines those topics under one cover. Two books are devoted to the topic covered in Chapter 4; we include the most relevant material between these covers as well.

[1] This student's thoughts as he takes the first course are recorded in Chapter 0, entitled "A Thoughtful Student's Retrospective on Statistics 101".

Next the book presents two chapters of *prescriptions*. Chapter 5 answers the questions:

- How do I fix up my data so that we can treat them as if they were normally distributed?
- How do I fix up my regression variables so that the residuals are homoscedastic?
- How do I concoct an appropriate set of independent variables for a regression?

Chapter 6 answers the most frequently asked questions by the student in that "first course":

- How do I know how many independent variables to put into a regression model?
- What do I do about multicollinearity?

The material in Chapter 5 is, to my knowledge, not to be found in any other book. Answers to the questions covered in my Chapter 6 are well presented in Chapter 6 of Draper and Smith [1966] or Chatterjee and Price [1977]; again my justification for writing this chapter is to include it between the covers of a single book of the timbre of this one.

Finally, the book contains two chapters on *techniques* for making inferences from specialized data, mixed categorical and measured data (Chapter 7), and cross-classified data (Chapter 8). Though the material in these chapters can be found in a number of other books, our approach to this material may be a little different from the usual fare, and this slant may give the reader added insight into the techniques covered in these chapters—in the main, into analysis of variance and into log r linear modeling of contingency tables. Chapter 7 was motivated by the observation that a first course teaches the student both the two-sample t test and multiple regression. What would then be more natural in a second course than to use the multiple regression model with dummy independent variables as the vehicle for teaching students the generalization of the t test, the analysis of variance? Once dummy variables are introduced in regression, a natural further topic is the use of a "dummy" variable as a dependent variable. This, too, is included in Chapter 7.

The last topic, along with the technology of the analysis of variance, leads naturally to a topic not treated in a first course but which is probably as important as t tests and regression, namely the analysis of counted data. Chapter 8 gives a brief summary of log r linear model analysis for counted data, more to give the reader an acquaintance with these methods, and whet his appetite for more, than replace the many excellent books devoted to this subject.

In summary, I have written a book that is ostensibly for a "second course" in applied statistics but actually may have merit as a ready handbook for the practitioner. To this end, it is self-contained in that all necessary tables are included in this book. When there was a number of competitive procedures for the same problem, I was faced with an editorial choice: Should I only include the best procedure and disregard the rest, or include them all? My answer to that question was the latter, for a few reasons. First, in some cases we don't yet know which procedure is best. (Of course, for situations in which something is known about the properties of the procedure, I include that information.) Second, sometimes a "better" procedure may be more complex to implement than a "worse" procedure; by describing both the reader can select the one he chooses to use based not merely on "goodness" but on practicality. Finally, some of the "not-so-good" procedures are so well known and well ingrained in statistical practice, or are intellectual forerunners of the "better" procedures, that their absence would be noticed (and have been, in earlier drafts).

One institutional idiosyncrasy which creeps into this book, but in a minor way and can be brushed aside by the disinterested reader, is my occasional use of IDA, an interactive data analysis system available to students at the University of Chicago Graduate School of Business (see Roberts and Ling [1982]). These uses have purposely not been edited out, partly to enhance this book's direct value to my students, but also partly to stimulate the reader's interest in interactive data analysis systems as the "way of life" in current statistical practice. Moreover, as MINITAB (see Ryan, Joiner, and Ryan [1985]) embodies commands which implement some of the procedures described herein, I include some examples on the use of that data analysis system as well.

As I was writing these chapters, I had to grapple with the problem as to what level of mathematical and statistical background to presuppose of the reader. Some of the material requires no more mathematics than high school algebra; others require a good knowledge of matrix algebra. To cope with this range of prerequisites, I have adopted a number of expository ground rules:

(1) describe the procedures in as clear a manner as I can consonant with the maturity level necessary for the user of the procedure;
(2) provide motivation for the procedure, but stop short of a mathematical derivation;
(3) provide an example of the use and computation of each procedure;
(4) provide a FORTRAN computer program for implementing the more mathematically complex procedures.

We require three items of statistics background:

(1) familiarity with multiple regression modeling and at least a nodding acquaintance with the matricial representation of the regression parameter estimates;
(2) exposure to the t, F, chi-square, beta, gamma, binomial, Poisson, and uniform densities, at least to the point of knowing their mean and variance and how to read tables of their distributions; and
(3) understanding of the workings of the expectation ($\mathscr{E}$), variance ($\mathscr{V}$), and covariance ($\mathscr{C}$) operators.

These are for the most part covered in a good first course in statistics. Any gap between what the reader has learned in such a course and a particular fact needed to read this book can be bridged by the reader with a minimum amount of individual effort.

Thus, the book is not written toward a single prototypical reader with a specific background. The applied statistician with a background in matrix algebra and a prior background in regression will find this book, at even its most difficult points, right at his level and in its easiest parts quite tedious. The reader whose background is merely high school algebra, and a first course in statistics which treated only the normal distribution, will find some of the material opaque. To him, I can only recommend that he read the opaque material, at least to ascertain the general drift of the described material, in order that, knowing of its existence, he can invoke it at appropriate occasions with the help of a more mathematically trained statistician. I have helped the reader along by segregating the material in each section of the book into two categories, labeled "Prescription" and "Theoretical Background", respectively. I have tried to make the "Prescription" material more intuitive, readable to the practitioner, and "cook-bookish"; and even in the "Theoretical Background" most theoretical results are stated without proof. Of course, one can pursue any topic in the book in greater depth by going back to the cited references.

The reader should be forewarned that I am a bit idiosyncratic in my notation. I use lowercase letters to denote scalar variables and uppercase letters to denote matrix (and vector) variables. I use regular type to denote nonrandom variables and boldface type to denote random variables. Thus we can look at a symbol and know whether or not it is a matrix and whether or not it is random. (No longer need we write X for a random variable and x for its value, or equations like $X = x$, and no longer will we be confused about what symbols to use for a random matrix and its value.) Schematically, the notation is the following:

	Nonrandom	Random
Scalar	x	$\mathbf{x}$
Matrix	X	$\mathbf{X}$

At the blackboard one can easily denote randomness by using the printer's symbol ~ under the letter to indicate randomness, e.g., what appears in this book as **u** would be written on the blackboard as $\underset{\sim}{u}$. (This is by contrast to those who are already denoting randomness by using the ~ above the letter to indicate randomness.) Finally, I have attempted to name all parameters with Greek letters, and here again the case of the letter indicates whether it is a scalar or a matrix.

To make this book as usable as possible, I have appended to each chapter all the specialized tables necessary to perform the statistical procedures described therein, though in most cases only for the 5% significance level. Since tables of the χ^2 and F distribution are necessary for almost all the chapters of this book, I append them for ready reference in this Preface.

References

Chatterjee, S. and Price, B. 1977. *Regression Analysis by Example.* New York: Wiley.

Draper, N. R. and Smith, H. 1966. *Applied Regression Analysis.* New York: Wiley.

Pearson, E. S. and Hartley, H. O., 1966 *Biometrika Tables for Statisticians*, Vol. I, 3rd ed. Cambridge: Cambridge University Press.

Ryan, B. F., Joiner, B. I., and Ryan, T. R., Jr. 1985. *Minitab Handbook.* Florence, KY: Wadsworth.

Roberts, H. V. and Ling, R. F. 1982. *Conversational Statistics with IDA.* New York: Scientific Press and McGraw-Hill.

Percentage Points of the χ^2 Distribution.

ν \ Q	0·995	0·990	0·975	0·950	0·900	0·750	0·500
1	392704.10^{-10}	157088.10^{-9}	982069.10^{-9}	393214.10^{-8}	0·0157908	0·1015308	0·454937
2	0·0100251	0·0201007	0·0506356	0·102587	0·210720	0·575364	1·38629
3	0·0717212	0·114832	0·215795	0·351846	0·584375	1·212534	2·36597
4	0·206990	0·297110	0·484419	0·710721	1·063623	1·92255	3·35670
5	0·411740	0·554300	0·831211	1·145476	1·61031	2·67460	4·35146
6	0·675727	0·872085	1·237347	1·63539	2·20413	3·45460	5·34812
7	0·989265	1·239043	1·68987	2·16735	2·83311	4·25485	6·34581
8	1·344419	1·646482	2·17973	2·73264	3·48954	5·07064	7·34412
9	1·734926	2·087912	2·70039	3·32511	4·16816	5·89883	8·34283
10	2·15585	2·55821	3·24697	3·94030	4·86518	6·73720	9·34182
11	2·60321	3·05347	3·81575	4·57481	5·57779	7·58412	10·3410
12	3·07382	3·57056	4·40379	5·22603	6·30380	8·43842	11·3403
13	3·56503	4·10691	5·00874	5·89186	7·04150	9·29906	12·3398
14	4·07468	4·66043	5·62872	6·57063	7·78953	10·1653	13·3393
15	4·60094	5·22935	6·26214	7·26094	8·54675	11·0365	14·3389
16	5·14224	5·81221	6·90766	7·96164	9·31223	11·9122	15·3385
17	5·69724	6·40776	7·56418	8·67176	10·0852	12·7919	16·3381
18	6·26481	7·01491	8·23075	9·39046	10·8649	13·6753	17·3379
19	6·84398	7·63273	8·90655	10·1170	11·6509	14·5620	18·3376
20	7·43386	8·26040	9·59083	10·8508	12·4426	15·4518	19·3374
21	8·03366	8·89720	10·28293	11·5913	13·2396	16·3444	20·3372
22	8·64272	9·54249	10·9823	12·3380	14·0415	17·2396	21·3370
23	9·26042	10·19567	11·6885	13·0905	14·8479	18·1373	22·3369
24	9·88623	10·8564	12·4011	13·8484	15·6587	19·0372	23·3367
25	10·5197	11·5240	13·1197	14·6114	16·4734	19·9393	24·3366
26	11·1603	12·1981	13·8439	15·3791	17·2919	20·8434	25·3364
27	11·8076	12·8786	14·5733	16·1513	18·1138	21·7494	26·3363
28	12·4613	13·5648	15·3079	16·9279	18·9392	22·6572	27·3363
29	13·1211	14·2565	16·0471	17·7083	19·7677	23·5666	28·3362
30	13·7867	14·9535	16·7908	18·4926	20·5992	24·4776	29·3360
40	20·7065	22·1643	24·4331	26·5093	29·0505	33·6603	39·3354
50	27·9907	29·7067	32·3574	34·7642	37·6886	42·9421	49·3349
60	35·5346	37·4848	40·4817	43·1879	46·4589	52·2938	59·3347
70	43·2752	45·4418	48·7576	51·7393	55·3290	61·6983	69·3344
80	51·1720	53·5400	57·1532	60·3915	64·2778	71·1445	79·3343
90	59·1963	61·7541	65·6466	69·1260	73·2912	80·6247	89·3342
100	67·3276	70·0648	74·2219	77·9295	82·3581	90·1332	99·3341
X	−2·5758	−2·3263	−1·9600	−1·6449	−1·2816	−0·6745	0·0000

SOURCE: Pearson and Hartley [1958]. Reprinted from *Biometrika Tables for Statisticians*, Vol. I, with the permission of the Biometrika Trustees.

Percentage Points of the χ^2 Distribution (*continued*)

ν \ Q	0·250	0·100	0·050	0·025	0·010	0·005	0·001
1	1·32330	2·70554	3·84146	5·02389	6·63490	7·87944	10·828
2	2·77259	4·60517	5·99147	7·37776	9·21034	10·5966	13·816
3	4·10835	6·25139	7·81473	9·34840	11·3449	12·8381	16·266
4	5·38527	7·77944	9·48773	11·1433	13·2767	14·8602	18·467
5	6·62568	9·23635	11·0705	12·8325	15·0863	16·7496	20·515
6	7·84080	10·6446	12·5916	14·4494	16·8119	18·5476	22·458
7	9·03715	12·0170	14·0671	16·0128	18·4753	20·2777	24·322
8	10·2188	13·3616	15·5073	17·5346	20·0902	21·9550	26·125
9	11·3887	14·6837	16·9190	19·0228	21·6660	23·5893	27·877
10	12·5489	15·9871	18·3070	20·4831	23·2093	25·1882	29·588
11	13·7007	17·2750	19·6751	21·9200	24·7250	26·7569	31·264
12	14·8454	18·5494	21·0261	23·3367	26·2170	28·2995	32·909
13	15·9839	19·8119	22·3621	24·7356	27·6883	29·8194	34·528
14	17·1170	21·0642	23·6848	26·1190	29·1413	31·3193	36·123
15	18·2451	22·3072	24·9958	27·4884	30·5779	32·8013	37·697
16	19·3688	23·5418	26·2962	28·8454	31·9999	34·2672	39·252
17	20·4887	24·7690	27·5871	30·1910	33·4087	35·7185	40·790
18	21·6049	25·9894	28·8693	31·5264	34·8053	37·1564	42·312
19	22·7178	27·2036	30·1435	32·8523	36·1908	38·5822	43·820
20	23·8277	28·4120	31·4104	34·1696	37·5662	39·9968	45·315
21	24·9348	29·6151	32·6705	35·4789	38·9321	41·4010	46·797
22	26·0393	30·8133	33·9244	36·7807	40·2894	42·7956	48·268
23	27·1413	32·0069	35·1725	38·0757	41·6384	44·1813	49·728
24	28·2412	33·1963	36·4151	39·3641	42·9798	45·5585	51·179
25	29·3389	34·3816	37·6525	40·6465	44·3141	46·9278	52·620
26	30·4345	35·5631	38·8852	41·9232	45·6417	48·2899	54·052
27	31·5284	36·7412	40·1133	43·1944	46·9630	49·6449	55·476
28	32·6205	37·9159	41·3372	44·4607	48·2782	50·9933	56·892
29	33·7109	39·0875	42·5569	45·7222	49·5879	52·3356	58·302
30	34·7998	40·2560	43·7729	46·9792	50·8922	53·6720	59·703
40	45·6160	51·8050	55·7585	59·3417	63·6907	66·7659	73·402
50	56·3336	63·1671	67·5048	71·4202	76·1539	79·4900	86·661
60	66·9814	74·3970	79·0819	83·2976	88·3794	91·9517	99·607
70	77·5766	85·5271	90·5312	95·0231	100·425	104·215	112·317
80	88·1303	96·5782	101·879	106·629	112·329	116·321	124·839
90	98·6499	107·565	113·145	118·136	124·116	128·299	137·208
100	109·141	118·498	124·342	129·561	135·807	140·169	149·449
X	+0·6745	+1·2816	+1·6449	+1·9600	+2·3263	+2·5758	+3·0902

SOURCE: Pearson and Hartley [1958]. Reprinted from *Biometrika Tables for Statisticians*, Vol. I, with the permission of the Biometrika Trustees.

Percentage Points of the F Distribution.
Upper 2.5% Points.

ν_2 \ ν_1	1	2	3	4	5	6	7	8	9	10	12	15	20	24	30	40	60	120	∞
1	647·8	799·5	864·2	899·6	921·8	937·1	948·2	956·7	963·3	968·6	976·7	984·9	993·1	997·2	1001	1006	1010	1014	1018
2	38·51	39·00	39·17	39·25	39·30	39·33	39·36	39·37	39·39	39·40	39·41	39·43	39·45	39·46	39·46	39·47	39·48	39·49	39·50
3	17·44	16·04	15·44	15·10	14·88	14·73	14·62	14·54	14·47	14·42	14·34	14·25	14·17	14·12	14·08	14·04	13·99	13·95	13·90
4	12·22	10·65	9·98	9·60	9·36	9·20	9·07	8·98	8·90	8·84	8·75	8·66	8·56	8·51	8·46	8·41	8·36	8·31	8·26
5	10·01	8·43	7·76	7·39	7·15	6·98	6·85	6·76	6·68	6·62	6·52	6·43	6·33	6·28	6·23	6·18	6·12	6·07	6·02
6	8·81	7·26	6·60	6·23	5·99	5·82	5·70	5·60	5·52	5·46	5·37	5·27	5·17	5·12	5·07	5·01	4·96	4·90	4·85
7	8·07	6·54	5·89	5·52	5·29	5·12	4·99	4·90	4·82	4·76	4·67	4·57	4·47	4·42	4·36	4·31	4·25	4·20	4·14
8	7·57	6·06	5·42	5·05	4·82	4·65	4·53	4·43	4·36	4·30	4·20	4·10	4·00	3·95	3·89	3·84	3·78	3·73	3·67
9	7·21	5·71	5·08	4·72	4·48	4·32	4·20	4·10	4·03	3·96	3·87	3·77	3·67	3·61	3·56	3·51	3·45	3·39	3·33
10	6·94	5·46	4·83	4·47	4·24	4·07	3·95	3·85	3·78	3·72	3·62	3·52	3·42	3·37	3·31	3·26	3·20	3·14	3·08
11	6·72	5·26	4·63	4·28	4·04	3·88	3·76	3·66	3·59	3·53	3·43	3·33	3·23	3·17	3·12	3·06	3·00	2·94	2·88
12	6·55	5·10	4·47	4·12	3·89	3·73	3·61	3·51	3·44	3·37	3·28	3·18	3·07	3·02	2·96	2·91	2·85	2·79	2·72
13	6·41	4·97	4·35	4·00	3·77	3·60	3·48	3·39	3·31	3·25	3·15	3·05	2·95	2·89	2·84	2·78	2·72	2·66	2·60
14	6·30	4·86	4·24	3·89	3·66	3·50	3·38	3·29	3·21	3·15	3·05	2·95	2·84	2·79	2·73	2·67	2·61	2·55	2·49
15	6·20	4·77	4·15	3·80	3·58	3·41	3·29	3·20	3·12	3·06	2·96	2·86	2·76	2·70	2·64	2·59	2·52	2·46	2·40
16	6·12	4·69	4·08	3·73	3·50	3·34	3·22	3·12	3·05	2·99	2·89	2·79	2·68	2·63	2·57	2·51	2·45	2·38	2·32
17	6·04	4·62	4·01	3·66	3·44	3·28	3·16	3·06	2·98	2·92	2·82	2·72	2·62	2·56	2·50	2·44	2·38	2·32	2·25
18	5·98	4·56	3·95	3·61	3·38	3·22	3·10	3·01	2·93	2·87	2·77	2·67	2·56	2·50	2·44	2·38	2·32	2·26	2·19
19	5·92	4·51	3·90	3·56	3·33	3·17	3·05	2·96	2·88	2·82	2·72	2·62	2·51	2·45	2·39	2·33	2·27	2·20	2·13
20	5·87	4·46	3·86	3·51	3·29	3·13	3·01	2·91	2·84	2·77	2·68	2·57	2·46	2·41	2·35	2·29	2·22	2·16	2·09
21	5·83	4·42	3·82	3·48	3·25	3·09	2·97	2·87	2·80	2·73	2·64	2·53	2·42	2·37	2·31	2·25	2·18	2·11	2·04
22	5·79	4·38	3·78	3·44	3·22	3·05	2·93	2·84	2·76	2·70	2·60	2·50	2·39	2·33	2·27	2·21	2·14	2·08	2·00
23	5·75	4·35	3·75	3·41	3·18	3·02	2·90	2·81	2·73	2·67	2·57	2·47	2·36	2·30	2·24	2·18	2·11	2·04	1·97
24	5·72	4·32	3·72	3·38	3·15	2·99	2·87	2·78	2·70	2·64	2·54	2·44	2·33	2·27	2·21	2·15	2·08	2·01	1·94
25	5·69	4·29	3·69	3·35	3·13	2·97	2·85	2·75	2·68	2·61	2·51	2·41	2·30	2·24	2·18	2·12	2·05	1·98	1·91
26	5·66	4·27	3·67	3·33	3·10	2·94	2·82	2·73	2·65	2·59	2·49	2·39	2·28	2·22	2·16	2·09	2·03	1·95	1·88
27	5·63	4·24	3·65	3·31	3·08	2·92	2·80	2·71	2·63	2·57	2·47	2·36	2·25	2·19	2·13	2·07	2·00	1·93	1·85
28	5·61	4·22	3·63	3·29	3·06	2·90	2·78	2·69	2·61	2·55	2·45	2·34	2·23	2·17	2·11	2·05	1·98	1·91	1·83
29	5·59	4·20	3·61	3·27	3·04	2·88	2·76	2·67	2·59	2·53	2·43	2·32	2·21	2·15	2·09	2·03	1·96	1·89	1·81
30	5·57	4·18	3·59	3·25	3·03	2·87	2·75	2·65	2·57	2·51	2·41	2·31	2·20	2·14	2·07	2·01	1·94	1·87	1·79
40	5·42	4·05	3·46	3·13	2·90	2·74	2·62	2·53	2·45	2·39	2·29	2·18	2·07	2·01	1·94	1·88	1·80	1·72	1·64
60	5·29	3·93	3·34	3·01	2·79	2·63	2·51	2·41	2·33	2·27	2·17	2·06	1·94	1·88	1·82	1·74	1·67	1·58	1·48
120	5·15	3·80	3·23	2·89	2·67	2·52	2·39	2·30	2·22	2·16	2·05	1·94	1·82	1·76	1·69	1·61	1·53	1·43	1·31
∞	5·02	3·69	3·12	2·79	2·57	2·41	2·29	2·19	2·11	2·05	1·94	1·83	1·71	1·64	1·57	1·48	1·39	1·27	1·00

SOURCE: Pearson and Hartley [1958]. Reprinted from *Biometrika Tables for Statisticians*, Vol. I, with the permission of the Biometrika Trustees.

Percentage Points of the F Distribution (*continued*)
Upper 5% Points.

ν_2 \ ν_1	1	2	3	4	5	6	7	8	9	10	12	15	20	24	30	40	60	120	∞
1	161·4	199·5	215·7	224·6	230·2	234·0	236·8	238·9	240·5	241·9	243·9	245·9	248·0	249·1	250·1	251·1	252·2	253·3	254·3
2	18·51	19·00	19·16	19·25	19·30	19·33	19·35	19·37	19·38	19·40	19·41	19·43	19·45	19·45	19·46	19·47	19·48	19·49	19·50
3	10·13	9·55	9·28	9·12	9·01	8·94	8·89	8·85	8·81	8·79	8·74	8·70	8·66	8·64	8·62	8·59	8·57	8·55	8·53
4	7·71	6·94	6·59	6·39	6·26	6·16	6·09	6·04	6·00	5·96	5·91	5·86	5·80	5·77	5·75	5·72	5·69	5·66	5·63
5	6·61	5·79	5·41	5·19	5·05	4·95	4·88	4·82	4·77	4·74	4·68	4·62	4·56	4·53	4·50	4·46	4·43	4·40	4·36
6	5·99	5·14	4·76	4·53	4·39	4·28	4·21	4·15	4·10	4·06	4·00	3·94	3·87	3·84	3·81	3·77	3·74	3·70	3·67
7	5·59	4·74	4·35	4·12	3·97	3·87	3·79	3·73	3·68	3·64	3·57	3·51	3·44	3·41	3·38	3·34	3·30	3·27	3·23
8	5·32	4·46	4·07	3·84	3·69	3·58	3·50	3·44	3·39	3·35	3·28	3·22	3·15	3·12	3·08	3·04	3·01	2·97	2·93
9	5·12	4·26	3·86	3·63	3·48	3·37	3·29	3·23	3·18	3·14	3·07	3·01	2·94	2·90	2·86	2·83	2·79	2·75	2·71
10	4·96	4·10	3·71	3·48	3·33	3·22	3·14	3·07	3·02	2·98	2·91	2·85	2·77	2·74	2·70	2·66	2·62	2·58	2·54
11	4·84	3·98	3·59	3·36	3·20	3·09	3·01	2·95	2·90	2·85	2·79	2·72	2·65	2·61	2·57	2·53	2·49	2·45	2·40
12	4·75	3·89	3·49	3·26	3·11	3·00	2·91	2·85	2·80	2·75	2·69	2·62	2·54	2·51	2·47	2·43	2·38	2·34	2·30
13	4·67	3·81	3·41	3·18	3·03	2·92	2·83	2·77	2·71	2·67	2·60	2·53	2·46	2·42	2·38	2·34	2·30	2·25	2·21
14	4·60	3·74	3·34	3·11	2·96	2·85	2·76	2·70	2·65	2·60	2·53	2·46	2·39	2·35	2·31	2·27	2·22	2·18	2·13
15	4·54	3·68	3·29	3·06	2·90	2·79	2·71	2·64	2·59	2·54	2·48	2·40	2·33	2·29	2·25	2·20	2·16	2·11	2·07
16	4·49	3·63	3·24	3·01	2·85	2·74	2·66	2·59	2·54	2·49	2·42	2·35	2·28	2·24	2·19	2·15	2·11	2·06	2·01
17	4·45	3·59	3·20	2·96	2·81	2·70	2·61	2·55	2·49	2·45	2·38	2·31	2·23	2·19	2·15	2·10	2·06	2·01	1·96
18	4·41	3·55	3·16	2·93	2·77	2·66	2·58	2·51	2·46	2·41	2·34	2·27	2·19	2·15	2·11	2·06	2·02	1·97	1·92
19	4·38	3·52	3·13	2·90	2·74	2·63	2·54	2·48	2·42	2·38	2·31	2·23	2·16	2·11	2·07	2·03	1·98	1·93	1·88
20	4·35	3·49	3·10	2·87	2·71	2·60	2·51	2·45	2·39	2·35	2·28	2·20	2·12	2·08	2·04	1·99	1·95	1·90	1·84
21	4·32	3·47	3·07	2·84	2·68	2·57	2·49	2·42	2·37	2·32	2·25	2·18	2·10	2·05	2·01	1·96	1·92	1·87	1·81
22	4·30	3·44	3·05	2·82	2·66	2·55	2·46	2·40	2·34	2·30	2·23	2·15	2·07	2·03	1·98	1·94	1·89	1·84	1·78
23	4·28	3·42	3·03	2·80	2·64	2·53	2·44	2·37	2·32	2·27	2·20	2·13	2·05	2·01	1·96	1·91	1·86	1·81	1·76
24	4·26	3·40	3·01	2·78	2·62	2·51	2·42	2·36	2·30	2·25	2·18	2·11	2·03	1·98	1·94	1·89	1·84	1·79	1·73
25	4·24	3·39	2·99	2·76	2·60	2·49	2·40	2·34	2·28	2·24	2·16	2·09	2·01	1·96	1·92	1·87	1·82	1·77	1·71
26	4·23	3·37	2·98	2·74	2·59	2·47	2·39	2·32	2·27	2·22	2·15	2·07	1·99	1·95	1·90	1·85	1·80	1·75	1·69
27	4·21	3·35	2·96	2·73	2·57	2·46	2·37	2·31	2·25	2·20	2·13	2·06	1·97	1·93	1·88	1·84	1·79	1·73	1·67
28	4·20	3·34	2·95	2·71	2·56	2·45	2·36	2·29	2·24	2·19	2·12	2·04	1·96	1·91	1·87	1·82	1·77	1·71	1·65
29	4·18	3·33	2·93	2·70	2·55	2·43	2·35	2·28	2·22	2·18	2·10	2·03	1·94	1·90	1·85	1·81	1·75	1·70	1·64
30	4·17	3·32	2·92	2·69	2·53	2·42	2·33	2·27	2·21	2·16	2·09	2·01	1·93	1·89	1·84	1·79	1·74	1·68	1·62
40	4·08	3·23	2·84	2·61	2·45	2·34	2·25	2·18	2·12	2·08	2·00	1·92	1·84	1·79	1·74	1·69	1·64	1·58	1·51
60	4·00	3·15	2·76	2·53	2·37	2·25	2·17	2·10	2·04	1·99	1·92	1·84	1·75	1·70	1·65	1·59	1·53	1·47	1·39
120	3·92	3·07	2·68	2·45	2·29	2·17	2·09	2·02	1·96	1·91	1·83	1·75	1·66	1·61	1·55	1·50	1·43	1·35	1·25
∞	3·84	3·00	2·60	2·37	2·21	2·10	2·01	1·94	1·88	1·83	1·75	1·67	1·57	1·52	1·46	1·39	1·32	1·22	1·00

SOURCE: Pearson and Hartley [1958]. Reprinted from *Biometrika Tables for Statisticians*, Vol. I, with the permission of the Biometrika Trustees.

Acknowledgments

My editors, Steve Fienberg and Ingram Olkin, are to be thanked for not only their useful comments on earlier drafts of this book, but also for their encouragement in seeing that this enterprise reached its conclusion. Michael Cohen and Diane Lambert contributed immeasurably with their detailed critical reading of an earlier draft of this book. The extent to which they do not recognize this version is almost wholely attributable to my response to their comments. Chapter 8 benefitted from the extensive comments of the "four most" experts in the field, Steve Fienberg, Zvi Gilula, Leo Goodman, and Shelby Haberman. Evelyn Shropshire typed each of the drafts of this book, always with a smile and a willingness to push on and get it out, and this made my job so much easier. Finally, I would like to express my appreciation to each of my students who suffered through the various versions of this book and my effort to "debug" it.

Chicago, Illinois A. M.
June 1987

CHAPTER 0

A Thoughtful Student's Retrospective on Statistics 101

"Take it upon yourselves now, for all beginnings are difficult."
Chapter II, Tractate Bahodesh,
Mekilta de Rabbi Ishmael, Lauterbach [1933]

"If it ain't broken, don't fix it!"
Motto attributed to Western Electric

0. Introduction

The first course in applied statistics, dubbed here Statistics 101, has a standardized syllabus, as exemplified by both classical elementary textbooks such as Dixon and Massey [1957], and modern elementary textbooks such as Freedman, Pisani, and Purves [1978] or Mosteller, Fienberg, and Rourke [1983]. In the main, the course covers two fundamental topics, that of inference about parameters of a single normal distribution based on independent observations, and that of inference about parameters of a (simple or multiple) linear regression model with independent, identically normally distributed residuals. This book is designed to be used as a next course, either in universities or that "school of hard knocks", the "real world". To motivate some of the issues covered in this book, this chapter reviews some of the material from Statistics 101 from the point of view of a very thoughtful student, who asks herein all the questions you, the reader, had thought of when you took that course but, for some reason or other, didn't ask.

1. The Introductory Model

In the very first course in statistics, the standard situation presented is one where the data are a set of n independent observations from some (usually unspecified) distribution. The student is then taught to make inferences about one of the distribution's parameters, its mean μ, based on the sample mean, and either the population standard deviation σ or the sample standard deviation $\mathbf{s}$. He learns that the inferential procedure is based on a far-ranging theorem, the Central Limit Theorem, which says that the sampling distribution of the sample mean is approximately normal, with mean μ and standard deviation $\sigma/\sqrt{n}$, almost regardless of the nature and shape of the distribution from which the data were drawn.

It is rare, though, that the student is taught how to verify the independence assumption. Usually, this matter is deferred until he is exposed to the regression model and the concepts of correlation and autocorrelation. Unfortunately, the procedures derived from those concepts have a built-in assumption of their own underlying their validity, namely normality of the data, which may not be valid here. Thus the deferred solution may be of only limited value. A few texts do introduce various distribution-free tests, such as variants of the runs test, at this juncture, mainly to provide a handy tool to help check the validity of the independence assumption.[1]

The next situation presented is usually that in which one wishes to make inferences about some other parameter, say the population median or population standard deviation, from its sample counterpart. Let us assume that, just as with the use of the Central Limit Theorem to obtain the sampling distribution of the sample mean, one can use "large sample approximations" to the sampling distributions of these statistics. One then learns that these approximations are sensitive to more details about the data distribution than merely its mean and standard deviation.[2] For pedagogical ease, one then restricts oneself to making some particular assumption about the distribution from which the data were drawn, usually that the data come from the same normal distribution.

With respect to checking the assumption of normality of the data, the student is told a few salient facts about the percent of observations encompassed within intervals of the form $\mu \pm k\sigma$, where $k = 1$, 2, and 3, and asked to look at a histogram of the sample data and compare it visually with the familiar bell-shaped curve. If the histogram looks bell-shaped and if the requisite fractions of the data fall within 1, 2, and 3 sample standard deviations from the sample mean, then he should feel comfortable with the distributional assumption upon which the inferential procedures are based.

To make what I have described concrete, let us consider the random sample of 20 sixth grade verbal achievement scores from data of the Coleman report, dubbed the Northeast achievement data and given as Table 14-1 of Mosteller, Fienberg, and Rourke [1983]. Because the data were collected by randomly sampling 20 schools, the data satisfy the independence assumption. But what

[1] Throughout this chapter I use the footnote as a vehicle for forward referencing the reader to the appropriate chapter in which to find these answers. Chapter 3 presents methods for independence testing, including both runs tests and tests based on the autocorrelations of the data.

[2] In particular, if ν is the population median and $f_{\mathbf{x}}(x)$ is the probability density function of the data, then the sample median is approximately normally distributed with mean ν and standard deviation $0.5/[f_{\mathbf{x}}(\nu)\sqrt{n}]$. With respect to the sample standard deviation, for large samples it is approximately normally distributed with mean σ and variance $(\mu_4 - \sigma^4)/4\sigma^2(n-1)$, where μ_4 is the population fourth central moment.

of normality? We find that the sample mean is 35.082, the sample standard deviation is 5.8171, and following is a histogram of the 20 scores:

Sample Frequency

```
5                             *
 -                         * **
 -                         ****
 -                      *  ****
 -                      ** *****
  ------:---------:---------:---------:---------:------
      -25         5         35        65        95
```

Score

The histogram does not appear very bell-shaped, but since there are only 20 observations, we can look at a frequency distribution, viz.,

Frequency Distribution of Score

Class midpoint	Percentage	Cumulative percentage
23.0	10.0	10.0
26.0	5.0	15.0
29.0	0.0	15.0
32.0	20.0	35.0
35.0	15.0	50.0
38.0	20.0	70.0
41.0	25.0	95.0
44.0	5.0	100.0
Total	100.0	
(Number)	(20)	

We note that 85% of the sample (i.e., 17 observations) is in the interval 35.052 ± 5.8171, that is, within one standard deviation of the sample mean, in contrast to the 67% expected. There are 19 observations, or 95% of the sample, within two standard deviations, and all 20 observations are within three standard deviations of the sample mean. We are at a loss as to what to make of this possibly conflicting visual evidence about the normality of the distribution of the verbal achievement scores.

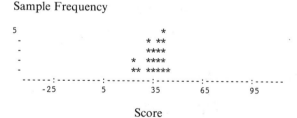

"If things don't 'look right', though, then what? Can I trust these elementary tools (and my eyesight) to tell me when our data aren't drawn from a normal distribution? And if I do trust these tools and decide that things really don't look normal, now what should I do? Is the deviation from normality due to outliers? And if so, should I just delete the ones I don't like and proceed with the remaining data? Or is the deviation truly due the fact that the data come from a nonnormal distribution? And if the distribution is nonnormal, how can I rectify the situation so that I can use the procedures I have just learned in my statistics course? And what about that independence assumption? How can I check it, and what effect does its violation have on the checking of normality?" All these are reasonable questions which come to the mind of the thoughtful student, and it is the purpose of this book to provide some answers to them.[3]

2. The Regression Model

The next standard situation introduced in that first course is the simple linear regression model, followed by the multiple linear regression model. A "dependent variable" **y** is introduced, along with a set of "independent variables" $x_1, \ldots, x_p$, and the student is told that a reasonable model relating **y** to the x's is

$$\mathbf{y} = \beta_0 + \beta_1 x_1 + \cdots + \beta_p x_p + \mathbf{u}.$$

Again to make matters concrete, let us consider the complete data set given in Table 14-1 of Mosteller, Fienberg, and Rourke [1983], as reproduced in Table 1. There are five variables suggested as candidates for independent

Table 1. Northeast Achievement Data.

	y	x_1	x_2	x_3	x_4	x_5
1	37.01	3.83	28.87	7.20	26.60	6.19
2	26.51	2.89	20.10	-11.71	24.40	5.17
3	36.51	2.86	69.05	12.32	25.70	7.04
4	40.70	2.92	65.40	14.28	25.70	7.10
5	37.10	3.06	29.59	6.31	25.40	6.15
6	33.90	2.07	44.82	6.16	21.60	6.41
7	41.80	2.52	77.37	12.70	24.90	6.86
8	33.40	2.45	24.67	-0.17	25.01	5.78
9	41.01	3.13	65.01	9.85	26.60	6.51
10	37.20	2.44	9.99	-0.05	28.01	5.57
11	23.30	2.09	12.20	-12.86	23.51	5.62
12	35.20	2.52	22.55	0.92	23.60	5.34
13	34.90	2.22	14.30	4.77	24.51	5.80
14	33.10	2.67	31.79	-0.96	25.80	6.19
15	22.70	2.71	11.60	-16.04	25.20	5.62
16	39.70	3.14	68.47	10.62	25.01	6.94
17	31.80	3.54	42.64	2.66	25.01	6.33
18	31.70	2.52	16.70	-10.99	24.80	6.01
19	43.10	2.68	86.27	15.03	25.51	7.51
20	41.01	2.37	76.73	12.77	24.51	6.96

[3] Chapter 1 covers tests for normality, Chapter 4 covers tests for outliers from a normally distributed population, Chapter 5 describes methods for transforming nonnormally distributed data to approximate normality, and, as pointed out earlier, Chapter 3 shows how to check for independence.

variables in a multiple linear regression with the aforementioned sixth grade verbal achievement scores as the dependent variable **y**. These are:

x_1 = staff salaries per pupil,

x_2 = sixth grade percentage white-collar fathers,

x_3 = sixth grade composite socioeconomic status,

x_4 = mean teacher's verbal test score,

x_5 = mean sixth grade mother's educational level.

Immediately questions come to the mind of the student. "Why these particular independent variables? Do we need all of them? Are there others we should be considering? I see that y is a random quantity, presumably drawn from some probability distribution, but what about the x's? They do not appear to be random quantities in this formulation! And why are we confined to this linear form for the model? And finally, what is that meaningless word 'regression' supposed to convey?"

In answer to his first question, our thoughtful student trusts to his intuition and agrees that each of these variables may be relevant in contributing to the level of the sixth grade verbal test scores. But he may not be that positive about treating these variables as continuously measured quantities. He is dubious, for example, about whether any small increase in per pupil staff salaries will have a direct proportionate effect in increasing student verbal achievement scores. Rather, he suspects that the effect of per pupil staff salaries is that of a categorical variable, i.e., one should categorize levels of x_1 as "low", "medium", and "high", and use only these categorical levels as independent variables in the multiple linear regression model.[4] Indeed, he may even be dubious about using the continuously measured dependent variable y as the quantity to be modeled. He may, for example, believe that it is not the specific numeric value of y that is effected by the x's, but rather the level of y, i.e., whether it is "low", "medium", or "high".[5] But he is willing to proceed first with a model which uses the data to its fullest, knowing that he can subsequently proceed to categorizing the variable(s) and developing an alternate model if this model is unsatisfactory due to the continuously measured aspect of the variables in the model.

He may not have the same feeling of reasonableness about the form of the model, but is willing to go along with it as a first-order approximation to whatever is the true function relating the dependent variable to these five independent variables. And so he is willing to proceed.

In response to some of his other questions, the student may be told that the adjective "linear", describing the form of the model, does not refer to the

[4] Chapter 7 treats special aspects of regression models of this variety.

[5] Chapter 7 discusses a simple case in which y is a dichotomous categorical variable and the x's are continuously measured quantities. Chapter 8 is devoted to modeling the situation in which both y and the x's are categorical variables.

way the independent variables appear in the model, but rather to the way the parameters appear in the model. That is, the model

$$\mathbf{y} = \beta_0 + \beta_1 x + \beta_2 x^2 + \mathbf{u}$$

is a "linear" multiple regression, in that it is linear in the parameters β_0, β_1, and β_2, though quadratic in x (but also linear in the two independent variables $x_1 = x$ and $x_2 = x^2$). He may also be told that the term "linear regression" is perhaps a misnomer for this model, whose full name is "linear functional relation". He may be told that the x's are either nonrandom variables observed perfectly (i.e., without any measurement error), or are values of random quantities but where we are modeling the relationship of $\mathbf{y}$ to the $\mathbf{x}$'s conditionally on the observed values of the $\mathbf{x}$'s.

If and when he absorbs (or at the very least accepts) all this, he is told about three more requirements that the model must satisfy, namely that the $\mathbf{u}$'s, sometimes called the "true residuals", must be independent and identically normally distributed variables with mean 0 and standard deviation σ. These three requirements, independence, normality, and homoscedasticity of the true residuals, the student is told, are necessary for the validity of all the statistical inferential procedures that he is about to learn. When he then asks, "How do I check on these assumptions when the only data I have are observations on y and the x's, and not on the u's?", he is told to fit the model anyway, regardless of whether the assumptions about the u's are true or not, then calculate estimates $\hat{u}$ of the u's, and test the model assumptions about the "true residuals" using these "estimated residuals".

It is more compact to express our estimates of the u's using matrix notation. Let $\mathbf{Y}$ be the $n \times 1$ vector of our observations, i.e., $\mathbf{Y}' = [\mathbf{y}_1, \ldots, \mathbf{y}_n]$, let X be the $n \times p + 1$ matrix of observations on the independent variables (including a column associated with the intercept), namely

$$X = \begin{bmatrix} 1 & x_{11} & \ldots & x_{p1} \\ \vdots & \vdots & & \vdots \\ 1 & x_{1n} & \ldots & x_{pn} \end{bmatrix} = \begin{bmatrix} X_1' \\ \vdots \\ X_n' \end{bmatrix},$$

where X_i is the $p + 1 \times 1$ vector of observations on the ith set of independent variables, $i = 1, \ldots, n$, let B be the $p + 1 \times 1$ vector of regression coefficients

$$B' = [\beta_0 \quad \beta_1 \quad \ldots \quad \beta_p],$$

and let $\mathbf{U}$ be the $n \times 1$ vector of the $\mathbf{u}_i$'s. Then the model is expressible as

$$\mathbf{y}_i = X_i'B + \mathbf{u}_i, \qquad i = 1, \ldots, n,$$

or, more compactly, as

$$\mathbf{Y} = XB + \mathbf{U},$$

and the usual[6] estimator of B is given by

$$\hat{\mathbf{B}} = (X'X)^{-1}X'\mathbf{Y}.$$

The usual estimator of the true residual vector is given by the estimated residual vector

$$\hat{\mathbf{U}} = \mathbf{Y} - X\hat{\mathbf{B}} = [I - X(X'X)^{-1}X']\mathbf{Y}.$$

Applying these computations to our example, the student obtains the following results:

Variable	$\hat{\mathbf{B}}$	
x_1	−1.7933	$\hat{\sigma} = 2.0743$
x_2	0.043602	
x_3	0.55576	
x_4	1.1102	
x_5	−1.8109	
Constant	19.949	

and finds that the estimated residuals are as given in Table 2.

How is he to check these assumptions about the true residuals? The independence and normality assumptions are the same kind of assumptions underlying the standard situation first encountered by the student and described earlier. So, with whatever tools that were made available to him, the

Table 2. Regression Residuals.

	$\hat{u}$		$\hat{u}$
1	0.34878	11	-2.20804
2	-0.34992	12	1.74647
3	-3.94972	13	-1.04872
4	-0.47361	14	-0.34567
5	0.78091	15	-1.77888
6	-0.08570	16	1.29742
7	0.71876	17	-1.43983
8	-0.43425	18	5.00176
9	0.62442	19	1.12262
10	0.21039	20	0.26282

[6] This estimator is "usually" not presented explicitly in a first course, primarily because of the unfamiliarity of matrices to the student. Rather, this estimator is embodied in the statistical package used as an adjunct to the text in this course, and alluded to as necessary by the instructor. In a few courses where the estimator is presented, the fact that it is the ordinary least squares estimator of B and the maximum likelihood estimator for this model that is stated, and some properties of $\hat{\mathbf{B}}$ (e.g., unbiasedness, minimum variance among estimators linear in $\mathbf{Y}$) are stated.

student can go ahead and check these two assumptions ... right? Wrong! He would be shocked to learn that the $\hat{\mathbf{u}}$'s are themselves correlated, so that the procedures taught to him relating to the standard situation are technically inapplicable in dealing with the $\hat{\mathbf{u}}$'s.[7] But he is told to use these procedures anyway.

So he plots a histogram of the $\hat{\mathbf{u}}_i$'s and finds the following:

Sample Frequency

```
  -                          *
  -                          *
  5                          **
  -                          **
  -                        * **
  -                        ****
  -                      * *****    *
   ------:---------:---------:---------:---------:------
       -18        -9         0         9        18
```

Residual

and the frequency distribution

Class midpoint	Percentage	Cumulative percentage
−3.6	5.0	5.0
−2.7	0.0	5.0
−1.8	15.0	20.0
−0.9	10.0	30.0
0.0	35.0	65.0
0.9	25.0	90.0
1.8	5.0	95.0
2.7	0.0	95.0
3.6	0.0	95.0
4.5	0.0	95.0
5.4	5.0	100.0
Total	100.0	
(Number)	(20)	

He notes that 90% of the sample (i.e., 18 residuals) is in the interval 0.0 ± 2.0743, that is, within one standard deviation of 0, in contrast to the 67% expected. There are 19 observations, or 95% of the sample, within two standard deviations, and all 20 observations are within three standard deviations of the sample mean. Thus, though there may be "too many" residuals within one standard deviation of 0, the histogram "looks" like a bell and so the

[7] See Chapter 2 for a proof of this.

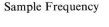

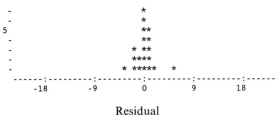

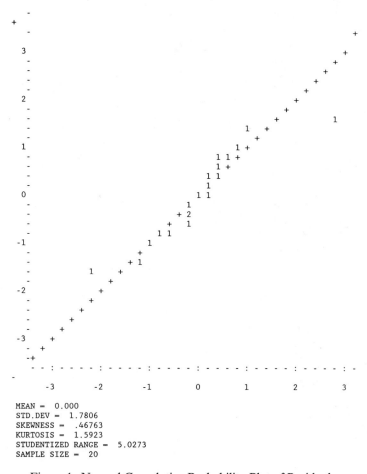

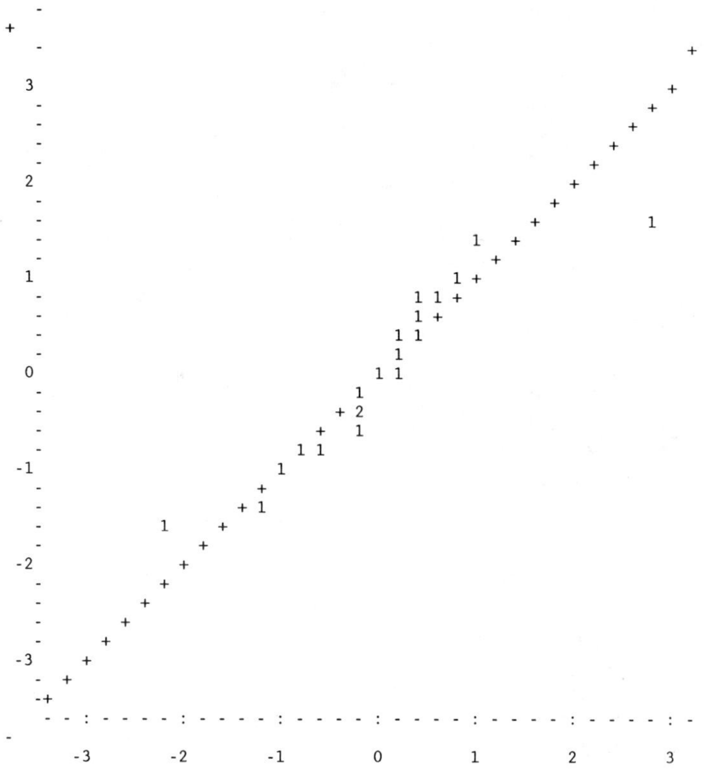

```
MEAN =   0.000
STD.DEV =   1.7806
SKEWNESS =   .46763
KURTOSIS =   1.5923
STUDENTIZED RANGE =   5.0273
SAMPLE SIZE =   20
```

Figure 1. Normal Cumulative Probability Plot of Residuals.

student may go away feeling that the residuals are normally distributed.[8] A look at Table 2, though, shows that one residual, $\hat{\mathbf{u}}_3$, is -2.21 and another, $\hat{\mathbf{u}}_{18}$, is 2.81 standard deviations from the mean, and so one is concerned about whether these are outliers.[9]

As to the third assumption, at best the student is taught a neat *ad hoc* procedure, dubbed the "tic-tac-toe method" by Roberts and Ling [1982]. This consists of plotting the residuals versus the fitted values of the dependent variable, demarcating the ± 1 standard deviation values of both these vari-

[8] For the benefit of the reader, a normal probability paper plot of the residuals is given in Figure 1, along with values of other statistics described in Chapter 1 and used in testing for normality. After reading Chapter 1, the reader can then definitely conclude that the residuals can be considered as coming from a normal distribution.

[9] Methods presented in Chapter 4 will allay this concern.

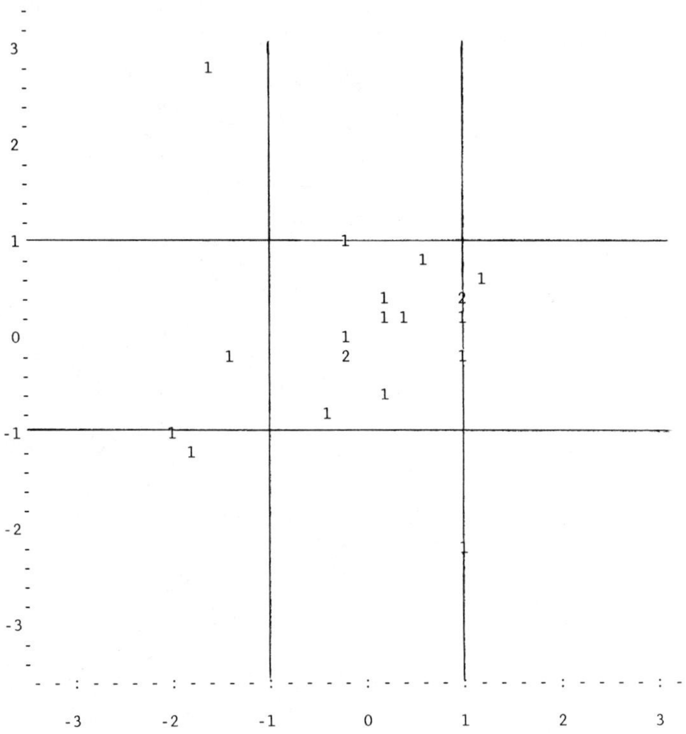

	MEAN	STD.DEV.
VERT. VAR.	0.00000	1.7806
HORIZ. VAR.	35.082	5.5378
SAMPLE SIZE =	20	

Figure 2. Scatter Plot of 20 Standardized Values of $\hat{u}$ versus $\hat{y}$.

ables, and counting the proportion of residuals within one standard deviation in each of the three regions demarcated on the fitted variable. A graph of our data implementing this procedure is given as Figure 2. If the residuals have constant variance, we would expect the same proportion of residuals, namely 67%, to be within ± 1 standard deviation in each of the regions demarcated on the fitted variable. A glance at Figure 2 leads the student to conclude that the constant variance assumption appears not to be violated.[10]

Finally, the student is warned about another danger lurking in such models, that of multicollinearity. In brief, he is told that the independent variables used in the model must be a linearly independent set.[11] How to check on this

[10] The reader can apply the methods of Chapter 2 to these data to reach a formal conclusion about this assumption.

[11] He now has to keep straight three different uses of the word "independent", the statistical use relating to the **u**'s, the general mathematical use relating to the x's as arguments of a function y, and the linear algebraic use relating to the linear interrelationship amongst the x's.

is another story. At a minimum he is told to avoid x's that are "highly" correlated with each other. Sometimes he is told that the computer regression package will check for this phenomenon for him. But it is rare that he is given a full explanation of this phenomenon and how to guard against it.

For the Northeast achievement data the intercorrelations of the independent variables are:

	x_1	x_2	x_3	x_4
x_1	1.00000			
x_2	0.18114	1.00000		
x_3	0.22963	0.82718	1.00000	
x_4	0.50266	0.05106	0.18333	1.00000
x_5	0.19677	0.92710	0.81906	0.12381

It appears that x_2 and x_5 are highly correlated, as are x_2 and x_3, and as are x_3 and x_5. Is this sufficient to cause the student to eliminate one or two of these three independent variables from consideration?[12]

Having survived all this, the student has still more questions. "How do I know if I have a good model? How do I know I can't do better? Isn't it the case that the more independent variables I use the better off I will be?" He is then introduced to some time-honored criteria for goodness of regression models, namely R^2 and adjusted R^2, which measure the overall quality of the fitted model. But if he is merely interested in using the model to predict **y** associated with *one* new set of x's, the student may rightly say, "I am not really interested in overall measures of quality, but only in a measure of how good the model will be for my purposes." At this juncture the course moves on to the topic of prediction intervals, followed by the use of the standard deviation of the prediction, rather than the traditional general measures, as the figure of merit for the regression.

Let us denote the set of independent variables to be used in the prediction of y by the vector $Z' = (1, z_1, \ldots, z_p)$. Then the prediction of y using this model is $Z'\hat{\mathbf{B}}$. Since $\hat{\mathbf{B}}$ is a random matrix, in order to determine the variance of the prediction one needs to know the covariance matrix of $\hat{\mathbf{B}}$. The student is told[13] that the covariance matrix of $\hat{\mathbf{B}}$ is given by

$$\sigma^2(X'X)^{-1}$$

so that the variance of the prediction is $v_Z = \sigma^2(1 + Z'(X'X)^{-1}Z)$. Thus his figure of merit in assessing the model is the computation of this quantity v_Z. Of course, this figure of merit changes with variations in any of the components of Z. But when, as stated earlier, the model is to be used to make a

[12] The reader is invited to apply the material of Chapter 6 to these data to resolve this issue.

[13] Alternatively, he is referred to the appropriate command of the statistical program used in the course which generates this covariance matrix.

Table 3. Values of R^2, Adjusted R^2, and Standard Deviation of Prediction.

Independent variables	R^2	Adjusted R^2	v_Z
x_3	.8597	.8519	2.3457
x_2, x_3, x_4, x_5	.8921	.8633	2.3854
x_2, x_3, x_4	.8874	.8662	6.1712
x_3, x_4, x_5	.8889	.8680	2.2397
x_1, x_2, x_3, x_4, x_5	.9063	.8728	2.4278
x_3, x_4	.8874	.8741	11.832
x_1, x_2, x_3, x_4	.9010	.8746	2.3064
x_1, x_3, x_4, x_5	.9017	.8755	2.3215
x_1, x_3, x_4	.9008	.8822	2.2336

prediction based on one particular value of Z, this figure of merit is the appropriate one, more so than the general measures of goodness of the model.

Table 3 presents the values of R^2, adjusted R^2, and v_Z for the Northeast achievement data, assuming that $Z' = (1, 2, 15, -6, 25, 6)$, for the nine combinations of independent variables associated with the largest values of adjusted R^2. One finds that the best set of independent variables is x_1, x_3, x_4, which happen to be the set associated with the highest adjusted R^2 as well. Thus in this case one would use the model incorporating only these variables as a predictive model.

But from a perusal of this table the student sees that, although the usual measures tell him that all nine of these models are quite good, at least two of them, the one using x_3 and x_4 and the one using x_2, x_3, and x_4, will not lead to good predictions given the Z vector of independent variable values in which he is interested. He is left with a quandary about the use of v_Z as a criterion of goodness of a regression model, in that this figure of merit may depend so heavily on the set of independent variables in which one is interested. He is stranded with no further handle on how to create a general measure of goodness of a regression model, that incorporates the notion of how good the regression model is as a predictor given some reasonable sets of independent variables, for the first course has just ended.[14]

References

Dixon, W. J. and Massey, F. J. 1957. *Introduction to Statistical Analysis*. New York: McGraw-Hill.

[14] Chapter 6 presents a set of criteria for goodness of regression models, including the PRESS criterion which will satisfy this zealous student's desire for an overall criterion which reflects the usability of the model as a general predictor (at least over the range of the independent variables used in developing the regression equation).

Freedman, D., Pisani, R., and Purves, R. 1978. *Statistics.* New York: Norton.

Lauterbach, J. Z. 1933. *Mekilta de Rabbi Ishmael,* Vol. II. Philadelphia: Jewish Publication Society.

Mosteller, F., Fienberg, S. E., and Rourke, R. E. K. 1983. *Beginning Statistics with Data Analysis.* Reading, MA: Addison-Wesley.

Roberts, H. V. and Ling, R. F. 1982. *Conversational Statistics with IDA.* New York: Scientific Press and McGraw-Hill.

CHAPTER 1

Testing for Normality

"Normality is a myth; there never was, and never will be, a normal distribution."

Geary [1947]

0. Introduction

The Central Limit Theorem and other limit theorems of statistics provide the basis for the centrality of the normal distribution in statistical inference. These theorems tell us that, regardless of the underlying distribution of the data, as the sample size on which they are based gets very large the sampling distribution of statistics calculated from the sample becomes closer to that of some normally distributed variable. But there are many inferential situations which require more than merely the normality of the sampling distribution of a statistic; they require normality of the data distribution itself. For example, standard inference about regression parameters requires that the residuals are normally distributed. In this chapter we describe methods for testing for the normality of the distribution of data. In a subsequent chapter we will describe methods for transforming data which fail the normality test so that the transformed data has a normal distribution.

This chapter contains a mixture of modern and classical methods of testing for normality. We begin in Section 1 with an old approach which has recently become very popular; namely, plotting the data in some special manner to facilitate an eyeball assessment of whether the data could come from a normally distributed population. Though this does not provide a formal procedure for testing for normality, normal plotting is what I would advise as the first thing to do. The formal procedure that I recommend is the Shapiro–Wilk procedure (or approximations thereto) given in Section 2. So why bother with the material in the other sections of this chapter, you may well ask. The reasons are partly historical, partly because they are still used, and partly to provide a context for and a contrast to the modern procedures of Section 2. Section 3 describes an extremely effective test for normality, the use of the studentized range statistic. As will be seen in Section 6, except when the data come from

Table 1. One Hundred Independent Gamma Observations (x) and the Cumulative Normal $N(25, 125)$ Distribution ($F(x)$) Associated with Each Observation.

Row	x	$F(x)$	Row	x	$F(x)$
1	6.48875	0.03208	51	24.25881	0.47045
2	7.36181	0.03889	52	24.73173	0.48929
3	7.70804	0.04189	53	24.96917	0.49877
4	8.11865	0.04570	54	25.24047	0.50960
5	8.46150	0.04909	55	25.24985	0.50997
6	12.75674	0.11042	56	25.37080	0.51480
7	13.08055	0.11664	57	26.35537	0.55391
8	13.43281	0.12369	58	26.47865	0.55878
9	13.93881	0.13433	59	26.48096	0.55887
10	14.22067	0.14053	60	26.99213	0.57895
11	14.25665	0.14133	61	27.52403	0.59963
12	14.69054	0.15128	62	28.05616	0.62005
13	15.02005	0.15913	63	28.09613	0.62157
14	15.03517	0.15950	64	28.29106	0.62896
15	15.14141	0.16209	65	28.46719	0.63559
16	15.31778	0.16646	66	28.71782	0.64496
17	15.44101	0.16955	67	29.20037	0.66276
18	15.52946	0.17180	68	29.56230	0.67588
19	15.75355	0.17757	69	29.73896	0.68221
20	15.76179	0.17778	70	30.30640	0.70216
21	15.78501	0.17839	71	30.32239	0.70271
22	16.24040	0.19052	72	30.57425	0.71138
23	16.62110	0.20104	73	31.49794	0.74209
24	16.65918	0.20211	74	31.84531	0.75318
25	16.71427	0.20367	75	31.85266	0.75341
26	17.42898	0.22449	76	32.70566	0.77952
27	18.76179	0.26637	77	32.93997	0.78641
28	18.79385	0.26743	78	33.31204	0.79708
29	19.07174	0.27665	79	33.35007	0.79815
30	19.32876	0.28532	80	33.43215	0.80045
31	19.62610	0.29550	81	33.71671	0.80831
32	19.78951	0.30117	82	34.05206	0.81733
33	19.96001	0.30714	83	34.25993	0.82278
34	20.04533	0.31014	84	34.44229	0.82748
35	20.14921	0.31382	85	35.43620	0.85167
36	20.62760	0.33098	86	35.90700	0.86230
37	20.99996	0.34459	87	36.77312	0.88046
38	21.00931	0.34493	88	36.86588	0.88231
39	21.24029	0.35348	89	37.67851	0.89757
40	21.47477	0.36223	90	37.84356	0.90049
41	21.81918	0.37522	91	38.00723	0.90332
42	22.05309	0.38412	92	40.46576	0.93901
43	22.33860	0.39507	93	40.72330	0.94205
44	22.87242	0.41576	94	43.38677	0.96702
45	22.94075	0.41842	95	45.75185	0.98102
46	23.23921	0.43011	96	45.79899	0.98124
47	23.45028	0.43842	97	46.53636	0.98437
48	23.74068	0.44989	98	47.08834	0.98641
49	23.81423	0.45280	99	52.63478	0.99714
50	24.15124	0.46617	100	52.84011	0.99731

asymmetric leptokurtic distributions (in which case this test performs very poorly), this test is as good as the recommended Shapiro–Wilk procedure and is much easier to use. Section 4 describes the classical analysis of sample skewness and kurtosis coefficients and their generalizations as a test for normality.

Section 5 is included for those readers who would naturally think of using a standard test of goodness-of-fit to test for normality, such as the chi-square test or the Kolgomorov test.[1] We point out the problems of employing these tests, especially when the parameters of the normal distribution are unknown, and finally recommend a preferred procedure of this goodness-of-fit genre, the Durbin test. Finally, Section 6 summarizes the evaluative research on all the procedures given in this chapter.

We will illustrate each of the tests for normality using a random sample of 100 observations[2] from the gamma distribution

$$f_{\mathbf{x}}(x) = \frac{x^{\alpha-1}}{\Gamma(\alpha)\beta^{\alpha}} e^{-x/\beta}.$$

In this distribution, $\mathscr{E}\mathbf{x} = \alpha\beta$ and $\mathscr{V}\mathbf{x} = \alpha\beta^2$, and so we will take as our null hypothesis that the underlying distribution is $N(\alpha\beta, \alpha\beta^2)$. The data are given in Table 1, based on $\alpha = \beta = 5$, so that $\mathscr{E}\mathbf{x} = 25$ and $\mathscr{V}\mathbf{x} = 125$. For these data, $\bar{x} = 25.4439$ and $s^2 = 102.47918$.

1. Normal Plots

Prescription

The normal cumulative distribution function (sometimes called the normal ogive) has the characteristic S-like shape given in Figure 1. If we were to plot this curve on the special graph paper given in Figure 2, this curve would appear as a straight line. Figure 3, which plots the cumulative distribution of 1,000 randomly selected observations from a normal distribution with $\mu = 25$ and $\sigma^2 = 125$, illustrates this. A visual check of normality then would consist of plotting the sample cumulative distribution function on such paper and eyeballing the plot to see how deviant it is from a straight line. If one standardizes the observations by subtracting μ from each observation and dividing the result by σ, then a plot of these standardized data on such graph paper will fall on the 45° line. Such a plot of our 1,000 normal observations, where μ is replaced by $\bar{x} = 24.63$ and σ by $s = 11.305$, is given as Figure 4.

Let the observations be ordered so that $\mathbf{x}_1 \leq \mathbf{x}_2 \leq \cdots \leq \mathbf{x}_n$. Then the sample cumulative distribution function, sometimes called the empirical dis-

[1] This test is sometimes mistakenly called the Kolgomorov–Smirnov test—that test compares two empirical distributions and tests whether they came from the same underlying population whose distribution function is unspecified, whereas here we test whether an empirical distribution comes from a population with a specific distribution function, namely the normal distribution function.

[2] In Section 1 we select 50 observations to illustrate the various procedures therein, due to limitations of tables for facilitating these procedures.

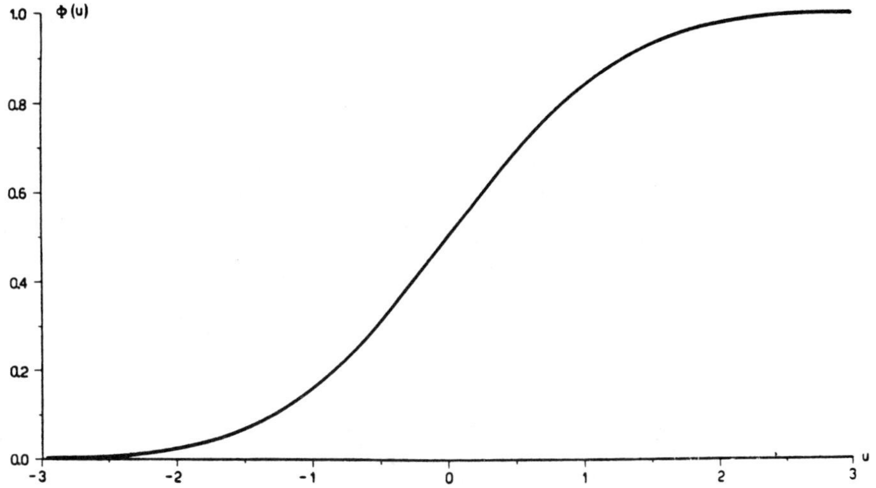

Figure 1. Cumulative Normal Distribution.

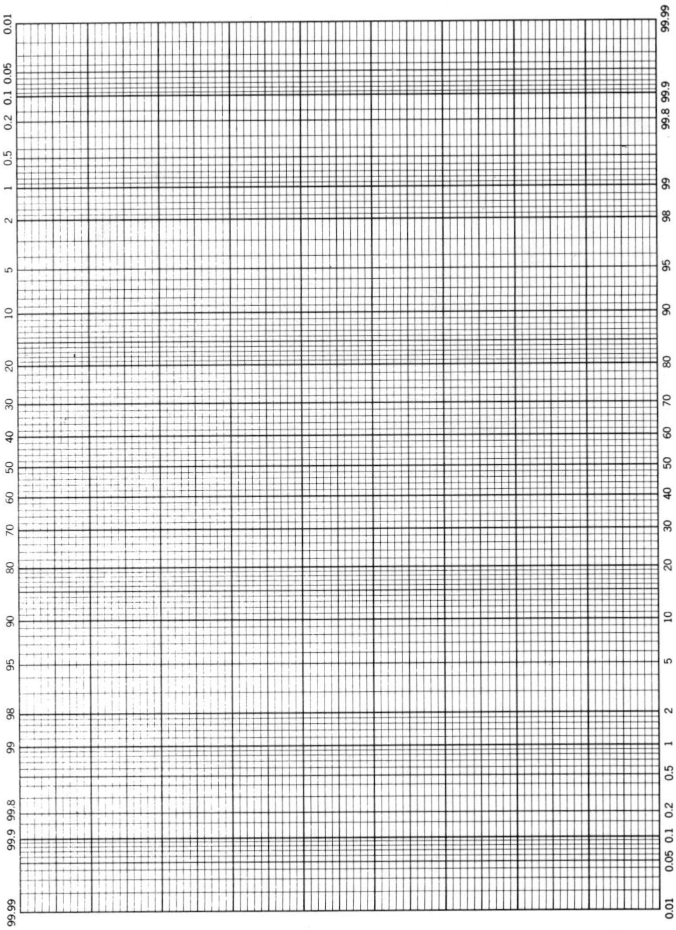

Figure 2. Normal Probability Paper.

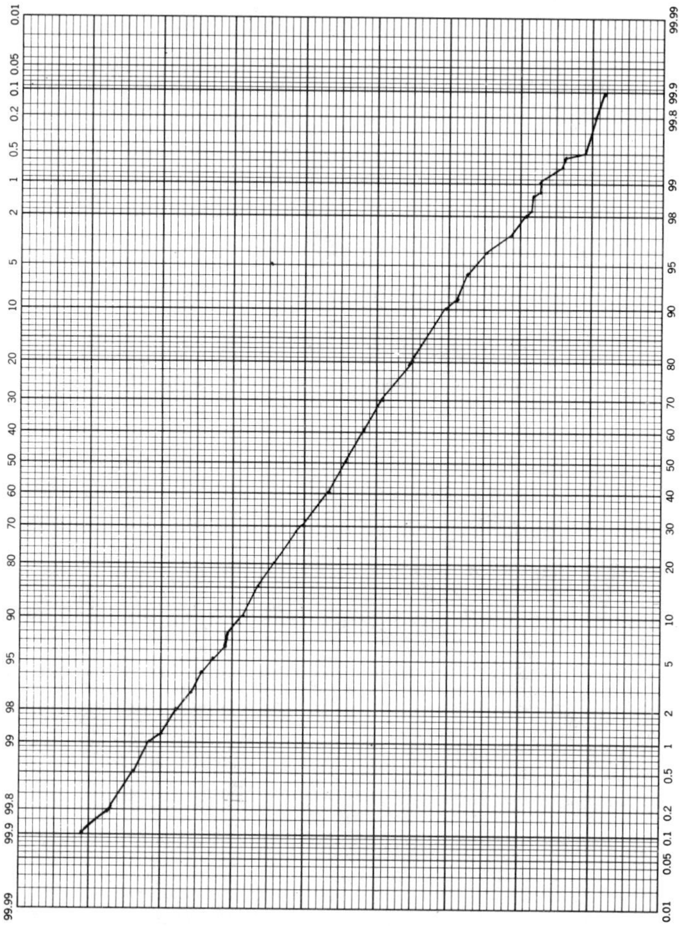

Figure 3. Plot of 1,000 Independent $N(25, 125)$ Observations on Normal Probability Paper.

tribution function $\mathbf{F}_n(x)$, is a step function with values

$$\mathbf{F}_n(x) = \begin{cases} 0, & x < \mathbf{x}_1, \\ i/n, & \mathbf{x}_i \le x < \mathbf{x}_{i+1}, \qquad i = 1, \ldots, n-1. \\ 1, & \mathbf{x}_n \le x, \end{cases}$$

Thus $\mathbf{x}_i$ is the (i/n)th quantile of the empirical distribution function. Let q_i denote the (i/n)th quantile of the $N(0, 1)$ distribution. Then a plot of the pairs $(\mathbf{x}_i, q_i)$ for $i = 1, \ldots, n$ is what is accomplished by using this graph paper; that is, the virtue of the paper is that it is unnecessary to look up the n values of the q_i in order to create the plot.

One can intuitively understand what this graphical procedure tries to accomplish by the following argument. First, suppose we had only one obser-

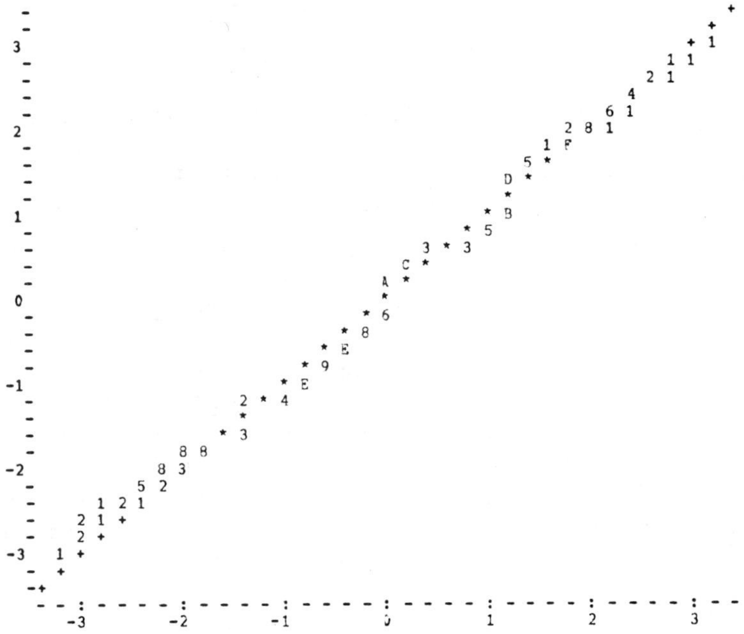

Figure 4. IDA-Generated Standardized Probability Plot of 1,000 Independent $N(25, 125)$ Observations.

vation from a $N(0, 1)$ distribution. Then we would expect that observation to be $x = 0$, i.e., $x = \Phi^{-1}(0.5)$, the value of x associated with cumulative normal probability equal to 1/2. Now suppose we had three observations from a $N(0, 1)$ distribution. What values of x do we expect them to be? Intuitively, they should be $\Phi^{-1}(0.25)$, $\Phi^{-1}(0.5)$, and $\Phi^{-1}(0.75)$, in that we expect the cumulative probabilities associated with these three observations to be three uniformly spaced probabilities. In general, with n observations from a $N(0, 1)$ distribution, we expect the observations to equal $\Phi^{-1}(i/(n + 1))$. So if we plot the pairs $(x_i, \Phi^{-1}(i/(n + 1))$ we would expect the n pairs of points to lie on a straight line.

One can formalize and generalize the eyeball procedure described above as follows:

(1) select a set of p_i;
(2) determine the values of the $q_i = \Phi^{-1}(p_i)$;
(3) calculate the slope of the regression of the q_i on the x_i; and
(4) test whether the slope is significantly different from 1.

Section 3, and in particular the Filliben procedure described therein, does this. Critical to this generalization is the definition of the set of p_i. If the empirical distribution function is used, $p_i = i/n$; if the intuitive argument given above is used, $p_i = i/(n + 1)$. Other choices of p_i are given below. But, whatever the choice of the set of p_i, the eyeball procedure is the same—a graphic comparison of the x_i with $q_i = \Phi^{-1}(p_i)$.

Theoretical Background

There are two ways of using a set of points $(x, F_{\mathbf{x}}(x))$ for comparison purposes. The usual way of viewing the pairs is to treat x as a *quantile* and $F_{\mathbf{x}}(x)$ as the *percentile* associated with the given x. Thus, given two distribution functions $F(x)$ and $G(x)$, one can prespecify the set of quantiles $x_1, \ldots, x_n$ at which $F(x)$ and $G(x)$ are to be evaluated and then plot the pairs $(F(x_i), G(x_i))$. Usually, the prespecified x's are the observations themselves, and $F(x)$ is the empirical distribution function, so that $F(x_i) = i/n$ (if the x's are ordered), and the pairs plotted are $(i/n, G(x_i))$. Such a graph is called a $p - p$ *plot* (see Wilk and Gnanadesikan [1968]) and is used in the Kolmogorov test described in Section 5(b).

Another way is to fix $F_{\mathbf{x}}(x)$ at some value p and find the quantile $x = F_{\mathbf{x}}^{-1}(p)$ associated with the percentile p. Then we can prespecify the set of percentiles $p_1, \ldots, p_n$ at which $F^{-1}(p)$ and $G^{-1}(p)$ are to be evaluated and plot the pairs $(F^{-1}(p_i), G^{-1}(p_i))$. Usually, $p_i = i/n$ and F is the empirical distribution function, so that $F^{-1}(p_i) = x_i$ and the pairs plotted are, as described in our use of normal probability paper, $(x_i, G^{-1}(i/n))$. Such a graph is called a $q - q$ *plot* (Wilk and Gnanadesikan [1968]). A good discussion of $q - q$ plots is Chambers *et al.* [1983].

One problem with this $q - q$ plot is that the point $(F^{-1}(1), G^{-1}(1))$ will not fit on the graph paper if G has infinite range. Many suggestions for alternative choices of p_i have been made to circumvent this difficulty, the most commonly cited being:

(1) $p_i = i/(n + 1)$; (2) $p_i = (i - 1/2)/n$; (3) $p_i = (i - 3/8)/(n + 1/4)$.

If, as is typical, $F(x)$ is the empirical distribution function, then $F^{-1}(p_i)$ will equal $F^{-1}(i/n) = x_i$ for each of these choices of p_i.

Kimball [1960] has suggested that the best choice of p_i, to meet the objective of testing whether sample data indicate that the underlying population distribution is normal, is alternative (3) above, due to Blom [1958]. Thus it is suggested that the pair of points to be plotted is $(x_i, \Phi^{-1}((i - 3/8)/(n + 1/4)))$ rather than the pair $(x_i, \Phi^{-1}(i/n))$ as plotted on the graph paper of Figure 1.

Chambers *et al.* [1983] provide a method of estimating a confidence interval for each point in a $q - q$ plot. Essentially, it is based on the asymptotic distribution of the order statistics from a normal distribution. Michael [1983] uses variance stabilizing transformations to obtain a new kind of plot, the "stabilized probability plot", which he uses to draw acceptance regions for the hypothesis that the data came from the postulated distribution.

2. Regression Procedures

(a) Shapiro–Wilk Statistic

Prescription

As noted in Section 1, one can formalize the eyeballing of the plot of the x_i's against the $q_i = \Phi^{-1}(i/n)$, i.e., against the (i/n)th quantiles of the $N(0, 1)$ distribution by regressing the x_i against the q_i. Since the ordered x_i are not independent, standard regression computations are invalid. Shapiro and Wilk [1965] have developed the appropriate regression-based statistic for testing for normality, and that procedure consists of calculating

$$\mathbf{w} = \frac{(A'\mathbf{X})^2}{(n-1)\mathbf{s}^2},$$

where $\mathbf{X}$ is the vector of ordered $\mathbf{x}_i$'s, $\mathbf{s}^2$ is the usual estimator of σ^2, namely $\mathbf{s}^2 = \sum_{i=1}^{n} (\bar{\mathbf{x}}_i - \bar{\mathbf{x}})^2/(n-1)$, and A is a vector of tabulated coefficients especially derived to be used in computing this statistic.

To use this procedure, one needs both tables of the elements of A and also tables of the sampling distribution of $\mathbf{w}$. These are reproduced from Shapiro and Wilk [1965] as Tables 2 and 3 herein. We note by symmetry that $a_i = -a_{n-i+1}$, the elements tabulated in Table 2.

Table 2. Values of a_i for the Shapiro–Wilk Procedure.

i \ n	2	3	4	5	6	7	8	9	10
1	0·7071	0·7071	0·6872	0·6646	0·6431	0·6233	0·6052	0·5888	0·5739
2	—	·0000	·1677	·2413	·2806	·3031	·3164	·3244	·3291
3	—	—	—	·0000	·0875	·1401	·1743	·1976	·2141
4	—	—	—	—	—	·0000	·0561	·0947	·1224
5	—	—	—	—	—	—	—	·0000	·0399

i \ n	11	12	13	14	15	16	17	18	19	20
1	0·5601	0·5475	0·5359	0·5251	0·5150	0·5056	0·4968	0·4886	0·4808	0·4734
2	·3315	·3325	·3325	·3318	·3306	·3290	·3273	·3253	·3232	·3211
3	·2260	·2347	·2412	·2460	·2495	·2521	·2540	·2553	·2561	·2565
4	·1429	·1586	·1707	·1802	·1878	·1939	·1988	·2027	·2059	·2085
5	·0695	·0922	·1099	·1240	·1353	·1447	·1524	·1587	·1641	·1686
6	0·0000	0·0303	0·0539	0·0727	0·0880	0·1005	0·1109	0·1197	0·1271	0·1334
7	—	—	·0000	·0240	·0433	·0593	·0725	·0837	·0932	·1013
8	—	—	—	—	·0000	·0196	·0359	·0496	·0612	·0711
9	—	—	—	—	—	—	·0000	·0163	·0303	·0422
10	—	—	—	—	—	—	—	—	·0000	·0140

i \ n	21	22	23	24	25	26	27	28	29	30
1	0·4643	0·4590	0·4542	0·4493	0·4450	0·4407	0·4366	0·4328	0·4291	0·4254
2	·3185	·3156	·3126	·3098	·3069	·3043	·3018	·2992	·2968	·2944
3	·2578	·2571	·2563	·2554	·2543	·2533	·2522	·2510	·2499	·2487
4	·2119	·2131	·2139	·2145	·2148	·2151	·2152	·2151	·2150	·2148
5	·1736	·1764	·1787	·1807	·1822	·1836	·1848	·1857	·1864	·1870
6	0·1399	0·1443	0·1480	0·1512	0·1539	0·1563	0·1584	0·1601	0·1616	0·1630
7	·1092	·1150	·1201	·1245	·1283	·1316	·1346	·1372	·1395	·1415
8	·0804	·0878	·0941	·0997	·1046	·1089	·1128	·1162	·1192	·1219
9	·0530	·0618	·0696	·0764	·0823	·0876	·0923	·0965	·1002	·1036
10	·0263	·0368	·0459	·0539	·0610	·0672	·0728	·0778	·0822	·0862
11	0·0000	0·0122	0·0228	0·0321	0·0403	0·0476	0·0540	0·0598	0·0650	0·0697
12	—	—	·0000	·0107	·0200	·0284	·0358	·0424	·0483	·0537
13	—	—	—	—	·0000	·0094	·0178	·0253	·0320	·0381
14	—	—	—	—	—	—	·0000	·0084	·0159	·0227
15	—	—	—	—	—	—	—	—	·0000	·0076

(continued)

Table 2 (*continued*)

i \ n	31	32	33	34	35	36	37	38	39	40
1	0·4220	0·4188	0·4156	0·4127	0·4096	0·4068	0·4040	0·4015	0·3989	0·3964
2	·2921	·2898	·2876	·2854	·2834	·2813	·2794	·2774	·2755	·2737
3	·2475	·2463	·2451	·2439	·2427	·2415	·2403	·2391	·2380	·2368
4	·2145	·2141	·2137	·2132	·2127	·2121	·2116	·2110	·2104	·2098
5	·1874	·1878	·1880	·1882	·1883	·1883	·1883	·1881	·1880	·1878
6	0·1641	0·1651	0·1660	0·1667	0·1673	0·1678	0·1683	0·1686	0·1689	0·1691
7	·1433	·1449	·1463	·1475	·1487	·1496	·1505	·1513	·1520	·1526
8	·1243	·1265	·1284	·1301	·1317	·1331	·1344	·1356	·1366	·1376
9	·1066	·1093	·1118	·1140	·1160	·1179	·1196	·1211	·1225	·1237
10	·0899	·0931	·0961	·0988	·1013	·1036	·1056	·1075	·1092	·1108
11	0·0739	0·0777	0·0812	0·0844	0·0873	0·0900	0·0924	0·0947	0·0967	0·0986
12	·0585	·0629	·0669	·0706	·0739	·0770	·0798	·0824	·0848	·0870
13	·0435	·0485	·0530	·0572	·0610	·0645	·0677	·0706	·0733	·0759
14	·0289	·0344	·0395	·0441	·0484	·0523	·0559	·0592	·0622	·0651
15	·0144	·0206	·0262	·0314	·0361	·0404	·0444	·0481	·0515	·0546
16	0·0000	0·0068	0·0131	0·0187	0·0239	0·0287	0·0331	0·0372	0·0409	0·0444
17	—	—	·0000	·0062	·0119	·0172	·0220	·0264	·0305	·0343
18	—	—	—	—	·0000	·0057	·0110	·0158	·0203	·0244
19	—	—	—	—	—	—	·0000	·0053	·0101	·0146
20	—	—	—	—	—	—	—	—	·0000	·0049

i \ n	41	42	43	44	45	46	47	48	49	50
1	0·3940	0·3917	0·3894	0·3872	0·3850	0·3830	0·3808	0·3789	0·3770	0·3751
2	·2719	·2701	·2684	·2667	·2651	·2635	·2620	·2604	·2589	·2574
3	·2357	·2345	·2334	·2323	·2313	·2302	·2291	·2281	·2271	·2260
4	·2091	·2085	·2078	·2072	·2065	·2058	·2052	·2045	·2038	·2032
5	·1876	·1874	·1871	·1868	·1865	·1862	·1859	·1855	·1851	·1847
6	0·1693	0·1694	0·1695	0·1695	0·1695	0·1695	0·1695	0·1693	0·1692	0·1691
7	·1531	·1535	·1539	·1542	·1545	·1548	·1550	·1551	·1553	·1554
8	·1384	·1392	·1398	·1405	·1410	·1415	·1420	·1423	·1427	·1430
9	·1249	·1259	·1269	·1278	·1286	·1293	·1300	·1306	·1312	·1317
10	·1123	·1136	·1149	·1160	·1170	·1180	·1189	·1197	·1205	·1212
11	0·1004	0·1020	0·1035	0·1049	0·1062	0·1073	0·1085	0·1095	0·1105	0·1113
12	·0891	·0909	·0927	·0943	·0959	·0972	·0986	·0998	·1010	·1020
13	·0782	·0804	·0824	·0842	·0860	·0876	·0892	·0906	·0919	·0932
14	·0677	·0701	·0724	·0745	·0765	·0783	·0801	·0817	·0832	·0846
15	·0575	·0602	·0628	·0651	·0673	·0694	·0713	·0731	·0748	·0764
16	0·0476	0·0506	0·0534	0·0560	0·0584	0·0607	0·0628	0·0648	0·0667	0·0685
17	·0379	·0411	·0442	·0471	·0497	·0522	·0546	·0568	·0588	·0608
18	·0283	·0318	·0352	·0383	·0412	·0439	·0465	·0489	·0511	·0532
19	·0188	·0227	·0263	·0296	·0328	·0357	·0385	·0411	·0436	·0459
20	·0094	·0136	·0175	·0211	·0245	·0277	·0307	·0335	·0361	·0386
21	0·0000	0·0045	0·0087	0·0126	0·0163	0·0197	0·0229	0·0259	0·0288	0·0314
22		—	·0000	·0042	·0081	·0118	·0153	·0185	·0215	·0244
23	—	—	—	—	·0000	·0039	·0076	·0111	·0143	·0174
24	—	—	—	—	—	—	·0000	·0037	·0071	·0104
25	—	—	—	—	—	—	—	—	·0000	·0035

SOURCE: Shapiro and Wilk [1965]. Reprinted from *Biometrika*, Vol. 52, with the permission of the Biometrika Trustees.

Table 3. Percentage Points of the Shapiro–Wilk Statistic.

n	**0·01**	**0·02**	**0·05**	**0·10**	**0·50**	**0·90**	**0·95**	**0·98**	**0·99**
3	0·753	0·756	0·767	0·789	0·959	0·998	0·999	1·000	1·000
4	·687	·707	·748	·792	·935	·987	·992	·996	·997
5	·686	·715	·762	·806	·927	·979	·986	·991	·993
6	0·713	0·743	0·788	0·826	0·927	0·974	0·981	0·986	0·989
7	·730	·760	·803	·838	·928	·972	·979	·985	·988
8	·749	·778	·818	·851	·932	·972	·978	·984	·987
9	·764	·791	·829	·859	·935	·972	·978	·984	·986
10	·781	·806	·842	·869	·938	·972	·978	·983	·986
11	0·792	0·817	0·850	0·876	0·940	0·973	0·979	0·984	0·986
12	·805	·828	·859	·883	·943	·973	·979	·984	·986
13	·814	·837	·866	·889	·945	·974	·979	·984	·986
14	·825	·846	·874	·895	·947	·975	·980	·984	·986
15	·835	·855	·881	·901	·950	·975	·980	·984	·987
16	0·844	0·863	0·887	0·906	0·952	0·976	0·981	0·985	0·987
17	·851	·869	·892	·910	·954	·977	·981	·985	·987
18	·858	·874	·897	·914	·956	·978	·982	·986	·988
19	·863	·879	·901	·917	·957	·978	·982	·986	·988
20	·868	·884	·905	·920	·959	·979	·983	·986	·988
21	0·873	0·888	0·908	0·923	0·960	0·980	0·983	0·987	0·989
22	·878	·892	·911	·926	·961	·980	·984	·987	·989
23	·881	·895	·914	·928	·962	·981	·984	·987	·989
24	·884	·898	·916	·930	·963	·981	·984	·987	·989
25	·888	·901	·918	·931	·964	·981	·985	·988	·989
26	0·891	0·904	0·920	0·933	0·965	0·982	0·985	0·988	0·989
27	·894	·906	·923	·935	·965	·982	·985	·988	·990
28	·896	·908	·924	·936	·966	·982	·985	·988	·990
29	·898	·910	·926	·937	·966	·982	·985	·988	·990
30	·900	·912	·927	·939	·967	·983	·985	·988	·900
31	0·902	0·914	0·929	0·940	0·967	0·983	0·986	0·988	0·990
32	·904	·915	·930	·941	·968	·983	·986	·988	·990
33	·906	·917	·931	·942	·968	·983	·986	·989	·990
34	·908	·919	·933	·943	·969	·983	·986	·989	·990
35	·910	·920	·934	·944	·969	·984	·986	·989	·990
36	0·912	0·922	0·935	0·945	0·970	0·984	0·986	0·989	0·990
37	·914	·924	·936	·946	·970	·984	·987	·989	·990
38	·916	·925	·938	·947	·971	·984	·987	·989	·990
39	·917	·927	·939	·948	·971	·984	·987	·989	·991
40	·919	·928	·940	·949	·972	·985	·987	·989	·991
41	0·920	0·929	0·941	0·950	0·972	0·985	0·987	0·989	0·991
42	·922	·930	·942	·951	·972	·985	·987	·989	·991
43	·923	·932	·943	·951	·973	·985	·987	·990	·991
44	·924	·933	·944	·952	·973	·985	·987	·990	·991
45	·926	·934	·945	·953	·973	·985	·988	·990	·991
46	0·927	0·935	0·945	0·953	0·974	0·985	0·988	0·990	0·991
47	·928	·936	·946	·954	·974	·985	·988	·990	·991
48	·929	·937	·947	·954	·974	·985	·988	·990	·991
49	·929	·937	·947	·955	·974	·985	·988	·990	·991
50	·930	·938	·947	·955	·974	·985	·988	·990	·991

SOURCE: Shapiro and Wilk [1965]. Reprinted from *Biometrika*, Vol. 52, with the permission of the Biometrika Trustees.

To illustrate the use of this procedure, consider as a random sample 50 of our 100 gamma distributed observations, as given in the first column of Table 4, labeled X. The second column contains the appropriate vector A for $n = 50$ taken from Table 2, and the third column contains the elements $a_i x_i$. Thus, using the IDA MEAN command we find (see Table 4A) that $s = 10.9707$ (so that $s^2 = 120.356$), and that $A'X/50 = 1.48449$, so that $A'X = 74.2245$ and thus $\mathbf{w} = 0.934$. Comparing this with the 95% point for $n = 50$ given in Table 3 we find that it is less than 0.947 and so we reject the hypothesis of normality.

This statistic is highly recommended, based on the Shapiro–Wilk–Chen study (see Section 6), as it had best power for testing normality against all classes of alternative distributions.

Shapiro and Wilk [1968] provide an approximation for the distribution of $\mathbf{w}$, namely a set of parameters γ, δ, ε such that $\gamma + \delta \log[(\mathbf{w} - \varepsilon)/(1 - \mathbf{w})]$ has a $N(0, 1)$ distribution. These parameters are tabulated for $n = 3(1)50$ in that paper.

Because of the difficulty in developing tables of A, and because tables of the q_i are readily available for an extensive set of values of n,[3] Shapiro and Francia [1972] suggested the statistic

$$\mathbf{w}' = \frac{(Q'\mathbf{X})^2}{(n-1)Q'Q\mathbf{s}^2}.$$

A table of the sampling distribution of $\mathbf{w}'$ is given as Table 5 herein.

Royston [1982a] provides a good approximation for $A^* = Q'V^{-1}$ (and hence for A) for $7 \le n \le 2{,}000$ given that one can calculate (or look up in tables) the values of the q_i. His approximations are:

$$a_i^* = \begin{cases} 2q_i, & i = 2, 3, \ldots, n-1, \\ \left[\dfrac{a_1^2}{1 - 2a_1^2} \displaystyle\sum_{j=2}^{n-1} a_j^{*2}\right]^{1/2}, & i = 1 \text{ and } n, \end{cases}$$

where

$$a_1^2 = a_n^2 = \begin{cases} g(n-1), & n \le 20, \\ g(n), & n > 20, \end{cases}$$

and

$$g(n) = \frac{\Gamma(\frac{1}{2}(n+1))}{\sqrt{2}\Gamma\left(\dfrac{n}{2}+1\right)} \approx \left[\frac{6n+7}{6n+13}\right]\left\{\frac{e}{n+2}\left[\frac{n+1}{n+2}\right]^{n-2}\right\}^{1/2}.$$

[3] See Harter [1969] for tables for $n = 2(1)100$, 105, 108, 112, 120, 126, 128, 135, 140, 144, 147, 150, 160, 162, 168, 175, 180, 189, 192, 196, 200, 210, 216, 224, 225, 240, 243, 245, 250, 252, 256, 270, 280, 288, 294, 300, 315, 320, 324, 336, 343, 350, 360, 375, 378, 384, 392, 400.

Table 4. Computations for Regression Procedures.

Row	X	A	AX	QH	QHX
1	6.48875	-0.37510	-2.43393	-2.24907	-14.59365
2	7.70804	-0.25740	-1.98405	-1.85487	-14.29741
3	8.46150	-0.22600	-1.91230	-1.62863	-13.78065
4	12.75674	-0.20320	-2.59217	-1.46374	-18.67255
5	13.08055	-0.18470	-2.41598	-1.33109	-17.41139
6	13.43281	-0.16910	-2.27149	-1.21846	-16.36734
7	13.93881	-0.15540	-2.16609	-1.11948	-15.60422
8	15.03517	-0.14300	-2.15003	-1.03042	-15.49254
9	15.44101	-0.13170	-2.03358	-0.94887	-14.65151
10	15.75355	-0.12120	-1.90933	-0.87321	-13.75616
11	15.78501	-0.11130	-1.75687	-0.80225	-12.66352
12	16.65918	-0.10200	-1.69924	-0.73513	-12.24666
13	16.71427	-0.09320	-1.55777	-0.67117	-11.21812
14	17.42898	-0.08460	-1.47449	-0.60986	-10.62924
15	19.62610	-0.07640	-1.49943	-0.55077	-10.80947
16	19.78951	-0.06850	-1.35558	-0.49354	-9.76691
17	19.96001	-0.06080	-1.21357	-0.43789	-8.74029
18	20.04533	-0.05320	-1.06641	-0.38357	-7.68879
19	20.14921	-0.04590	-0.92485	-0.33036	-6.65649
20	20.62760	-0.03860	-0.79623	-0.27807	-5.73592
21	20.99996	-0.03140	-0.65940	-0.22653	-4.75712
22	21.00931	-0.02440	-0.51263	-0.17559	-3.68902
23	22.33860	-0.01740	-0.38869	-0.12511	-2.79478
24	22.87242	-0.01040	-0.23787	-0.07494	-1.71406
25	22.94075	-0.00350	-0.08029	-0.02496	-0.57260
26	23.23921	0.00350	0.08134	0.02496	0.58005
27	23.81423	0.01040	0.24767	0.07494	1.78464
28	24.15124	0.01740	0.42023	0.12511	3.02156
29	24.25881	0.02440	0.59191	0.17559	4.25960
30	25.24047	0.03140	0.79255	0.22653	5.71772
31	25.37080	0.03860	0.97931	0.27807	7.05486
32	26.48096	0.04590	1.21548	0.33036	8.74825
33	26.99213	0.05320	1.43598	0.38357	10.35337
34	28.29106	0.06080	1.72010	0.43789	12.38837
35	29.20037	0.06850	2.00023	0.49354	14.41155
36	29.56230	0.07640	2.25856	0.55077	16.28203
37	30.30640	0.08460	2.56392	0.60986	18.48266
38	30.57425	0.09320	2.84952	0.67117	20.52052
39	31.85266	0.10200	3.24897	0.73513	23.41585
40	32.70566	0.11130	3.64014	0.80225	26.23812
41	33.31204	0.12120	4.03742	0.87321	29.08841
42	33.35007	0.13170	4.39220	0.94887	31.64488
43	33.43215	0.14300	4.78080	1.03042	34.44916
44	34.05206	0.15540	5.29169	1.11948	38.12060
45	40.46576	0.16910	6.84276	1.21846	49.30591
46	45.79899	0.18470	8.45907	1.33109	60.96258
47	46.53636	0.20320	9.45619	1.46374	68.11713
48	47.08834	0.22600	10.64196	1.62863	76.68948
49	52.63478	0.25740	13.54819	1.85487	97.63067
50	52.84011	0.37510	19.82033	2.24907	118.84111

(continued)

Table 4 (*continued*)

	QHQH	*U*	*Q*	*QX*	*QQ*
1	5.05832	0.01244	-2.24380	-14.55944	5.03463
2	3.44054	0.03234	-1.84792	-14.24386	3.41482
3	2.65244	0.05224	-1.62389	-13.74058	2.63703
4	2.14253	0.07214	-1.46034	-18.62918	2.13259
5	1.77180	0.09204	-1.32852	-17.37780	1.76497
6	1.48464	0.11194	-1.21642	-16.33999	1.47969
7	1.25324	0.13184	-1.11781	-15.58093	1.24950
8	1.06177	0.15174	-1.02900	-15.47120	1.05884
9	0.90035	0.17164	-0.94764	-14.63246	0.89801
10	0.76250	0.19154	-0.87210	-13.73870	0.76056
11	0.64361	0.21144	-0.80124	-12.64757	0.64198
12	0.54042	0.23134	-0.73419	-12.23098	0.53903
13	0.45047	0.25124	-0.67029	-11.20339	0.44929
14	0.37193	0.27114	-0.60902	-10.61459	0.37090
15	0.30335	0.29104	-0.54996	-10.79359	0.30246
16	0.24358	0.31095	-0.49277	-9.75166	0.24282
17	0.19175	0.33085	-0.43715	-8.72562	0.19110
18	0.14713	0.35075	-0.38287	-7.67480	0.14659
19	0.10914	0.37065	-0.32971	-6.64336	0.10871
20	0.07732	0.39055	-0.27748	-5.72365	0.07699
21	0.05132	0.41045	-0.22601	-4.74611	0.05108
22	0.03083	0.43035	-0.17515	-3.67973	0.03068
23	0.01565	0.45025	-0.12476	-2.78698	0.01557
24	0.00562	0.47015	-0.07471	-1.70888	0.00558
25	0.00062	0.49005	-0.02488	-0.57078	0.00062
26	0.00062	0.50995	0.02488	0.57821	0.00062
27	0.00562	0.52985	0.07471	1.77925	0.00558
28	0.01565	0.54975	0.12476	3.01313	0.01557
29	0.03083	0.56965	0.17515	4.24888	0.03068
30	0.05132	0.58955	0.22601	5.70449	0.05108
31	0.07732	0.60945	0.27748	7.03977	0.07699
32	0.10914	0.62935	0.32971	8.73099	0.10871
33	0.14713	0.64925	0.38287	10.33454	0.14659
34	0.19175	0.66915	0.43715	12.36758	0.19110
35	0.24358	0.68905	0.49277	14.38905	0.24282
36	0.30335	0.70896	0.54996	16.25812	0.30246
37	0.37193	0.72886	0.60902	18.45719	0.37090
38	0.45047	0.74876	0.67029	20.49358	0.44929
39	0.54042	0.76866	0.73419	23.38587	0.53903
40	0.64361	0.78856	0.80124	26.20507	0.64198
41	0.76250	0.80846	0.87210	29.05149	0.76056
42	0.90035	0.82836	0.94764	31.60374	0.89801
43	1.06177	0.84826	1.02900	34.40171	1.05884
44	1.25324	0.86816	1.11781	38.06371	1.24950
45	1.48464	0.88806	1.21642	49.22352	1.47969
46	1.77180	0.90796	1.32852	60.84498	1.76497
47	2.14253	0.92786	1.46034	67.95891	2.13259
48	2.65244	0.94776	1.62389	76.46647	2.63703
49	3.44054	0.96766	1.84792	97.26500	3.41482
50	5.05832	0.98756	2.24380	118.56252	5.03463

(*continued*)

Table 4 (*continued*)

	U^*	Q^*	D	DX
1	0.01377	-2.20432	-24.50000	-158.97437
2	0.03341	-1.83337	-23.50000	-181.13894
3	0.05326	-1.61439	-22.50000	-190.38375
4	0.07312	-1.45326	-21.50000	-274.26991
5	0.09297	-1.32290	-20.50000	-268.15128
6	0.11283	-1.21178	-19.50000	-261.93979
7	0.13268	-1.11388	-18.50000	-257.86798
8	0.15254	-1.02562	-17.50000	-263.11547
9	0.17239	-0.94469	-16.50000	-254.77666
10	0.19225	-0.86952	-15.50000	-244.18002
11	0.21210	-0.79896	-14.50000	-228.88264
12	0.23196	-0.73218	-13.50000	-224.89893
13	0.25181	-0.66851	-12.50000	-208.92837
14	0.27167	-0.60744	-11.50000	-200.43327
15	0.29152	-0.54857	-10.50000	-206.07405
16	0.31138	-0.49155	-9.50000	-188.00035
17	0.33123	-0.43609	-8.50000	-169.66009
18	0.35109	-0.38195	-7.50000	-150.33998
19	0.37094	-0.32893	-6.50000	-130.96987
20	0.39080	-0.27682	-5.50000	-113.45180
21	0.41065	-0.22548	-4.50000	-94.49982
22	0.43051	-0.17474	-3.50000	-73.53259
23	0.45036	-0.12447	-2.50000	-55.84650
24	0.47022	-0.07454	-1.50000	-34.30863
25	0.49007	-0.02482	-0.50000	-11.47037
26	0.50993	0.02482	0.50000	11.61960
27	0.52978	0.07454	1.50000	35.72134
28	0.54964	0.12447	2.50000	60.37810
29	0.56949	0.17474	3.50000	84.90584
30	0.58935	0.22548	4.50000	113.58212
31	0.60920	0.27682	5.50000	139.53940
32	0.62906	0.32893	6.50000	172.12624
33	0.64891	0.38195	7.50000	202.44098
34	0.66877	0.43609	8.50000	240.47401
35	0.68862	0.49155	9.50000	277.40351
36	0.70848	0.54857	10.50000	310.40415
37	0.72833	0.60744	11.50000	348.52360
38	0.74819	0.66851	12.50000	382.17812
39	0.76804	0.73218	13.50000	430.01091
40	0.78790	0.79896	14.50000	474.23207
41	0.80775	0.86952	15.50000	516.33662
42	0.82761	0.94469	16.50000	550.27615
43	0.84746	1.02562	17.50000	585.06262
44	0.86732	1.11388	18.50000	629.96311
45	0.88717	1.21178	19.50000	789.08231
46	0.90703	1.32290	20.50000	938.87929
47	0.92688	1.45326	21.50000	1000.53174
48	0.94674	1.61439	22.50000	1059.48764
49	0.96659	1.83337	23.50000	1236.91733
50	0.98623	2.20432	24.50000	1294.58269

Table 4A. Mean and Standard Deviation of Table 4 Entries.

Variable	Mean	Standard deviation
X	25.0119	10.9707
A	0.000000	0.142851
AX	1.48449	4.51301
QH	0.000000	0.983763
QHX	10.2760	30.6028
$QHQH$	0.948434	1.23345
U	0.500000	0.290097
Q	0.000000	0.981545
QX	10.2522	30.5277
QQ	0.944162	1.22672
U^*	0.500000	0.289420
Q^*	2.980232E − 08	0.974258
D	0.000000	14.5774
DX	148.771	428.437

Note:

X	Gamma data.
A	Coefficient for the Shapiro–Wilk statistics (see Table 2).
AX	$A * X$.
QH	Expected value of the normal order statistics.
QHX	$QH * X$.
$QHQH$	$QH * QH$.
U	$(I - \frac{3}{8})(N + \frac{1}{4})$.
Q	Blom approximation of $\Phi^{-1}(U)$: Expected value of normal order statistic.
QX	$Q * X$.
U^*	Median of the normal order statistic.
Q^*	$\Phi^{-1}(U^*)$.
D	Coefficient for the D'Agostino statistic (see Section 2(c)).
DX	$D * X$.

He also provides (Royston [1982b]) a computer program for calculating the q_i.

To continue with our example, the appropriate vector Q for $n = 50$, taken from Harter [1961], is in the fourth column of Table 4. The fifth column contains the $q_i x_i$, the sixth column contains the q_i^2's and so the MEAN command (see Table 4A) shows us that $Q'X/50 = 10.2760$ and $Q'Q/50 = 0.948434$, so that $w' = 0.9439$.

Weisberg and Bingham [1975] suggest an approximation to Q which obviates the necessity for special tables, namely

$$\tilde{q}_i = \Phi^{-1}\left(\frac{i - 3/8}{n + 1/4}\right),$$

Table 5. Empirical Percentage Points of the Approximate W' Test.

n	p										
	.01	.05	.10	.15	.20	.50	.80	.85	.90	.95	.99
35	0.919	0.943	0.952	0.956	0.964	0.976	0.982	0.985	0.987	0.989	0.992
50	.935	.953	.963	.968	.971	.981	.987	.988	.990	.991	.994
51	0.935	0.954	0.964	0.968	0.971	0.981	0.988	0.989	0.990	0.992	0.994
53	.938	.957	.964	.969	.972	.982	.988	.989	.990	.992	.994
55	.940	.958	.965	.971	.973	.983	.988	.990	.991	.992	.994
57	.944	.961	.966	.971	.974	.983	.989	.990	.991	.992	.994
59	.945	.962	.967	.972	.975	.983	.989	.990	.991	.992	.994
61	0.947	0.963	0.968	0.973	0.975	0.984	0.990	0.990	0.991	0.992	0.994
63	.947	.964	.970	.973	.976	.984	.990	.991	.992	.993	.994
65	.948	.965	.971	.974	.976	.985	.990	.991	.992	.993	.995
67	.950	.966	.971	.974	.977	.985	.990	.991	.992	.993	.995
69	.951	.966	.972	.976	.978	.986	.990	.991	.992	.993	.995
71	0.953	0.967	0.972	0.976	0.978	0.986	0.990	0.991	0.992	0.994	0.995
73	.956	.968	.973	.976	.979	.986	.991	.992	.993	.994	.995
75	.956	.969	.973	.976	.979	.986	.991	.992	.993	.994	.995
77	.957	.969	.974	.977	.980	.987	.991	.992	.993	.994	.996
79	.957	.970	.975	.978	.980	.987	.991	.992	.993	.994	.996
81	0.958	0.970	0.975	0.979	0.981	0.987	0.992	0.992	0.993	0.994	0.996
83	.960	.971	.976	.979	.981	.988	.992	.992	.993	.994	.996
85	.961	.972	.977	.980	.981	.988	.992	.992	.993	.994	.996
87	.961	.972	.977	.980	.982	.988	.992	.993	.994	.994	.996
89	.961	.972	.977	.981	.982	.988	.992	.993	.994	.995	.986
91	0.962	0.973	0.978	0.981	0.983	0.989	0.992	0.993	0.994	0.995	0.996
93	.963	.973	.979	.981	.983	.989	.992	.993	.994	.995	.996
95	.965	.974	.979	.981	.983	.989	.993	.993	.994	.995	.996
97	.965	.975	.979	.982	.984	.989	.993	.993	.994	.995	.996
99	.967	.976	.980	.982	.984	.989	.993	.994	.994	.995	.996

SOURCE: Shapiro and Francia [1972]. Reprinted from the *Journal of the American Statistical Association*, Vol. 67, with the permission of the ASA.

based on the approximation due to Blom [1958] cited earlier. They note[4] that the sampling distribution of the resulting **w**$'$ is quite close to that given in Table 3. The seventh column of Table 4 provides the values of $(i - 3/8)/(n + 1/4)$, and the Blom approximation values of the $\tilde{q}_i$ are given in the eighth column of Table 4. The ninth and tenth columns contain the $\tilde{q}_i x_i$ and $\tilde{q}_i^2$, respectively, and so the MEAN command shows us that $\tilde{Q}'X/50 = 10.2522$ and $\tilde{Q}'\tilde{Q}/50 = 0.944162$, so that the approximate value of w' is 0.9438.

Theoretical Background

Shapiro and Wilk [1965] recognized that the ordered $\mathbf{x}_i$ are not independent so that one cannot simply regress the x_i onto the q_i to estimate the slope of that regression. If the **x**'s are drawn independently from a $N(\mu, \sigma^2)$ distribu-

[4] As noted earlier, the q_i can be calculated using the computer program given in Royston [1982b].

tion, then the expected value of the ordered $\mathbf{x}_i$ is

$$\mathscr{E}\mathbf{x}_i = \mu + \sigma q_i.$$

Let v_{ij} denote the covariance between x_i and x_j, and V be the matrix of v_{ij}'s. Let Q be the vector of q_i's. Then in general the minimum variance linear (in the $\mathbf{x}$'s) unbiased estimators of μ and σ are

$$\hat{\boldsymbol{\mu}} = \frac{Q'V^{-1}[QE' - EQ']V^{-1}\mathbf{X}}{(E'V^{-1}E)(Q'V^{-1}Q) - (E'V^{-1}Q)^2},$$

$$\hat{\boldsymbol{\sigma}} = \frac{E'V^{-1}[EQ' - QE']V^{-1}\mathbf{X}}{(E'V^{-1}E)(Q'V^{-1}Q) - (E'V^{-1}Q)^2},$$

where $\mathbf{X}$ is the n vector of the $\mathbf{x}$'s and E is an n vector all of whose elements equal to 1 (see Lloyd [1952]). If the underlying distribution is symmetric, $E'V^{-1}Q = 0$, so that

$$\hat{\boldsymbol{\mu}} = \frac{(Q'V^{-1}Q)E'V^{-1}\mathbf{X}}{(E'V^{-1}E)(Q'V^{-1}Q)} = \frac{E'V^{-1}\mathbf{X}}{E'V^{-1}E},$$

$$\hat{\boldsymbol{\sigma}} = \frac{(E'V^{-1}E)Q'V^{-1}\mathbf{X}}{(E'V^{-1}E)(Q'V^{-1}Q)} = \frac{Q'V^{-1}\mathbf{X}}{Q'V^{-1}Q}.$$

Since we know that for the normal distribution the best linear unbiased estimator of μ is $\bar{\mathbf{x}} = \sum_{i=1}^{n} \mathbf{x}_i/n$, we see that $E'V^{-1}$ must equal E' and so $\hat{\boldsymbol{\mu}} = E'\mathbf{X}/n = \bar{\mathbf{x}}$.

Under normality the expected slope of the regression of the $\mathbf{x}_i$ on the q_i is σ, as seen above, and an unbiased estimator of σ is $\hat{\boldsymbol{\sigma}}$. The usual unbiased estimator of σ^2 is $\mathbf{s}^2 = \sum_{i=1}^{n} (\mathbf{x}_i - \bar{\mathbf{x}})^2/(n-1)$. Shapiro and Wilk defined as a test of normality the statistic

$$\mathbf{w} = \frac{(Q'V^{-1}Q)^2\hat{\boldsymbol{\sigma}}^2}{(n-1)(Q'V^{-1}V^{-1}Q)\mathbf{s}^2} = \frac{(A'\mathbf{X})^2}{(n-1)\mathbf{s}^2},$$

where the vector A is given by

$$A = \frac{V^{-1}Q}{(Q'V^{-1}V^{-1}Q)^{1/2}}$$

and is defined so that $A'A = 1$. In effect, this test statistic, except for a constant, compares two different estimators of σ^2, one valid only under normality and the other generally valid.

(b) Filliben Statistic

Prescription

Filliben [1975] suggested directly calculating $\mathbf{r}$, the correlation of the $\mathbf{x}$'s with the q_i^*, where q_i^* is the median of the sampling distribution of the ith order statistic from the normal distribution (in contrast to q_i, the mean of the sampling distribution of $\mathbf{x}_i$). To obtain the q_i^*, one must first obtain u_i^*, medians of the sampling distribution of the ith order statistic from the uniform distribution, for then it is easy to show that $q_i^* = \Phi^{-1}(u_i^*)$. There are no tables available of the u_i^*, but Filliben suggests the following approximation:

$$u_i^* = \begin{cases} 1 - 0.5^{1/n}, & i = 1, \\ (i - 0.3175)/(n + 0.365), & i = 2, \ldots, n-1, \\ 0.5^{1/n}, & i = n. \end{cases}$$

Table 6 gives the sampling distribution of **r**.

Continuing with our example, the eleventh column of Table 4 contains the u_i^2, and the q_i^2 are given in the twelfth column. We find that the correlation between the q_i^* and the x_i is 0.97151, which falls between the 1% and 2.5% point of the sampling distribution of **r**.

(c) D'Agostino Statistic

Prescription

All the above-described statistics are based on a numerator of the form

$$\sum_{i=1}^{h} d_{in}(\mathbf{x}_{n-i+1} - \mathbf{x}_i),$$

where $h = n/2$ or $h = (n-1)/2$ according to whether n is even or odd. D'Agostino [1971, 1972] developed a procedure of the same form, where d_{in} is simpler to calculate, namely,

$$d_{in} = (1/2)(n + 1) - i.$$

His statistic is

$$\mathbf{d} = \frac{\sum_{i=1}^{h} d_{in}(\mathbf{x}_{n-i+1} - \mathbf{x}_i)}{n^{3/2}\sqrt{n - 1}\mathbf{s}}.$$

He shows that $\mathbf{y} = \sqrt{n}(\mathbf{d} - 0.282095)/0.029986$ is asymptotically distributed as a $N(0, 1)$ variable. Tables of selected percentage points of the distribution of **y** are given in Table 7. Once again continuing our example, column 13 of Table 4 contains the d_{in}, and the $d_{in}x_i$ are given in column 7. From the MEAN command we see (Table 4A) that $\sum_{i=1}^{n} d_{in}x_i/50 = 148.771$, so that $\mathbf{d} = 0.27397$ and $\mathbf{y} = -1.916$.

Table 6. Percentage Points of the Filliben Statistic.

n	0.000	0.005	0.01	0.025	0.05	0.10	0.25	0.50	0.75	0.90	0.95	0.975	0.99	0.995
3	0.866	0.867	0.869	0.872	0.879	0.891	0.924	0.966	0.991	0.999	1.000	1.000	1.000	1.000
4	0.784	0.813	0.822	0.845	0.868	0.894	0.931	0.958	0.979	0.992	0.996	0.998	0.999	1.000
5	0.726	0.803	0.822	0.855	0.879	0.902	0.935	0.960	0.977	0.988	0.992	0.995	0.997	0.998
6	0.683	0.818	0.835	0.868	0.890	0.911	0.940	0.962	0.977	0.986	0.990	0.993	0.996	0.997
7	0.648	0.828	0.847	0.876	0.899	0.916	0.944	0.965	0.978	0.986	0.990	0.992	0.995	0.996
8	0.619	0.841	0.859	0.886	0.905	0.924	0.948	0.967	0.979	0.986	0.990	0.992	0.995	0.996
9	0.595	0.851	0.868	0.893	0.912	0.929	0.951	0.968	0.980	0.987	0.990	0.992	0.994	0.995
10	0.574	0.860	0.876	0.900	0.917	0.934	0.954	0.970	0.981	0.987	0.990	0.992	0.994	0.995
11	0.556	0.868	0.883	0.906	0.922	0.938	0.957	0.972	0.982	0.988	0.990	0.992	0.994	0.995
12	0.539	0.875	0.889	0.912	0.926	0.941	0.959	0.973	0.982	0.988	0.990	0.992	0.994	0.995
13	0.525	0.882	0.895	0.917	0.931	0.944	0.962	0.975	0.983	0.988	0.991	0.993	0.994	0.995
14	0.512	0.888	0.901	0.921	0.934	0.947	0.964	0.976	0.984	0.989	0.991	0.993	0.994	0.995
15	0.500	0.894	0.907	0.925	0.937	0.950	0.965	0.977	0.984	0.989	0.991	0.993	0.994	0.995
16	0.489	0.899	0.912	0.928	0.940	0.952	0.967	0.978	0.985	0.989	0.991	0.993	0.994	0.995
17	0.478	0.903	0.916	0.931	0.942	0.954	0.968	0.979	0.986	0.990	0.992	0.993	0.994	0.995
18	0.469	0.907	0.919	0.934	0.945	0.956	0.969	0.979	0.986	0.990	0.992	0.993	0.995	0.995
19	0.460	0.909	0.923	0.937	0.947	0.958	0.971	0.980	0.987	0.990	0.992	0.993	0.995	0.995
20	0.452	0.912	0.925	0.939	0.950	0.960	0.972	0.981	0.987	0.991	0.992	0.994	0.995	0.995
21	0.445	0.914	0.928	0.942	0.952	0.961	0.973	0.981	0.987	0.991	0.993	0.994	0.995	0.996
22	0.437	0.918	0.930	0.944	0.954	0.962	0.974	0.982	0.988	0.991	0.993	0.994	0.995	0.996
23	0.431	0.922	0.933	0.947	0.955	0.964	0.975	0.983	0.988	0.991	0.993	0.994	0.995	0.996
24	0.424	0.926	0.936	0.949	0.957	0.965	0.975	0.983	0.988	0.992	0.993	0.994	0.995	0.996
25	0.418	0.928	0.937	0.950	0.958	0.966	0.976	0.984	0.989	0.992	0.993	0.994	0.995	0.996
26	0.412	0.930	0.939	0.952	0.959	0.967	0.977	0.984	0.989	0.992	0.993	0.994	0.995	0.996
27	0.407	0.932	0.941	0.953	0.960	0.968	0.977	0.984	0.989	0.992	0.994	0.995	0.995	0.996
28	0.402	0.934	0.943	0.955	0.962	0.969	0.978	0.985	0.990	0.992	0.994	0.995	0.995	0.996
29	0.397	0.937	0.945	0.956	0.962	0.969	0.979	0.985	0.990	0.992	0.994	0.995	0.995	0.996
30	0.392	0.938	0.947	0.957	0.964	0.970	0.979	0.986	0.990	0.993	0.994	0.995	0.996	0.996

31	0.388	0.939	0.948	0.958	0.965	0.971	0.980	0.986	0.990	0.993	0.994	0.995	0.996	0.996
32	0.383	0.939	0.949	0.959	0.966	0.972	0.980	0.986	0.990	0.993	0.994	0.995	0.996	0.996
33	0.379	0.940	0.950	0.960	0.967	0.973	0.981	0.987	0.991	0.993	0.994	0.995	0.996	0.996
34	0.375	0.941	0.951	0.960	0.967	0.973	0.981	0.987	0.991	0.993	0.994	0.995	0.996	0.996
35	0.371	0.943	0.952	0.961	0.968	0.974	0.982	0.987	0.991	0.993	0.995	0.995	0.996	0.997
36	0.367	0.945	0.953	0.962	0.968	0.974	0.982	0.987	0.991	0.994	0.995	0.996	0.996	0.997
37	0.364	0.947	0.955	0.962	0.969	0.975	0.982	0.988	0.991	0.994	0.995	0.996	0.996	0.997
38	0.360	0.948	0.956	0.964	0.970	0.975	0.983	0.988	0.992	0.994	0.995	0.996	0.996	0.997
39	0.357	0.949	0.957	0.965	0.971	0.976	0.983	0.988	0.992	0.994	0.995	0.996	0.996	0.997
40	0.354	0.949	0.958	0.966	0.972	0.977	0.983	0.988	0.992	0.994	0.995	0.996	0.996	0.997
41	0.351	0.950	0.958	0.967	0.972	0.977	0.984	0.989	0.992	0.994	0.995	0.996	0.996	0.997
42	0.348	0.951	0.959	0.967	0.973	0.978	0.984	0.989	0.992	0.994	0.995	0.996	0.997	0.997
43	0.345	0.953	0.959	0.967	0.973	0.978	0.984	0.989	0.992	0.994	0.995	0.996	0.997	0.997
44	0.342	0.954	0.960	0.968	0.973	0.978	0.984	0.989	0.992	0.994	0.995	0.996	0.997	0.997
45	0.339	0.955	0.961	0.969	0.974	0.978	0.985	0.989	0.993	0.994	0.995	0.996	0.997	0.997
46	0.336	0.956	0.962	0.969	0.974	0.979	0.985	0.990	0.993	0.995	0.995	0.996	0.997	0.997
47	0.334	0.956	0.963	0.970	0.974	0.979	0.985	0.990	0.993	0.995	0.995	0.996	0.997	0.997
48	0.331	0.957	0.963	0.970	0.975	0.980	0.985	0.990	0.993	0.995	0.996	0.996	0.997	0.997
49	0.329	0.957	0.964	0.971	0.975	0.980	0.986	0.990	0.993	0.995	0.996	0.996	0.997	0.997
50	0.326	0.959	0.965	0.972	0.977	0.981	0.986	0.990	0.993	0.995	0.996	0.996	0.997	0.997
55	0.315	0.962	0.967	0.974	0.978	0.982	0.987	0.991	0.994	0.995	0.996	0.997	0.997	0.997
60	0.305	0.965	0.970	0.976	0.980	0.983	0.988	0.991	0.994	0.995	0.996	0.997	0.997	0.998
65	0.296	0.967	0.972	0.977	0.981	0.984	0.989	0.992	0.994	0.996	0.996	0.997	0.997	0.998
70	0.288	0.969	0.974	0.978	0.982	0.985	0.989	0.993	0.995	0.996	0.997	0.997	0.998	0.998
75	0.281	0.971	0.975	0.979	0.983	0.986	0.990	0.993	0.995	0.996	0.997	0.997	0.998	0.998
80	0.274	0.973	0.976	0.980	0.984	0.987	0.991	0.993	0.995	0.996	0.997	0.997	0.998	0.998
85	0.268	0.974	0.977	0.981	0.985	0.987	0.991	0.994	0.995	0.997	0.997	0.997	0.998	0.998
90	0.263	0.976	0.978	0.982	0.985	0.988	0.991	0.994	0.996	0.997	0.997	0.998	0.998	0.998
95	0.257	0.977	0.979	0.983	0.986	0.989	0.992	0.994	0.996	0.997	0.997	0.998	0.998	0.998
100	0.252	0.979	0.981	0.984	0.987	0.989	0.992	0.994	0.996	0.997	0.998	0.998	0.998	0.998

SOURCE: Filliben [1975]. Reprinted from *Technometrics*, Vol. 17, with the permission of the ASA.

Table 7. Percentage Points of the D'Agostino Statistic.

n	0·5	1	2·5	5	10	90	95	97·5	99	99·5
50	−3·949	−3·442	−2·757	−2·220	−1·661	0·759	0·923	1·038	1·140	1·192
60	−3·846	−3·360	−2·699	−2·179	−1·634	0·807	0·986	1·115	1·236	1·301
70	−3·762	−3·293	−2·652	−2·146	−1·612	0·844	1·036	1·176	1·312	1·388
80	−3·693	−3·237	−2·613	−2·118	−1·594	0·874	1·076	1·226	1·374	1·459
90	−3·635	−3·100	−2·580	−2·095	−1·579	0·899	1·109	1·268	1·426	1·518
100	−3·584	−3·150	−2·552	−2·075	−1·566	0·920	1·137	1·303	1·470	1·569
150	−3·409	−3·009	−2·452	−2·004	−1·520	0·990	1·233	1·423	1·623	1·746
200	−3·302	−2·922	−2·391	−1·960	−1·491	1·032	1·290	1·496	1·715	1·853
250	−3·227	−2·861	−2·348	−1·926	−1·471	1·060	1·328	1·545	1·779	1·927
300	−3·172	−2·816	−2·316	−1·906	−1·456	1·080	1·357	1·528	1·826	1·983
350	−3·129	−2·781	−2·291	−1·888	−1·444	1·096	1·379	1·610	1·863	2·026
400	−3·094	−2·753	−2·270	−1·873	−1·434	1·108	1·396	1·633	1·893	2·061
450	−3·064	−2·729	−2·253	−1·861	−1·426	1·119	1·411	1·652	1·918	2·090
500	−3·040	−2·709	−2·239	−1·850	−1·419	1·127	1·423	1·668	1·938	2·114
550	−3·019	−2·691	−2·226	−1·841	−1·413	1·135	1·434	1·682	1·957	2·136
600	−3·000	−2·676	−2·215	−1·833	−1·408	1·141	1·443	1·694	1·972	2·154
650	−2·984	−2·663	−2·206	−1·826	−1·403	1·147	1·451	1·704	1·986	2·171
700	−2·969	−2·651	−2·197	−1·820	−1·399	1·152	1·458	1·714	1·999	2·185
750	−2·956	−2·640	−2·189	−1·814	−1·395	1·157	1·465	1·722	2·010	2·199
800	−2·944	−2·630	−2·182	−1·809	−1·392	1·161	1·471	1·730	2·020	2·211
850	−2·933	−2·621	−2·176	−1·804	−1·389	1·165	1·476	1·737	2·029	2·221
900	−2·923	−2·613	−2·170	−1·800	−1·386	1·168	1·481	1·743	2·037	2·231
950	−2·914	−2·605	−2·164	−1·796	−1·383	1·171	1·485	1·749	2·045	2·241
1000	−2·906	−2·599	−2·159	−1·792	−1·381	1·174	1·489	1·754	2·052	2·249

SOURCE: D'Agostino [1971]. Reprinted from *Biometrika*, Vol. 58, with the permission of the Biometrika Trustees.

3. Studentized Range

Prescription

We all learned in elementary statistics that 99.73% of the area of the normal curve lies within $\pm 3\sigma$ from the population mean μ, 99.9937% lies within $\pm 4\sigma$ from μ, and "for sure" (i.e., 99.999942%) lies within $\pm 5\sigma$ from μ. Thus a "quick-and-dirty" test for normality based on this last fact is to take the sample range, divide by 10, and compare this with **s**, the sample standard deviation. But the sample range will vary in size based on the sample size. For small samples it is unlikely to be very large. And only for very large samples will it approximately equal 10σ.

David, Hartley, and Pearson [1954] made this argument more precise by advocating this statistic, the "studentized range", as a test of normality. If the value of the ratio of sample range to sample standard deviation is too small or too large, this indicates a deviation from normality. All that is required is a precise determination of the sampling distribution of the studentized range as a function of n, to replace the rather loose standard described earlier. The best tables are given in Pearson and Stephens [1964] and reproduced here as Table 8.

Table 8. Percentage Points of the Studentized Range.

Size of sample *n*	Lower percentage points						Upper percentage points					
	0·0	0·5	1·0	2·5	5·0	10·0	10·0	5·0	2·5	1·0	0·5	0·0
3	1·732	1·735	1·737	1·745	1·758	1·782	1·997	1·999	2·000	2·000	2·000	2·000
4	1·732	1·83	1·87	1·93	1·98	2·04	2·409	2·429	2·439	2·445	2·447	2·449
5	1·826	1·98	2·02	2·09	2·15	2·22	2·712	2·753	2·782	2·803	2·813	2·828
6	1·826	2·11	2·15	2·22	2·28	2·37	2·949	3·012	3·056	3·095	3·115	3·162
7	1·871	2·22	2·26	2·33	2·40	2·49	3·143	3·222	3·282	3·338	3·369	3·464
8	1·871	2·31	2·35	2·43	2·50	2·59	3·308	3·399	3·471	3·543	3·585	3·742
9	1·897	2·39	2·44	2·51	2·59	2·68	3·449	3·552	3·634	3·720	3·772	4·000
10	1·897	2·46	2·51	2·59	2·67	2·76	3·57	3·685	3·777	3·875	3·935	4·243
11	1·915	2·53	2·58	2·66	2·74	2·84	3·68	3·80	3·903	4·012	4·079	4·472
12	1·915	2·59	2·64	2·72	2·80	2·90	3·78	3·91	4·02	4·134	4·208	4·690
13	1·927	2·64	2·70	2·78	2·86	2·96	3·87	4·00	4·12	4·244	4·325	4·899
14	1·927	2·70	2·75	2·83	2·92	3·02	3·95	4·09	4·21	4·34	4·431	5·099
15	1·936	2·74	2·80	2·88	2·97	3·07	4·02	4·17	4·29	4·44	4·53	5·292
16	1·936	2·79	2·84	2·93	3·01	3·12	4·09	4·24	4·37	4·52	4·62	5·477
17	1·944	2·83	2·88	2·97	3·06	3·17	4·15	4·31	4·44	4·60	4·70	5·657
18	1·944	2·87	2·92	3·01	3·10	3·21	4·21	4·37	4·51	4·67	4·78	5·831
19	1·949	2·90	2·96	3·05	3·14	3·25	4·27	4·43	4·57	4·74	4·85	6·000
20	1·949	2·94	2·99	3·09	3·18	3·29	4·32	4·49	4·63	4·80	4·91	6·164
25	1·961	3·09	3·15	3·24	3·34	3·45	4·53	4·71	4·87	5·06	5·19	6·93
30	1·966	3·21	3·27	3·37	3·47	3·59	4·70	4·89	5·06	5·26	5·40	7·62
35	1·972	3·32	3·38	3·48	3·58	3·70	4·84	5·04	5·21	5·42	5·57	8·25
40	1·975	3·41	3·47	3·57	3·67	3·79	4·96	5·16	5·34	5·56	5·71	8·83
45	1·978	3·49	3·55	3·66	3·75	3·88	5·06	5·26	5·45	5·67	5·83	9·38
50	1·980	3·56	3·62	3·73	3·83	3·95	5·14	5·35	5·54	5·77	5·93	9·90
55	1·982	3·62	3·69	3·80	3·90	4·02	5·22	5·43	5·63	5·86	6·02	10·39
60	1·983	3·68	3·75	3·86	3·96	4·08	5·29	5·51	5·70	5·94	6·10	10·86
65	1·985	3·74	3·80	3·91	4·01	4·14	5·35	5·57	5·77	6·01	6·17	11·31
70	1·986	3·79	3·85	3·96	4·06	4·19	5·41	5·63	5·83	6·07	6·24	11·75
75	1·987	3·83	3·90	4·01	4·11	4·24	5·46	5·68	5·88	6·13	6·30	12·17
80	1·987	3·88	3·94	4·05	4·16	4·28	5·51	5·73	5·93	6·18	6·35	12·57
85	1·988	3·92	3·99	4·09	4·20	4·33	5·56	5·78	5·98	6·23	6·40	12·96
90	1·989	3·96	4·02	4·13	4·24	4·36	5·60	5·82	6·03	6·27	6·45	13·34
95	1·990	3·99	4·06	4·17	4·27	4·40	5·64	5·86	6·07	6·32	6·49	13·71
100	1·990	4·03	4·10	4·21	4·31	4·44	5·68	5·90	6·11	6·36	6·53	14·07
150	1·993	4·32	4·38	4·48	4·59	4·72	5·96	6·18	6·39	6·64	6·82	17·26
200	1·995	4·53	4·59	4·68	4·78	4·90	6·15	6·39	6·60	6·84	7·01	19·95
500	1·998	5·06	5·13	5·25	5·37	5·49	6·72	6·94	7·15	7·42	7·60	31·59
1000	1·999	5·50	5·57	5·68	5·79	5·92	7·11	7·33	7·54	7·80	7·99	44·70

SOURCE: Pearson and Stephens [1964]. Reprinted from *Biometrika*, Vol. 51, with the permission of the Biometrika Trustees.

To illustrate their use in our example, we find that the sample range is 46.35, the sample standard deviation is 10.12, and so the studentized range is 4.5. Based on a comparison of this value with 4.21 and 6.11, the 2.5% values of the distribution of **u**, thus we would not reject the hypothesis of normality.

Theoretical Background

As will be seen in Section 6, this test has very good power for testing normality against many alternative distributions. In fact, Uthoff [1970] has shown that this test is the most powerful scale and location invariant test of normality against the alternative of a uniform distribution with the same scale and location parameters. The one situation in which it has lowest power is in testing against an asymmetric leptokurtic distribution. It is particularly bad for testing against a lognormal distribution.

A related statistic proposed by Geary [1935] is the ratio of the mean deviation to the standard deviation, i.e., $\sum_{i=1}^{n} |\mathbf{x}_i - \bar{\mathbf{x}}|/n\mathbf{s}$. It has been shown (Dumonceaux, Antle, and Haas [1973] and Uthoff [1973]) that this test statistic is both the likelihood ratio test and asymptotically the most powerful test for normality versus the alternative hypothesis of double exponentiality. However, a study conducted by D'Agostino and Rosman [1974] indicates that there is no situation in which this procedure's power dominates that of other procedures described in this chapter.

4. Moment Checking

Prescription

Two parameters that describe critical aspects of normality are the *skewness*

$$\gamma_1 = \frac{\mathscr{E}(\mathbf{x} - \mu)^3}{\sigma^3},$$

and the *kurtosis*

$$\gamma_2 = \frac{\mathscr{E}(\mathbf{x} - \mu)^4}{\sigma^4} - 3,$$

where $\mu = \mathscr{E}\mathbf{x}$ and $\sigma^2 = \mathscr{V}\mathbf{x}$. For a normal distribution, $\gamma_1 = \gamma_2 = 0$. For distributions not symmetric about their expected value γ_1 will be nonzero, being positive if the right tail is heavier than the left and negative if the left tail is heavier than the right. When the tails of the distribution have more mass than that of the normal, γ_2 is positive and the distribution is said to be *leptokurtic*. When the tails of the distribution are less heavy than even that of the normal, γ_2 is negative and the distribution is said to be *platykurtic*.

To check the values of these parameters the statistics $\mathbf{g}_1$ and $\mathbf{g}_2$, respectively, are calculated. The statistics are given by

$$\mathbf{g}_1 = \sqrt{n}\frac{\sum_{i=1}^{n}(\mathbf{x}_i - \bar{\mathbf{x}})^3}{\left[\sum_{j=1}^{n}(\mathbf{x}_j - \bar{\mathbf{x}})^2\right]^{3/2}},$$

and

$$\mathbf{g}_2 = \frac{n\sum_{i=1}^{n}(\mathbf{x}_i - \bar{\mathbf{x}})^4}{\left[\sum_{j=1}^{n}(\mathbf{x}_j - \bar{\mathbf{x}})^2\right]^{2}} - 3,$$

whose asymptotic distributions are $N(0, 6/n)$ and $N(0, 24/n)$ when $\mathbf{x}$ is normally distributed. The exact finite sample distribution of these statistics does, however, depend on the underlying data distribution, and has been tabulated for normally distributed data and reproduced as Tables 9 and 10.

In our example, $\mathbf{g}_1 = 0.504$ and $\mathbf{g}_2 = -0.117$. Since $0.504 > 0.389$, the 95% point of the distribution of $\mathbf{g}_1$ under the hypothesis of normality, we reject this hypothesis. Since $\mathbf{g}_2 + 3 = 2.88$ lies between 2.51 and 3.57, the 5% and 95% points of the null distribution of $\mathbf{g}_2 + 3$ when $n = 200$, we can conclude that the kurtosis of the data is consistent with that of a normal distribution.[5]

Theoretical Background

Ordinarily, statisticians summarize characteristics of a probability distribution by reference to its moments, either *raw moments* μ_l given by

$$\mu_l = \int_{-\infty}^{\infty} x^l f_{\mathbf{x}}(x)\,dx,$$

or *central moments* μ_l' given by

$$\mu_l' = \int_{-\infty}^{\infty} (x - \mu_1)^l f_{\mathbf{x}}(x)\,dx.$$

Indeed, a classical mathematical problem is the determination of when the sequence $\{\mu_k\}$ uniquely determines $f_{\mathbf{x}}(x)$ (see Shohat and Tamarkin [1943]). An alternative set of characteristic numbers of a probability distribution is the

[5] Since we know that the data came from a gamma distribution, and since we know that for the gamma distribution $\gamma_1 = 2/\sqrt{\alpha}$ and $\gamma_2 = 6/\alpha$, we ought to contrast the observed values of g_1 and g_2 with the theoretical values $\gamma_1 = 0.8944$ and $\gamma_2 = 1.2$. To do this properly, one must calculate the standard deviation of the sampling distributions of $\mathbf{g}_1$ and $\mathbf{g}_2$. These are given for large n in Cramér ([1946], p. 357). For this gamma distribution we calculate that $\mathscr{V}\mathbf{g}_1 = 0.1297$ and $\mathscr{V}\mathbf{g}_2 = 0.4891$, so that we expect to observe values of $\mathbf{g}_1$ in the interval $0.8944 \pm 2 \times 0.3576 = (0.18, 1.61)$ and values of $\mathbf{g}_2$ in the interval $1.2 \pm 2 \times 0.6994 = (-0.20, 2.60)$. Thus values of $\mathbf{g}_1 = 0.504$ and $\mathbf{g}_2 = -0.117$ are not surprising.

Table 9. Percentage Points of the Skewness Statistic $\mathbf{g}_1$.

Size of sample n	Percentage points 5%	1%	Standard deviation	Size of sample n	Percentage points 5%	1%	Standard deviation	Size of sample n	Percentage points 5%	1%	Standard deviation
25	·711	1·061	·4354	200	·280	·403	·1706	1000	·127	·180	·0772
30	·661	·982	·4052	250	·251	·360	·1531	1200	·116	·165	·0705
35	·621	·921	·3804	300	·230	·329	·1400	1400	·107	·152	·0653
40	·587	·869	·3596	350	·213	·305	·1298	1600	·100	·142	·0611
45	·558	·825	·3418	400	·200	·285	·1216	1800	·095	·134	·0576
50	·533	·787	·3264	450	·188	·269	·1147	2000	·090	·127	·0547
				500	·179	·255	·1089				
60	·492	·723	·3009	550	·171	·243	·1039	2500	·080	·114	·0489
70	·459	·673	·2806	600	·163	·233	·0995	3000	·073	·104	·0447
80	·432	·631	·2638	650	·157	·224	·0956	3500	·068	·096	·0414
90	·409	·596	·2498	700	·151	·215	·0922	4000	·064	·090	·0387
100	·389	·567	·2377	750	·146	·208	·0891	4500	·060	·085	·0365
				800	·142	·202	·0863	5000	·057	·081	·0346
125	·350	·508	·2139	850	·138	·196	·0837				
150	·321	·464	·1961	900	·134	·190	·0814				
175	·298	·430	·1820	950	·130	·185	·0792				
200	·280	·403	·1706	1000	·127	·180	·0772				

SOURCE: Pearson and Hartley [1958]. Reprinted from *Biometrika Tables for Statisticians*, Vol. I, with the permission of the Biometrika Trustees.

Table 10. Percentage Points of the Kurtosis Statistic $\mathbf{g}_2 + 3$.

Size of sample n	Percentage points Upper 1%	Upper 5%	Lower 5%	Lower 1%	Size of sample n	Percentage points Upper 1%	Upper 5%	Lower 5%	Lower 1%
200	3·98	3·57	2·51	2·37	1000	3·41	3·26	2·76	2·68
250	3·87	3·52	2·55	2·42	1200	3·37	3·24	2·78	2·71
300	3·79	3·47	2·59	2·46	1400	3·34	3·22	2·80	2·72
350	3·72	3·44	2·62	2·50	1600	3·32	3·21	2·81	2·74
400	3·67	3·41	2·64	2·52	1800	3·30	3·20	2·82	2·76
450	3·63	3·39	2·66	2·55	2000	3·28	3·18	2·83	2·77
500	3·60	3·37	2·67	2·57					
550	3·57	3·35	2·69	2·58	2500	3·25	3·16	2·85	2·79
600	3·54	3·34	2·70	2·60	3000	3·22	3·15	2·86	2·81
650	3·52	3·33	2·71	2·61	3500	3·21	3·14	2·87	2·82
700	3·50	3·31	2·72	2·62	4000	3·19	3·13	2·88	2·83
750	3·48	3·30	2·73	2·64	4500	3·18	3·12	2·88	2·84
800	3·46	3·29	2·74	2·65	5000	3·17	3·12	2·89	2·85
850	3·45	3·28	2·74	2·66					
900	3·43	3·28	2·75	2·66					
950	3·42	3·27	2·76	2·67					
1000	3·41	3·26	2·76	2·68					

SOURCE: Pearson and Hartley [1958]. Reprinted from *Biometrika Tables for Statisticians*, Vol. I, with the permission of the Biometrika Trustees.

set of *cumulants* κ_l of the distribution. The cumulants are defined in the following way. Let $\psi_{\mathbf{x}}(t)$, the logarithm of the characteristic function of the distribution of $\mathbf{x}$, be given by

$$\psi_{\mathbf{x}}(t) = \log \int_{-\infty}^{\infty} e^{itx} f_{\mathbf{x}}(x)\, dx.$$

Now expand $\psi_{\mathbf{x}}(t)$ in an infinite Taylor series in powers of t. Then κ_l is the coefficient of $t^l/l!$ in that series, i.e.,

$$\psi_{\mathbf{x}}(t) = \sum_{l=1}^{\infty} \frac{\kappa_l}{l!} t^l$$

The key property of cumulants for our purpose is that $\kappa_l = 0$ for $l \geq 3$ if and only if $f_{\mathbf{x}}(x)$ is a normal density function. Thus a test of normality could be based on sample estimates of the κ_l, $l \geq 3$, and a comparison of these estimates with a null hypothesis value of 0.

The most common cumulants under study are κ_3 and κ_4, unstandardized measures of skewness and kurtosis. In terms of central moments, these cumulants are expressible as

$$\kappa_3 = \mu_3',$$
$$\kappa_4 = \mu_4' - 3(\mu_2')^2.$$

The corresponding unbiased sample estimators of these cumulants are given by

$$\mathbf{k}_3 = n \sum_{i=1}^{n} (\mathbf{x}_i - \bar{\mathbf{x}})^3/(n-1)(n-2),$$

$$\mathbf{k}_4 = \frac{n\left[(n+1) \sum_{i=1}^{n} (\mathbf{x}_i - \bar{\mathbf{x}})^4 - 3(n-1)\left\{\sum_{i=1}^{n} (\mathbf{x}_i - \bar{\mathbf{x}})^2\right\}^2\right]}{(n-1)(n-2)(n-3)}.$$

The large sample distribution of the sample cumulants are known to be normal (see Kendall and Stuart [1977]), and in particular under the hypothesis of normality $\mathbf{k}_l$ has a $N(\kappa_l, l!\, \sigma^{2l}/n)$ distribution, so one can use these large sample results to test whether $\mathbf{k}_3$ and $\mathbf{k}_4$ are each significantly different from their null value of 0. Since the asymptotic variances of $\mathbf{k}_3$ and $\mathbf{k}_4$ involve functions of σ^2, statisticians have adopted standardized measures of skewness and kurtosis, respectively the parameters[6]

$$\gamma_1 = \frac{\kappa_3}{\sigma^3} = \frac{\mu_3'}{\sigma^3},$$

[6] The literature on this subject sometimes refers to γ_1 as $\sqrt{\beta_1}$ and uses the parameter $\beta_2 = \mu_4'/\sigma^4 = \gamma_2 + 3$ instead of γ_2. The associated statistics are called $\sqrt{b_1}$ and b_2 in some of the literature.

and

$$\gamma_2 = \frac{\kappa_4}{\sigma^4} = \frac{\mu_4'}{\sigma^4} - 3.$$

The exact expected values of $\mathbf{g}_1$ and $\mathbf{g}_2$ are 0 and $-6/(n + 1)$, respectively. The exact variance of $\mathbf{g}_1$ is

$$\mathscr{V}\mathbf{g}_1 = \frac{6(n-2)}{(n+1)(n+3)},$$

and that of $\mathbf{g}_2$ is

$$\mathscr{V}\mathbf{g}_2 = \frac{24n(n-2)(n-3)}{(n+1)^2(n+3)(n+5)}.$$

If instead we were to estimate γ_1 and γ_2 using $\mathbf{k}_3$ and $\mathbf{k}_4$, i.e., $\mathbf{g}_1^* = \mathbf{k}_3/\mathbf{k}_2^{3/2}$ and $\mathbf{g}_2^* = \mathbf{k}_4/\mathbf{k}_2^2$, we would find that $\mathscr{E}\mathbf{g}_1^* = \mathscr{E}\mathbf{g}_2^* = 0$.

The exact variance of $\mathbf{g}_1^*$ is

$$\mathscr{V}\mathbf{g}_1^* = \frac{6n(n-1)}{(n-2)(n+1)(n+3)},$$

and that of $\mathbf{g}_2^*$ is

$$\mathscr{V}\mathbf{g}_2^* = \frac{24n(n-1)^2}{(n-3)(n-2)(n+3)(n+5)}$$

under the hypothesis of normality, so that these expressions can be used to gain a little more precision in the asymptotic distributions of $\mathbf{g}_1$ and $\mathbf{g}_2$ (or $\mathbf{g}_1^*$ and $\mathbf{g}_2^*$).

5. Standard Tests of Goodness-of-Fit

An immediate reaction to the query of how to test for normality is to apply one of the tests of goodness-of-fit based on the sample histogram or cumulative distribution function. Unfortunately, those procedures are only applicable when testing whether the true distribution is a *specific* normal distribution, i.e., one wherein both μ and σ^2 are specified, and not a generic normal distribution with μ and σ^2 undetermined. In their comprehensive study of tests for normality, Shapiro, Wilk, and Chen [1968] addressed this problem and concluded that the errors produced by misspecification of values of μ and σ^2 are substantial even for relatively small departures. Of the procedures studied the three that fared best, i.e., were least subject to distortion by lack of precise knowledge of μ and σ^2, were the chi-square, Kolmogorov, and Durbin tests. We will therefore briefly present these procedures as they were derived, i.e., assuming μ and σ^2 known, and add some comments on their applicability when μ and σ^2 are estimated from the data.

(a) Chi-Square Test

This procedure, probably the oldest statistical procedure for testing "goodness-of-fit" of a theoretical distribution to data (see Pearson [1900]), consists of the following stages:

(1) divide the line into k mutually exclusive and exhaustive intervals $I_1, \ldots, I_k$;
(2) count the number of observations $\mathbf{n}_i$ which fall into interval I_i, $i = 1, \ldots, k$;
(3) calculate the expected number θ_i of observations in interval I_i if the hypothesized theoretical distribution were true, i.e.,

$$\theta_i = n \int_{I_i} f_{\mathbf{x}}(x)\,dx,$$

and finally
(4) calculate the statistic

$$\mathbf{u} = \sum_{i=1}^{k} (\mathbf{n}_i - \theta_i)^2/\theta_i.$$

It can be shown that $\mathbf{u}$ has approximately a $\chi^2(k-1)$ distribution. Values of $\mathbf{u}$ greater than $\chi^2_{0.95}(k-1)$, the upper 5% point of the $\chi^2(k-1)$ distribution, are inconsistent with the hypothesis that $f_{\mathbf{x}}(x)$ is the true underlying theoretical distribution.

With respect to the best choice of k and a rule for best choice of the size of the intervals $I_1, \ldots, I_k$, the following have been found to be practical rules of thumb: take k to be

$$k = 3.7653(n-1)^{2/5}$$

rounded up to the next integer (see Mann and Wald [1942]) and select the intervals such that $\theta_1 = \cdots = \theta_k = n/k$ (see Gumbel [1943]). For example, if $n = 100$, $k = 23.66$ so that, rounding k to 24, $\theta_i = 4.1667$ for $i = 1, \ldots, 24$. The intervals for the $N(0, 1)$ distribution in this case are given in Table 11, and a computer program to produce the endpoints of the intervals is given in Appendix 1.

If we use $k = 24$ and cutoff values given by $\mu + \sigma c_i = 25 + \sqrt{125}c_i$, where the c_i are taken from Table 11, we obtain the $\mathbf{n}_i$ also given in Table 11. From these we find that $\mathbf{u} = 34.4$ which, compared with $\chi^2_{0.95}(23) = 35.17$, leads us to accept the hypothesis of normality. Upon collapsing these cells into 12 cells by combining cells i and $i+1$, $i = 1, 3, 5, \ldots, 23$, we find that $\mathbf{u} = 14.96$ which, compared with $\chi^2_{0.95}(11) = 19.68$, also leads us to accept the hypothesis of normality.

Theoretical Background

Since each $\mathbf{n}_i$ defined above is approximately distributed as a Poisson variable with mean (and variance) equal to θ_i, the variable $(\mathbf{n}_i - \theta_i)/\sqrt{\theta_i}$ is approximately a $N(0, 1)$ variable, and so $(\mathbf{n}_i - \theta_i)^2/\theta_i$ is approximately distributed as

Table 11. Computation of Intervals and Frequency Counts for the Chi-Square Test.

Interval number	Unit normal upper limit of interval	Upper limit of interval	Frequency count
i	c_i	$25 + \sqrt{125}c_i$	n_i
1	-1.7321	5.6351	0
2	-1.3832	9.5350	5
3	-1.1504	12.1377	0
4	-0.9674	14.1846	4
5	-0.8120	15.9212	12
6	-0.6742	17.4623	5
7	-0.5481	18.8716	2
8	-0.4303	20.1892	7
9	-0.3182	21.4424	4
10	-0.2100	22.6516	4
11	-0.1044	23.8329	6
12	0.0	25.0000	4
13	0.1044	26.1671	3
14	0.2100	27.3484	4
15	0.3182	28.5576	5
16	0.4303	29.8108	4
17	0.5481	31.1284	3
18	0.6742	32.5377	3
19	0.8120	34.0788	7
20	0.9674	35.8154	3
21	1.1504	37.8623	5
22	1.3832	40.4650	1
23	1.7321	44.3649	3
24	∞	∞	6

$\chi^2(1)$. Since $\sum_{i=1}^{k} \mathbf{n}_i = n$, these $\chi^2(1)$ variables are not independent. But it can be shown that the sum of these k $\chi^2(1)$ variables has a $\chi^2(k-1)$ distribution.

One should note that the expected number of observations in each interval is $\theta_i = n/k \approx 0.26558n^{3/5}$. When $n = 100$, we expect only four observations per cell. The reader should not be concerned that this expected value is below some "minimum expectation level" cited in some textbooks as necessary for the validity of the chi-squared distribution as the sampling distribution of **u**. This minimum expectation level is probably a result of Fisher's comment ([1941], p. 82) that "it is desirable that the number expected should in no group be less than 5". Cramér ([1946], p. 420) recommends 10 as the minimum expectation level. Kendall ([1952], p. 292) recommends 20, though this recommendation is dropped in the Kendall and Stuart [1977] three-volume edition. Cochran [1942] has shown that there is little disturbance to the 5% level when a single expected value is as low as one-half or two expected values as low as 1 (a far contrast from the usual recommendations of 5 or even 10). Yarnold [1970] proposed an even more relaxed rule, namely that if $k \geq 3$ and if r denotes the number of expected values that are lower than 5, then the minimum expectation level can be as low as $5r/k$. (In our sample, $r = k$ and so the Yarnold minimum expectation level would be 5 here.) A further investigation by Tate and Hyer [1973], though, brings us back to Kendall's recommendation. Nonetheless, Williams [1950] has shown that the value of k given above can be halved with little loss of sensitivity of the test. Thus, for those concerned about minimum expected values, taking $k = 12$ and $\theta_i = 8.33$ leads to a successive combination of the intervals given in Table 11.

Of course, all of this is predicated on knowing the distribution fully, i.e., including all its parameters. But suppose the parameters are unknown but instead are estimated by the method of maximum likelihood. Then, let $\hat{\boldsymbol{\theta}}_i$ be the maximum likelihood estimate of θ_i $(i = 1, \ldots, k)$, and let

$$\hat{\mathbf{u}} = \sum_{i=1}^{k} (\mathbf{n}_i - \hat{\boldsymbol{\theta}}_i)^2/\hat{\boldsymbol{\theta}}_i.$$

Let s be the number of independent parameters estimated (e.g., for the normal distribution, $s = 2$). Chernoff and Lehmann [1954] have shown that $\hat{\mathbf{u}}$ is *not* distributed asymptotically as $\chi^2(k - s - 1)$, but that its asymptotic distribution is that of the sum of a $\chi^2(k - s - 1)$ variable and another positive random variable, independent of the $\chi^2(k - s - 1)$ variable and bounded from above by a $\chi^2(s)$ variable. Thus, in our example, if we had used $\bar{x}$ in place of μ and s in place of σ in calculating the θ_i, a comparison of **u** with the 95th percentile of the $\chi^2(k - 3)$ distribution will lead to a probability of rejection, when the null hypothesis is true, which is seriously greater than the desired level of significance 0.05.

We have heretofore only considered intervals with fixed endpoints. One might also consider situations in which the endpoints are determined from the sample. For example, Watson [1957] studied intervals around $\bar{x}$ with length some multiple of the sample standard deviation. He obtained results of the same form as that of Chernoff and Lehmann cited above. Upon analysis he concluded that with ten intervals one can be sure that the true significance level is between 0.05 and 0.06, taking care that none of the intervals have "small" expected frequencies. Dahiya and Gurland [1973] have studied the power of the chi-square test for normality with intervals with random endpoints against specific alternatives, and found the following recommended values of k:

Alternative	k
Exponential	7
Double exponential	3
Logistic	3
Pearson Type III	12
Power	3

(b) Kolmogorov Test

Prescription
Another measure of discrepancy between the observed data and a theoretical distribution $F(x)$ is to consider the statistic $\mathbf{d} = \max_x |\mathbf{F}_n(x) - F(x)|$, where $\mathbf{F}_n(x)$ is the empirical distribution function defined earlier. The sampling distribution of $\mathbf{d}$, the Kolmogorov statistic, is tabulated in Miller [1956] and reproduced as Table 12 herein. To apply this test, we must evaluate both $\mathbf{F}_n(x)$ and $F(x)$ for all possible values of x, not merely for the observed x's. One way to do this is to plot both curves and find the value of x which maximizes the absolute difference between $\mathbf{F}_n(x)$ and $F(x)$ graphically. Figure 5 depicts such a graph when $\mathbf{F}_n(x)$ is the empirical distribution function of the data, and $F(x)$ is the cumulative normal distribution function with mean 25 and standard deviation 15. Note that the largest deviation is at $x = 12.75674$, and is 0.06042.

Alternatively, one can find the value of d computationally as follows. Since $F_n(x) = i/n$ for $x_i \leq x < x_{i+1}$, we need only maximize $|(i-1)/n - F(x_i)|$ with respect to i, since the maximum must occur at the endpoint of an interval. Values of $F(x_i)$ for our example are given in Table 1, and $|(i-1)/n - F(x_i)|$ is maximized when $i = 6$, and its maximum value is 0.06042, which, when compared to the critical value in Table 12, 0.12067, is found not to be significant.

Theoretical Background
Lilliefors [1967] investigated properties of the Kolmogorov test when μ and σ^2 are unknown but instead are estimated by $\bar{\mathbf{x}}$ and $\mathbf{s}^2$. An improved version of his table of significance values of $\mathbf{d}$ under these circumstances is given as Table 13 herein. The values are about two-thirds the critical values for the Kolmogorov statistic when μ and σ^2 are known.

An argument for the use of the Kolmogorov statistic in place of the chi-square statistic as a test for normality is that, as shown by Kac, Kiefer, and Wolfowitz [1955], this test is asymptotically more powerful than the chi-square test.

Table 12. Percentage Points of the Kolmogorov Statistic.

n	$\alpha = .10$	$\alpha = .05$ ($P = .90$)	$\alpha = .025$ ($P = .95$)	$\alpha = .01$ ($P = .98$)	$\alpha = .005$ ($P = .99$)
1	.90000	.95000	.97500	.99000	.99500
2	.68377	.77639	.84189	.90000	.92929
3	.56481	.63604	.70760	.78456	.82900
4	.49265	.56522	.62394	.68887	.73424
5	.44698	.50945	.56328	.62718	.66853
6	.41037	.46799	.51926	.57741	.61661
7	.38148	.43607	.48342	.53844	.57581
8	.35831	.40962	.45427	.50654	.54179
9	.33910	.38746	.43001	.47960	.51332
10	.32260	.36866	.40925	.45662	.48893
11	.30829	.35242	.39122	.43670	.46770
12	.29577	.33815	.37543	.41918	.44905
13	.28470	.32549	.36143	.40362	.43247
14	.27481	.31417	.34890	.38970	.41762
15	.26588	.30397	.33760	.37713	.40420
16	.25778	.29472	.32733	.36571	.39201
17	.25039	.28627	.31796	.35528	.38086
18	.24360	.27851	.30936	.34569	.37062
19	.23735	.27136	.30143	.33685	.36117
20	.23156	.26473	.29408	.32866	.35241
21	.22617	.25858	.28724	.32104	.34427
22	.22115	.25283	.28087	.31394	.33666
23	.21645	.24746	.27490	.30728	.32954
24	.21205	.24242	.26931	.30104	.32286
25	.20790	.23768	.26404	.29516	.31657
26	.20399	.23320	.25907	.28962	.31064
27	.20030	.22898	.25438	.28438	.30502
28	.19680	.22497	.24993	.27942	.29971
29	.19348	.22117	.24571	.27471	.29466
30	.19032	.21756	.24170	.27023	.28987
31	.18732	.21412	.23788	.26596	.28530
32	.18445	.21085	.23424	.26189	.28094
33	.18171	.20771	.23076	.25801	.27677
34	.17909	.20472	.22743	.25429	.27279
35	.17659	.20185	.22425	.25073	.26897
36	.17418	.19910	.22119	.24732	.26532
37	.17188	.19646	.21826	.24404	.26180
38	.16966	.19392	.21544	.24089	.25843
39	.16753	.19148	.21273	.23786	.25518
40	.16547	.18913	.21012	.23494	.25205
41	.16349	.18687	.20760	.23213	.24904
42	.16158	.18468	.20517	.22941	.24613
43	.15974	.18257	.20283	.22679	.24332
44	.15796	.18053	.20056	.22426	.24060
45	.15623	.17856	.19837	.22181	.23798
46	.15457	.17665	.19625	.21944	.23544
47	.15295	.17481	.19420	.21715	.23298
48	.15139	.17302	.19221	.21493	.23059
49	.14987	.17128	.19028	.21277	.22828
50	.14840	.16959	.18841	.21068	.22604

(continued)

Table 12 (*continued*)

n	$\alpha=.10$	$\alpha=.05$ $(P=.90)$	$\alpha=.025$ $(P=.95)$	$\alpha=.01$ $(P=.98)$	$\alpha=.005$ $(P=.99)$
51	.14697	.16796	.18659	.20864	.22386
52	.14558	.16637	.18482	.20667	.22174
53	.14423	.16483	.18311	.20475	.21968
54	.14292	.16332	.18144	.20289	.21768
55	.14164	.16186	.17981	.20107	.21574
56	.14040	.16044	.17823	.19930	.21384
57	.13919	.15906	.17669	.19758	.21199
58	.13801	.15771	.17519	.19590	.21019
59	.13686	.15639	.17373	.19427	.20844
60	.13573	.15511	.17231	.19267	.20673
61	.13464	.15385	.17091	.19112	.20506
62	.13357	.15263	.16956	.18960	.20343
63	.13253	.15144	.16823	.18812	.20184
64	.13151	.15027	.16693	.18667	.20029
65	.13052	.14913	.16567	.18525	.19877
66	.12954	.14802	.16443	.18387	.19729
67	.12859	.14693	.16322	.18252	.19584
68	.12766	.14587	.16204	.18119	.19442
69	.12675	.14483	.16088	.17990	.19303
70	.12586	.14381	.15975	.17863	.19167
71	.12499	.14281	.15864	.17739	.19034
72	.12413	.14183	.15755	.17618	.18903
73	.12329	.14087	.15649	.17498	.18776
74	.12247	.13993	.15544	.17382	.18650
75	.12167	.13901	.15442	.17268	.18528
76	.12088	.13811	.15342	.17155	.18408
77	.12011	.13723	.15244	.17045	.18290
78	.11935	.13636	.15147	.16938	.18174
79	.11860	.13551	.15052	.16832	.18060
80	.11787	.13467	.14960	.16728	.17949
81	.11716	.13385	.14868	.16626	.17840
82	.11645	.13305	.14779	.16526	.17732
83	.11576	.13226	.14691	.16428	.17627
84	.11508	.13148	.14605	.16331	.17523
85	.11442	.13072	.14520	.16236	.17421
86	.11376	.12997	.14437	.16143	.17321
87	.11311	.12923	.14355	.16051	.17223
88	.11248	.12850	.14274	.15961	.17126
89	.11186	.12779	.14195	.15873	.17031
90	.11125	.12709	.14117	.15786	.16938
91	.11064	.12640	.14040	.15700	.16846
92	.11005	.12572	.13965	.15616	.16755
93	.10947	.12506	.13891	.15533	.16666
94	.10889	.12440	.13818	.15451	.16579
95	.10833	.12375	.13746	.15371	.16493
96	.10777	.12312	.13675	.15291	.16408
97	.10722	.12249	.13606	.15214	.16324
98	.10668	.12187	.13537	.15137	.16242
99	.10615	.12126	.13469	.15061	.16161
100	.10563	.12067	.13403	.14987	.16081

SOURCE: Miller [1956]. Reprinted from the *Journal of the American Statistical Association*, Vol. 51, with the permission of the ASA.

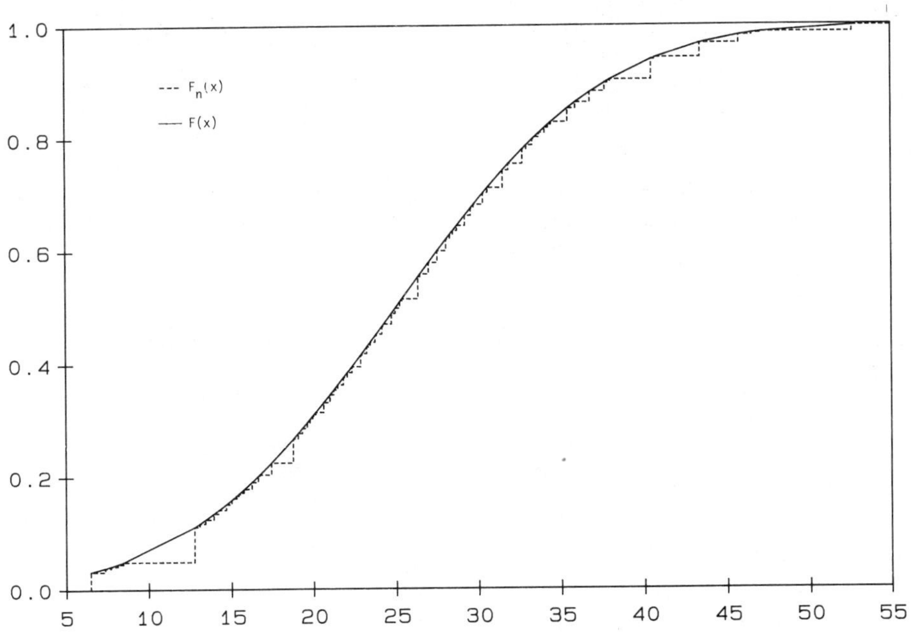

Figure 5. Empirical Distribution Function of 100 Independent Gamma Observations and Cumulative Normal $N(25, 125)$ Distribution.

Table 13. Levels of Significance for the Kolmogorov Statistic when Parameters Are Estimated from the Sample.

Sample size	*Upper tail percentiles*					
	20	*15*	*10*	*05*	*01*	*.1*
4	.303	.321	.346	.376	.413	.433
5	.289	.303	.319	.343	.397	.439
6	.269	.281	.297	.323	.371	.424
7	.252	.264	.280	.304	.351	.402
8	.239	.250	.265	.288	.333	.384
9	.227	.238	.252	.274	.317	.365
10	.217	.228	.241	.262	.304	.352
11	.208	.218	.231	.251	.291	.338
12	.200	.210	.222	.242	.281	.325
13	.193	.202	.215	.234	.271	.314
14	.187	.196	.208	.226	.262	.305
15	.181	.190	.201	.219	.254	.296
16	.176	.184	.195	.213	.247	.287
17	.171	.179	.190	.207	.240	.279
18	.167	.175	.185	.202	.234	.273
19	.163	.170	.181	.197	.228	.266
20	.159	.166	.176	.192	.223	.260
25	.143	.150	.159	.173	.201	.236
30	.131	.138	.146	.159	.185	.217
40	.115	.120	.128	.139	.162	.189
100	.074	.077	.082	.089	.104	.122
400	.037	.039	.041	.045	.052	.061
900	.025	.026	.028	.030	.035	.042

SOURCE: Dallal and Wilkinson [1986]. Reprinted from *The American Statistician*, Vol. 40, with the permission of the ASA.

Table 14. Computation of the Durbin Statistic.

X	U	C	G	$CUMG$	D
6.48875	0.03208	0.03208	0.00843	0.00843	0.00157
7.36181	0.03889	0.00681	0.00401	0.01245	0.00755
7.70804	0.04189	0.00301	0.00029	0.01274	0.01726
8.11865	0.04570	0.00381	0.00141	0.01414	0.02586
8.46150	0.04909	0.00339	0.01094	0.02508	0.02492
12.75674	0.11042	0.06133	0.00213	0.02721	0.03279
13.08055	0.11664	0.00623	0.00072	0.02793	0.04207
13.43281	0.12369	0.00705	0.01687	0.04480	0.03520
13.93881	0.13433	0.01064	0.00476	0.04957	0.04043
14.22067	0.14053	0.00619	0.01834	0.06791	0.03209
14.25665	0.14133	0.00080	0.02274	0.09065	0.01935
14.69054	0.15128	0.00995	0.00153	0.09218	0.02782
15.02005	0.15913	0.00786	0.00009	0.09227	0.03773
15.03517	0.15950	0.00037	0.03946	0.13173	0.00827
15.14141	0.16209	0.00259	0.00304	0.13477	0.01523
15.31778	0.16646	0.00437	0.02446	0.15923	0.00077
15.44101	0.16955	0.00309	0.01716	0.17639	−0.00639
15.52946	0.17180	0.00224	0.01697	0.19336	−0.01336
15.75355	0.17757	0.00577	0.00489	0.19825	−0.00825
15.76179	0.17778	0.00021	0.02380	0.22205	−0.02205
15.78501	0.17839	0.00061	0.00591	0.22797	−0.01797
16.24040	0.19052	0.01213	0.00191	0.22987	−0.00987
16.62110	0.20104	0.01052	0.01118	0.24105	−0.01105
16.65918	0.20211	0.00107	0.00625	0.24731	−0.00731
16.71427	0.20367	0.00156	0.00035	0.24765	0.00235
17.42898	0.22449	0.02082	0.00672	0.25437	0.00563
18.76179	0.26637	0.04188	0.00017	0.25454	0.01546
18.79385	0.26743	0.00105	0.00290	0.25744	0.02256
19.07174	0.27665	0.00922	0.00355	0.26099	0.02901
19.32876	0.28532	0.00867	0.00286	0.26385	0.03615
19.62610	0.29550	0.01019	0.01795	0.28180	0.02820
19.78951	0.30117	0.00567	0.02018	0.30197	0.01803
19.96001	0.30714	0.00597	0.00907	0.31104	0.01896
20.04533	0.31014	0.00300	0.03203	0.34307	−0.00307
20.14921	0.31382	0.00368	0.00586	0.34892	0.00108
20.62760	0.33098	0.01716	0.02198	0.37091	−0.01091
20.99996	0.34459	0.01361	0.00817	0.37908	−0.00908
21.00931	0.34493	0.00034	0.00283	0.38191	−0.00191
21.24029	0.35348	0.00855	0.03685	0.41876	−0.02876
21.47477	0.36223	0.00875	0.01332	0.43208	−0.03208
21.81918	0.37522	0.01299	0.00619	0.43827	−0.02827
22.05309	0.38412	0.00890	0.01175	0.45002	−0.03002
22.33860	0.39507	0.01095	0.01345	0.46347	−0.03347
22.87242	0.41576	0.02069	0.00187	0.46534	−0.02534
22.94075	0.41842	0.00267	0.00571	0.47105	−0.02105
23.23921	0.43011	0.01169	0.01739	0.48843	−0.02843
23.45028	0.43842	0.00830	0.00947	0.49791	−0.02791
23.74068	0.44989	0.01147	0.00410	0.50201	−0.02201
23.81423	0.45280	0.00291	0.00882	0.51083	−0.02083
23.15124	0.46617	0.01337	0.01758	0.52841	−0.02841
24.25881	0.47045	0.00428	0.02387	0.55228	−0.04228
24.73173	0.48929	0.01884	0.00018	0.55246	−0.03246

(continued)

Table 14 (*continued*).

X	U	C	G	$CUMG$	D
24.96917	0.49877	0.00948	0.02174	0.57420	−0.04420
25.24047	0.50960	0.01083	0.01169	0.58589	−0.04589
25.24985	0.50997	0.00037	0.00542	0.59131	−0.04131
25.37080	0.51480	0.00482	0.00020	0.59151	−0.03151
26.35537	0.55391	0.03912	0.00387	0.59538	−0.02538
26.47865	0.55878	0.00487	0.00655	0.60193	−0.02193
26.48096	0.55887	0.00009	0.00489	0.60682	−0.01682
26.99213	0.57895	0.02008	0.00873	0.61555	−0.01555
27.52403	0.59963	0.02068	0.00612	0.62167	−0.01167
28.05616	0.62005	0.02041	0.00409	0.62576	−0.00576
28.09613	0.62157	0.00152	0.01833	0.64409	−0.01409
28.29106	0.62896	0.00739	0.00913	0.65322	−0.01322
28.46719	0.63559	0.00664	0.01240	0.66562	−0.01562
28.71782	0.64496	0.00937	0.00381	0.66943	−0.00943
29.20037	0.66276	0.01780	0.00053	0.66996	0.00004
29.56230	0.67588	0.01312	0.00096	0.67093	0.00907
29.73896	0.68221	0.00633	0.00182	0.67275	0.01725
30.30640	0.70216	0.01996	0.00327	0.67602	0.02398
30.32239	0.70271	0.00055	0.00377	0.67979	0.03021
30.57425	0.71138	0.00866	0.00435	0.68415	0.03585
31.49794	0.74209	0.03071	0.01088	0.69502	0.03498
31.84531	0.75318	0.01109	0.00614	0.70116	0.03884
31.35266	0.75341	0.00023	0.01193	0.71310	0.03690
32.70566	0.77952	0.02611	0.02226	0.73536	0.02464
32.93997	0.78641	0.00688	0.00326	0.73862	0.03138
33.31204	0.79708	0.01067	0.00616	0.74478	0.03522
33.35007	0.79815	0.00107	0.00539	0.75017	0.03983
33.43215	0.80045	0.00230	0.00870	0.75887	0.04113
33.71671	0.80831	0.00786	0.02651	0.78538	0.02462
34.05206	0.81733	0.00902	0.03786	0.82324	−0.00324
34.25993	0.82278	0.00545	0.01214	0.83538	−0.00538
34.44229	0.82748	0.00470	0.00658	0.84196	−0.00196
35.43620	0.85167	0.02419	0.01158	0.85353	−0.00353
35.90700	0.86230	0.01063	0.01777	0.87130	−0.01130
36.77312	0.88046	0.01816	0.00190	0.87320	−0.00320
36.86588	0.88231	0.00184	0.00464	0.87784	0.00216
37.67851	0.89757	0.01527	0.00346	0.88130	0.00870
37.84356	0.90049	0.00292	0.00011	0.88141	0.01859
38.00723	0.90332	0.00283	0.00149	0.88289	0.02711
40.46576	0.93901	0.03569	0.03370	0.91659	0.00341
40.72330	0.94205	0.00305	0.00692	0.92352	0.00648
43.38677	0.96702	0.02496	0.00918	0.93269	0.00731
45.75185	0.98102	0.01400	0.03221	0.96490	−0.01490
45.79899	0.98124	0.00022	0.00822	0.97312	−0.01312
46.53636	0.98437	0.00313	0.01805	0.99117	−0.02117
47.08834	0.98641	0.00204	0.01371	1.00488	−0.02488
52.63478	0.99714	0.01073	0.00830	1.01318	−0.02318
52.84011	0.99731	0.00017	0.03889	1.05207	−0.05207
		0.00269			

Note:

X	Gamma data.	G	See Section 5(c).
U	Cumulative normal probability.	$CUMG$	Cumulative Sum of G.
C	See Section 5(c).	D	$I/N - CUMG$.

(c) Durbin Test

Prescription

Durbin [1961] proposed the following procedure. Let

$$\mathbf{u}_i = F(\mathbf{x}_i), \qquad i = 1, \ldots, n,$$

where $\mathbf{x}_1 \leq \mathbf{x}_2 \leq \cdots \leq \mathbf{x}_n$. Define

$$\mathbf{c}_1 = \mathbf{u}_1, \qquad \mathbf{c}_j = \mathbf{u}_j - \mathbf{u}_{j-1}, \quad j = 2, \ldots, n, \qquad \mathbf{c}_{n+1} = 1 - \mathbf{u}_n.$$

Let

$$\mathbf{c}_{(1)} \leq \mathbf{c}_{(2)} \leq \cdots \leq \mathbf{c}_{(n+1)}$$

denote the ordered set of $\mathbf{c}_i$. Define

$$\mathbf{g}_j = (n + 2 - j)(\mathbf{c}_{(j)} - \mathbf{c}_{(j-1)}), \qquad j = 1, \ldots, n.$$

The test statistic is given by

$$\mathbf{d} = \max_i \left[\frac{i}{n} - \sum_{j=1}^{i} \mathbf{g}_i \right], \qquad i = 1, \ldots, n,$$

and is distributed identically with that of the Kolmogorov statistic.

Table 14 provides the u_i, c_i, g_i, and the other elements necessary to calculate d from our gamma distributed data. From this we see that $d = 0.042067$. Since the critical value for the Kolmogorov statistic is $1.36/\sqrt{n}$ for $n > 35$, we compare $\mathbf{d}$ with 0.136 in this case and conclude that the data come from a normal distribution.

Theoretical Background

The main reason for including this test, although it is not as well known as either of the other two goodness-of-fit tests presented herein, is that it fared better than the other two in the Shapiro–Wilk–Chen [1968] study. Not only does it have better power but it is computationally simple to perform.

6. Evaluation

First let us look at a scorecard of what the various tests described in this chapter told us about our example of 100 observations from the gamma distribution.

Shapiro–Wilk	reject normality
Filliben	reject normality
D'Agostino	accept normality
Studentized range	accept normality
Skewness	reject normality
Kurtosis	accept normality
Chi-square	accept normality
Kolmogorov	accept normality
Durbin	accept normality

The gamma distribution had a theoretical skewness of $\gamma_1 = 2/\sqrt{\alpha} = 0.8944$ and a theoretical kurtosis of $\gamma_2 = 6/\alpha = 1.2$, so that tests which concentrate on the tails of the distribution would have a hard time picking up such mild kurtosis. One cannot generalize from one example, but it appears that this scorecard in this one instance matches what other studies have found about these various tests.

There have been a number of studies comparing the powers of many of these tests when the alternative hypothesis is one of a variety of nonnormal distributions. As noted earlier, D'Agostino and Rosman [1974] showed that for most alternative hypotheses the studentized range had better power than did the mean deviation to standard deviation ratio (see also Gastwirth and Owens [1977]). Pearson, D'Agostino, and Bowman [1977] showed that the one-tailed test based on $\mathbf{g}_2$ had for the most part better power than did the studentized range.

The most extensive power studies were those of Shapiro, Wilk, and Chen [1968] and the aforementioned study of Pearson, D'Agostino, and Bowman (hereafter referred to as the SWC and PDB studies, respectively). The SWC study compared (among others) all the test statistics described in the chapter except for the mean deviation to standard deviation ratio, the D'Agostino statistic, the Filliben statistic, and the various simplified variants of the Shapiro–Wilk statistic **w**. The PDC study compared (among others) the Shapiro–Wilk statistic **w** with the D'Agostino statistic **y**.

It is useful to classify the various alternatives by their skewness and kurtosis, since the power comparisons yield different "winning" and "losing" test procedures depending on the values of γ_1 and γ_2 of the alternative hypothesis. The classes are:

I. symmetric, platykurtic (short-tailed);
II. symmetric, leptokurtic (long-tailed);
III. asymmetric, platykurtic (short-tailed);
IV. asymmetric, leptokurtic (long-tailed).

For populations in group I the studentized range, $\mathbf{g}_2$, and **w** tests far outperformed the remaining procedures studied by SWC for the PDB study, **w** was more powerful than **y**, except for the case where the alternative was a beta distribution, sometims by a wide margin.

For populations in group II, the studentized range, $\mathbf{g}_2$, $\mathbf{g}_1$, and **w** tests were superior to the remaining procedures studied by SWC. In the PDB study **y** was always superior to **w**, though never by a wide margin.

For populations in group III, **w** had the highest power in the SWC study, and the studentized range also performed well. In the PDB study, **w** outperformed **d**. Among these are the symmetric stable distributions, and Fama and Roll [1971] confirm the good performance of the studentized range relative to **w**, and, because of the simplicity of the computation of the studentized range relative to that of **w**, they recommend it as the preferred test statistic.

For populations in group IV, in the SWC study **w** had highest power, and the studentized range had *lowest* power. Again in the PDB study, **w** outperformed **y**. Among these populations are those with a logrnormal distribution.

There is, however, a subset of population distributions, the "near normal", for which none of the procedures showed much sensitivity, i.e., could not discriminate between the true distribution and the normal distribution. These distributions are characterized by $|\gamma_1| \leq 0.3$ and $-0.5 \leq \gamma_2 \leq 1.5$; included among them is the logistic distribution with $\gamma_1 = 0$ and $\gamma_2 = 1.2$.

In summary, as an omnibus test of normality the Shapiro–Wilk statistic **w** comes out best regardless of alternatives. The studentized range is quite good, except for testing against asymmetric leptokurtic alternatives. The tests based on $\mathbf{g}_2$ are good for symmetric alternatives. And the goodness of fit tests are outperformed by the "winners" in all cases.

One observation made by Pearson, D'Agostino, and Bowman [1977] is that sometimes our alternative hypothesis is directional, e.g., we know that, if nonnormal, the distribution is skewed positively. In such cases, one can sharpen the use of $\mathbf{g}_1$, $\mathbf{g}_2$, and **y** by employing them in one-tailed tests. PDB considered four such directional alternatives:

(a) symmetric and platykurtic (group I);
(b) symmetric and leptokurtic (group II);
(c) asymmetric in a given direction (groups III and IV);
(d) positively skewed and leptokurtic (subset of group IV).

In these instances they studied the appropriate one-tailed tests based on the following statistics:

(a) $\mathbf{g}_2$, **y**;
(b) $\mathbf{g}_2$, **y**;
(c) $\mathbf{g}_1$;
(d) $\mathbf{g}_1$.

They found the following to be the case:

(a) $\mathbf{g}_2$ is more powerful than unidirectional **y**, as well as the omnibus **w**;
(b) there is no clear-cut winner between $\mathbf{g}_2$ and unidirectional **y**, but both surpass the omnibus tests;
(c) and (d) $\mathbf{g}_1$ surpasses all omnibus tests.

D'Agostino [1973] studied the power of the Shapiro–Wilk test (see also Chen [1971]) in contrast to the D'Agostino statistic when the alternative hypothesis is a contaminated normal distribution. When the contamination is scale contamination (i.e., the distribution is a mixture of $N(0, 1)$ and $N(0, \lambda^2)$), the D'Agostino statistic has greater power. When the contamination is location contamination (i.e., the distribution is a mixture of $N(0, 1)$ and $N(\lambda, 1)$), the Shapiro–Wilk test has greater power. In most of the cases studied, though, the difference in power was small.

Finally, one should note that if the distributional form of the alternative hypothesis is precisely specified (rather than generally characterized in terms of skewness and kurtosis), then one might be able to derive sharper tests appropriate for discerning normality from that particular alternative. As noted earlier, Dumonceaux, Antle, and Haas [1973] and Uthoff [1973] showed that the mean deviation to standard deviation ratio is best if the alternative is a double exponential (or LaPlace) distribution. Also, Uthoff [1970] has shown that the studentized range is best if the alternative is a uniform distribution, and that the ratio of $|\mathbf{x}_1 - \bar{\mathbf{x}}|$ to $\mathbf{s}$ (where $\mathbf{x}_1$ is the smallest value of the sample of $\mathbf{x}$'s) is best if the alternative is an exponential distribution.

Appendix I

```
      read(5,*)n
      do 10 i=1,n
      p1=.5+i/(2.*(n+1.))
      p=1.-p1
      t=sqrt(-2*alog(p))
      x=t-(2.515517+.802853*t+.010328*t*t)/(1.+1.432788*t
     +     +.189269*t*t+.001308*t**3)
      write(5,600)i,p1,x
600   format(1h i3,2(x,e12.6))
10    continue
      end
```

References

Blom, G. 1958. *Statistical Estimates and Transformed Beta-Variables.* New York: Wiley.

Chambers, J. M., Cleveland, W. S., Kleiner, B., and Tukey, P. A. 1983. *Graphical Methods for Data Analysis.* Boston: Duxbury Press.

Chen, E. H. 1971. The power of the Shapiro–Wilk *W* test for normality in samples from contaminated normal distributions. *Journal of the American Statistical Association* **66** (December): 760–62.

Chernoff, H. and Lehmann, E. L. 1954. The use of maximum likelihood estimates in χ^2 tests for goodness of fit. *Annals of Mathematical Statistics* **25** (September): 579–86.

Cochran, W. G. 1942. The χ^2 correction for continuity. *Iowa State Journal of Science* **16**: 421–36.

Cramér, H. 1946. *Mathematical Methods of Statistics.* Princeton: Princeton University Press.

D'Agostino, R. B. 1971. An omnibus test of normality for moderate and large size samples. *Biometrika* **58** (August): 341–48.

D'Agostino, R. B. 1972. Small sample probability points for the *D* test of normality. *Biometrika* **59** (April): 219–21.

D'Agostino, R. B. 1973. Monte Carlo power comparison of the *W'* and *D* tests of normality for $n = 100$. *Communications in Statistics* **1**: 545–51.

D'Agostino, R. B. and Rosman, B. 1974. The power of Geary's test of normality. *Biometrika* **61** (April): 181–84.

Dahiya, R. C. and Gurland, J. 1973. How many classes in the Pearson chi-square test? *Journal of the American Statistical Association* **68** (September): 707–12.

Dallal, G. E. and Wilkinson, L. 1986. An analytic approximation to the distribution of Lilliefors's test statistic for normality. *The American Statistician* **40** (November): 294–96.

David, H. A., Hartley, H. O., and Pearson, E. S. 1954. The distribution of the ratio, in a single normal sample, of range to standard deviation. *Biometrika* **41** (December): 482–93.

Dumonceaux, R., Antle, C. E., and Haas, G. 1973. Likelihood ratio test for discrimination between two models with unknown location and scale parameters. *Technometrics* **15** (February): 19–27.

Durbin, J. 1961. Some methods for constructing exact tests. *Biometrika* **48** (June): 41–55.

Fama, E. F. and Roll, R. 1971. Parameter estimates for symmetric stable distributions. *Journal of the American Statistical Association* **66** (June): 331–38.

Filliben, J. J. 1975. The probability plot correlation coefficient test for normality. *Technometrics* **17** (February): 111–18.

Fisher, R. A. 1941. *Statistical Methods for Research Workers*, 8th edn. Edinburgh: Oliver and Boyd.

Gastwirth, J. L. and Owens, M. E. B. 1977. On classical tests of normality. *Biometrika* **64** (April): 135–39.

Geary, R. C. 1935. The ratio of the mean deviation to the standard deviation as a test of normality. *Biometrika* **27**: 310–32.

Geary, R. C. 1947. Testing for normality. *Biometrika* **34**: 209–42.

Gumbel, E. J. 1943. On the reliability of the classical χ^2 test. *Journal of the American Statistical Association* **45** (March): 77–86.

Harter, H. L. 1969. *Order Statistics and Their Use in Testing and Estimation*, Vol. 2. Washington DC: U.S. Government Printing Office.

Kac, M., Kiefer, J., and Wolfowitz, J. 1955. On tests of normality and other tests of goodness of fit based on distance methods. *Annals of Mathematical Statistics* **26** (March): 189–211.

Kendall, M. 1952. *The Advanced Theory of Statistics*, Vol. 1, 5th edn. London: Griffin.

Kendall, M. and Stuart, A. 1977. *The Advanced Theory of Statistics*, Vol. 1. New York: Macmillan.

Kimball, B. F. 1960. On the choice of plotting positions on probability paper. *Journal of the American Statistical Association* **55** (September): 546–60.

Lilliefors, H. W. 1967. On the Kolmogorov–Smirnov test for normality with mean and variance unknown. *Journal of the American Statistical Association* **62** (June): 399–402.

Lloyd, E. H. 1952. Least squares estimation of location and scale parameters using order statistics. *Biometrika* **39**: 88–95.

Mann, H. B. and Wald, A. 1942. On the choice of the number of class intervals in the application of the chi-square test. *Annals of Mathematical Statistics* **13** (June): 306–17.

Michael, J. R. 1983. The stabilized probability plot. *Biometrika* **70** (April): 11–17.

Miller, L. H. 1956. Table of percentage points of Kolmogorov statistics. *Journal of the American Statistical Association* **51** (March): 111–21.

Pearson, E. S., D'Agostino, R. B., and Bowman, K. O. 1977. Tests for departure from normality: Comparison of powers. *Biometrika* **64** (August): 231–46.

Pearson, E. S. and Hartley, H. O. 1958. *Biometrika Tables for Statisticians*, Vol. I. Cambridge: Cambridge University Press.

Pearson, E. S. and Stephens, M. A. 1964. The ratio of range to standard deviation in the same normal sample. *Biometrika* **51**: 484–87.

Pearson, K. 1900. On the criterion that a given system of deviations from the probable in the case of a correlated system of variables is such that it can be reasonably supposed to have arisen from random sampling. *Philosophical Magazine, Series* 5 **50**: 157–72.

Royston, J. P. 1982a. An extension of Shapiro and Wilk's W test for normality to large samples. *Applied Statistics* **31** (No. 2): 115–24.

Royston, J. P. 1982b. Expected normal order statistics (exact and approximate). *Applied Statistics* **31** (No. 2): 161–65.

Shapiro, S. S. and Francia, R. S. 1972. An approximate analysis of variance test normality. *Journal of the American Statistical Association* **67** (March): 215–16.

Shapiro, S. S. and Wilk, M. B. 1965. An analysis of variance test for normality (complete samples). *Biometrika* **52** (December): 591–611.

Shapiro, S. S. and Wilk, M. B. 1968. Approximations for the null distribution of the W statistic. *Technometrics* **10** (November): 861–66.

Shapiro, S. S., Wilk, M. B., and Chen, H. J. 1968. A comparative study of various tests for normality. *Journal of the American Statistical Association* **63** (December): 1343–72.

Shohat, J. A. and Tamarkin, J. D. 1943. *The Problem of Moments.* New York: American Mathematical Society.

Tate, M. W. and Hyer, L. A. 1973. Inaccuracy of X^2 test of goodness of fit when expected frequencies are small. *Journal of the American Statistical Association* **68** (December): 836–41.

Uthoff, V. A. 1970. An optimum property of two well-known statistics. *Journal of the American Statistical Association* **65** (December): 1597–1600.

Uthoff, V. A. 1973. The most powerful scale and location invariant test of the normal versus the double exponential. *Annals of Statistics* **1** (January): 170–74.

Watson, G. S. 1957. The χ^2 goodness-of-fit for normal distributions. *Biometrika* **44** (December): 336–48.

Weisberg, S. and Bingham, C. 1975. An approximate analysis of variance test for non-normality suitable for machine calculation. *Technometrics* **17** (February): 133–34.

Wilk, M. B. and Gnanadesikan, R. 1968. Probability plotting methods for the analysis of data. *Biometrika* **55** (March): 1–17.

Williams, C. A. 1950. On the choice of the number and width of classes for the chi-square test of goodness of fit. *Journal of the American Statistical Association* **45** (March): 77–86.

Yarnold, J. K. 1970. The minimum expectation in X^2 goodness of fit tests. *Journal of the American Statistical Association* **65** (June): 864–86.

CHAPTER 2

Testing for Homoscedasticity

"To make a preliminary test on variances is rather like putting to sea in a rowing boat to find out whether conditions are sufficiently calm for an ocean liner to leave port!"

Box [1953]

0. Introduction

The problem statisticians have, when confronted with several populations with different variances, range from problems solved merely by a minor adjustment to problems for which no satisfactory solution exists. Many statistical procedures, based on the assumption of homoscedasticity of the populations under study, are highly sensitive to deviations of the population variances from equality. It is therefore critical to learn how to test for homoscedasticity. That is the goal of this chapter. But it may not be as critical to learn the appropriate modifications to each and every statistical procedure in the face of heteroscedasticity. It may be more worthwhile to learn portmanteau techniques, good for all occasions, for transforming the various population data sets into homoscedastic ones. That we shall do in the chapter on transformations.

We begin with the basic problem of testing whether two normal distributions have equal variances, followed by the Bartlett test, a classical treatment of the problem of testing whether k normally distributed populations are homoscedastic. It is well known that the Bartlett test is not robust against deviations from the assumed normality of the population. We therefore next study two procedures for testing for homoscedasticity, which are both robust and have good power even when the underlying data distributions are not normal, variants of procedures suggested by Levene and by Fligner and Killeen.

To illustrate the use of the various procedures for variance homogeneity testing described in this chapter, we will use a data set of operating cost per mile in 1951 for a fleet of automobiles, the fleet consisting of 17 Fords, 18

Table 1. Operating Costs per Mile.

Ford	Chevrolet	Plymouth
1.560	2.000	1.610
1.740	2.180	2.270
1.970	2.280	2.384
2.130	2.382	2.430
2.130	2.382	2.430
2.170	2.700	2.460
2.360	2.740	2.516
2.680	2.940	2.524
2.870	3.020	2.570
3.170	3.260	2.980
3.310	3.450	3.040
3.934	3.492	3.060
4.150	3.494	3.100
4.290	3.670	3.150
4.550	3.926	3.340
4.686	4.080	3.480
4.689	4.250	3.530
4.700	5.390	4.940
	5.900	5.940

Chevrolets, and 18 Plymouths. The complete data set is given as Table 1. We identify population 1 with Ford, population 2 with Chevrolet, and population 3 with Plymouth.

One of the more critical areas in which homoscedasticity is required is when our population is that of regression residuals. Unfortunately, the various test procedures referred to above require that the observations be independent. That this is not true for estimated residuals (even though the corresponding true residuals are independent) can be seen merely by noting that the sum of the estimated residuals is zero, so that even if they were homoscedastic (which, as we will see later, they are not) the covariance of any pair of observed residuals from among n is $-\sigma^2/[n-1]$, where σ^2 is the (assumed) common variance. Thus special techniques are required both to obtain independent estimates of the true residuals and to test for homoscedasticity of regression residuals. Section 3 discusses the statistical properties of estimated residuals and describes two procedures for producing independent residuals, the BLUS and the recursive residual method. Section 4 of this chapter describes four procedures for testing for homoscedasticity of residuals, one of which is known to be robust but not used much in econometric practice, and the other three being part of the standard toolkit in econometric practice. Finally, some nonparametric procedures are described.

To illustrate the various techniques for homogeneity testing of regression

Table 2. Netherlands Demand for Textiles 1923–39.

Row	Year	TEXDEM	LOGY/C	LOGDPI
1	1923	1.99651	1.98543	2.00432
2	1924	1.99564	1.99167	2.00043
3	1925	2.00000	2.00000	2.00000
4	1926	2.04766	2.02078	1.95713
5	1927	2.08707	2.02078	1.93702
6	1928	2.07041	2.03941	1.95279
7	1929	2.08314	2.04454	1.95713
8	1930	2.13354	2.05038	1.91803
9	1931	2.18808	2.03862	1.84572
10	1932	2.18639	2.02243	1.81558
11	1933	2.20003	2.00732	1.78746
12	1934	2.14799	1.97955	1.79588
13	1935	2.13418	1.98408	1.80346
14	1936	2.22531	1.98945	1.72099
15	1937	2.18837	2.01030	1.77597
16	1938	2.17319	2.00689	1.77452
17	1939	2.21880	2.01620	1.78746

SOURCE: Theil [1971]. Reprinted from *Principles of Econometrics* by permission of John Wiley & Sons, Inc.

residuals, we use a data set from Koerts and Abrahamse [1969], relating demand for textiles in the Netherlands to x_1 = log per capital income and x_2 = log deflated price index for 1923–39. These data are given in Table 2.

Why are we interested in testing for homoscedasticity? The reason usually given is that various standard statistical test procedures (e.g., testing equality of means, analysis of variance, and testing hypotheses about regression coefficients) all are based on the assumption of homoscedasticity. Though there is evidence that those procedures are quite robust when heteroscedasticity is present (see Box [1954a, b]), it is more reassuring to know that homoscedasticity prevails.

But what of the effect of heteroscedasticity on estimation? This question arises mostly in connection with estimation of regression coefficients. On the one hand, ordinary least squares produce a consistent but inefficient estimate of the set of regression coefficients. On the other hand, the most efficient estimator, based on weighted least squares, requires that we know the covariance matrix of the residuals (except for a scalar factor). Recent research has shown that there is something in between these two extremes to help the statistician interested in estimating regression coefficients efficiently in the face of heteroscedasticity without the requirement of knowledge of the covariance matrix. We do not discuss this problem in this book, but the reader is referred to Carroll [1982] and Carroll and Ruppert [1982] for an exposition of an adaptive robust procedure for estimating regression coefficients in heteroscedastic linear models. (Another approach to this problem involves the estimation of the variances themselves and using these estimates in weighted least squares. One such estimator, the MINQUE estimator, is given in Rao [1970].)

1. Comparing Variances of Two Normal Distributions

Prescription

Let $\mathbf{x}_{11}, \mathbf{x}_{12}, \ldots, \mathbf{x}_{1n_1}$ be n_1 independent observations from a $N(\mu_1, \sigma_1^2)$ distribution and let $\mathbf{x}_{21}, \mathbf{x}_{22}, \ldots, \mathbf{x}_{2n_2}$ be n_2 independent observations from a $N(\mu_2, \sigma_2^2)$ distribution. Also assume that the $\mathbf{x}_{1i}$'s and $\mathbf{x}_{2j}$'s are independent of each other for all i, j. Let μ_1 and μ_2 be unknown. Then the best statistical procedure for testing the hypothesis that $\sigma_1^2 = \sigma_2^2$ is based on the ratio

$$\mathbf{v} = \frac{\mathbf{s}_1^2}{\mathbf{s}_2^2} = \frac{\sum_{j=1}^{n_1} (\mathbf{x}_{1j} - \bar{\mathbf{x}}_1)^2/(n_1 - 1)}{\sum_{j=1}^{n_2} (\mathbf{x}_{2j} - \bar{\mathbf{x}}_2)^2/(n_2 - 1)},$$

where $\bar{\mathbf{x}}_i = \sum_{j=1}^{n_i} \mathbf{x}_{ij}/n_i$. We know that $(n_1 - 1)\mathbf{s}_1^2/\sigma_1^2$ has a $\chi^2(n_1 - 1)$ distribution and $(n_2 - 1)\mathbf{s}_2^2/\sigma_2^2$ has a $\chi^2(n_2 - 1)$ distribution, so, under the hypothesis $\sigma_1^2 = \sigma_2^2$, $\mathbf{v}$ has an $F(n_1 - 1, n_2 - 1)$ distribution.

If our alternative hypothesis is one-sided, then the best procedure to follow is to reject the homoscedasticity hypothesis if $\mathbf{v}$ falls into the appropriate tail of the F distribution. Thus if our alternative were $\sigma_1^2 > \sigma_2^2$ we would reject when $\mathbf{v} \geq F_{0.95}(n_1 - 1, n_2 - 1)$; if it were $\sigma_1^2 < \sigma_2^2$ we would reject when $\mathbf{v} \leq F_{0.05}(n_1 - 1, n_2 - 1)$. In either event, this procedure is best in the sense that its associated probability of a type II error is no larger than that associated with any other procedure, for all possible values of σ_1^2, σ_2^2, μ_1, and μ_2.

When our alternative is $\sigma_1^2 \neq \sigma_2^2$ we are confronted with a slight technical complication. We know that the best procedure for testing for homoscedasticity takes the form: reject if $\mathbf{v} \leq f_1$, or if $\mathbf{v} \geq f_2$, where f_1 and f_2 are so chosen that

$$0.05 = \Pr\{\mathbf{v} \leq f_1\} + \Pr\{\mathbf{v} \geq f_2\}.$$

Unfortunately, the selection of f_1 and f_2 is not completely arbitrary. An optimal procedure and tables to facilitate its implementation have been developed by Ramachandran [1958]. Table 3 contains the values of f_1 and f_2, labeled F_1' and F_2' therein, associated with the optimal two-tailed test procedure.

To begin our series of illustrations, let us assume that the observations in Table 1 are independent and normally distributed and that we wish to test whether the variance of operating costs per mile is identical for Fords and

Table 3. Critical Values for Two-Tailed Test About σ^2.
Values of F_1' and F_2' for $\alpha = 0.05$ and for Different Values of $n_1 - 1, n_2 - 1$.

n_2-1 \ n_1-1	2	3	4	6	8	10	12	16	20	24	30	40	60
2	39.0	33.2	30.5	28.0	26.8	26.1	25.6	25.1	24.8	24.6	24.4	24.2	24.0
3	18.4	15.4	14.0	12.6	11.96	11.58	11.33	11.05	10.83	10.68	10.54	10.40	10.26
4	12.9	10.7	9.60	8.56	8.05	7.75	7.55	7.30	7.16	7.06	6.97	6.88	6.79
6	9.14	7.46	6.64	5.82	5.42	5.17	5.01	4.81	4.69	4.61	4.53	4.45	4.37
8	7.73	6.26	5.53	4.80	4.43	4.21	4.07	3.88	3.77	3.69	3.62	3.54	3.47
10	7.00	5.63	4.95	4.27	3.93	3.72	3.58	3.40	3.29	3.22	3.14	3.06	2.99
12	6.56	5.26	4.61	3.95	3.62	3.41	3.28	3.10	3.00	2.93	2.85	2.78	2.71
16	6.05	4.83	4.21	3.58	3.26	3.06	2.93	2.75	2.65	2.58	2.51	2.44	2.37
20	5.76	4.58	3.98	3.37	3.06	2.87	2.74	2.56	2.46	2.39	2.32	2.25	2.18
24	5.58	4.42	3.84	3.24	2.93	2.74	2.61	2.43	2.34	2.27	2.20	2.13	2.06
30	5.41	4.26	3.70	3.12	2.81	2.62	2.49	2.31	2.22	2.15	2.07	2.00	1.91
40	5.24	4.11	3.56	2.99	2.69	2.50	2.36	2.20	2.11	2.04	1.96	1.87	1.77
60	5.07	3.98	3.44	2.88	2.58	2.39	2.27	2.09	2.00	1.93	1.89	1.79	1.67

* The values given in the table are F_2'. To obtain the value of F_1' for $n_1 - 1, n_2 - 1$, take the reciprocal of F_2' with $n_2 - 1, n_1 - 1$.

SOURCE: Ramachandran [1958]. Reprinted from the *Journal of the American Statistical Association*, Vol. 53, with the permission of the ASA.

Chevrolets. We record that $\mathbf{s}_1^2 = 1.29666$, $\mathbf{s}_2^2 = 1.10620$, $n_1 = 17$, $n_2 = 18$, and so $\mathbf{v} = 1.172$. If our alternative hypothesis were that $\sigma_1^2 > \sigma_2^2$, we would compare $\mathbf{v}$ with $F_{0.95}(16, 17) = 2.32$. If it were that $\sigma_1^2 < \sigma_2^2$, we would compare $\mathbf{v}$ with $F_{0.05}(16, 17) = .44$. And if it were that $\sigma_1^2 \neq \sigma_2^2$, we would compare $\mathbf{v}$ with $F_2'(16, 17) = 2.72$ and $F_1'(16, 17) = 1/F_2'(17, 16) = 1/2.270 = 0.37$. When compared with any of the three alternatives, the homoscedasticity hypothesis is accepted.

Theoretical Background

With regard to the problem of finding the best two-sided test procedure for this problem, suppose one took f_1 as the lower 2.5% point of the $F(n_1 - 1, n_2 - 1)$ distribution (the so-called "equal tail" procedure). It turns out that, for some values of (σ_1^2, σ_2^2), the probability of making a type II error will be greater than 0.95. This means that, at least for those values of (σ_1^2, σ_2^2), a superior procedure would entail disregarding the data, tossing a coin with probability 0.05 of heads, and rejecting the null hypothesis if heads came up. The probability of a type II error for this procedure is 0.95 for all values of σ_1^2, σ_2^2, and is thus superior to one whose probability of type II error is greater than 0.95.

To get around this difficulty statisticians restrict themselves to "unbiased" test procedures, ones whose probability of type II error is always at most one minus the probability of making a type I error, i.e.,

$$\Pr\{\text{type II error}\} \leq 1 - \alpha.$$

In our problem of finding a two-tailed test of the hypothesis that $\sigma_1^2 = \sigma_2^2$, such a condition rules out the "equal tail" test. But there are many choices of (f_1, f_2) which satisfy the "size" condition that $\Pr\{\text{type I error}\} = \alpha$. Amongst these choices we seek the one which, if possible, minimizes the probability of a type II error uniformly in (σ_1^2, σ_2^2). This was accomplished by Ramachandran [1958].

2. Testing Homoscedasticity of Many Populations

Generalizing the situation studied in the previous section, suppose we have k populations, the ith distributed as $N(\mu_i, \sigma_i^2)$, with both μ_i and σ_i^2 unknown for $i = 1, \ldots, k$. We observe $\mathbf{x}_{i1}, \ldots, \mathbf{x}_{in_i}$, a random sample from population i, and we assume that $\mathbf{x}_{ij}$, $\mathbf{x}_{i'j'}$ are independent for all (i, j), (i', j'). We wish to test the homoscedasticity hypothesis $\sigma_1^2 = \sigma_2^2 = \cdots = \sigma_k^2$.

(a) Bartlett Test

Prescription

The usual approach toward constructing a statistical hypothesis test is that of deriving the likelihood ratio procedure. For this problem the likelihood ratio procedure was derived by Neyman and Pearson in Neyman–Pearson [1931]. The exact distribution of the likelihood ratio test statistic was not derived; however, it was found that the test was biased. To correct this flaw, Bartlett [1937] developed the test statistic that bears his name,

$$\mathbf{b} = \frac{\prod_{i=1}^{k} (\mathbf{s}_i^2)^{f_i}}{\left(\sum_{i=1}^{k} f_i \mathbf{s}_i^2\right)},$$

where

$$f_i = \frac{n_i - 1}{\sum_{i=1}^{k} (n_i - 1)},$$

$$\mathbf{s}_i^2 = \frac{\sum_{j=1}^{n_i} (\mathbf{x}_{ij} - \overline{\mathbf{x}}_i)^2}{(n_i - 1)},$$

and

$$\overline{\mathbf{x}}_i = \frac{\sum_{i=1}^{n_i} \mathbf{x}_{ij}}{n_i},$$

Bartlett[1] approximated the critical value of $-\log \mathbf{b}$ as a constant times the $100(1-\alpha)\%$ point of the chi-square distribution with $k-1$ degrees of freedom, where the constant c is given by

$$c = \frac{1 + \left[\dfrac{1}{3(k-1)}\right] \displaystyle\sum_{i=1}^{k} \frac{1}{(n_i - 1)} - \frac{1}{\displaystyle\sum_{i=1}^{k} (n_i - 1)}}{\displaystyle\sum_{i=1}^{k} (n_i - 1)}.$$

Chao and Glaser [1978] found the exact distribution of b in the case $n_1 = n_2 = \cdots = n_k$. A table of critical values of $\mathbf{b}$ when $k = 0.05$ is given as Table 4 herein. Values of $\mathbf{b}$ smaller than the appropriate tabulated value lead to rejection of homoscedasticity. Dyer and Keating [1980] point out that a good approximation to the exact critical value of $\mathbf{b}$ when the n_i's are unequal is to use

$$\sum_{i=1}^{k} \frac{n_i b_k(\alpha; n_i)}{n},$$

where $b_k(\alpha; n_i)$ is the critical value of $\mathbf{b}$ when all the k sample sizes are equal to n_i, and where $n = \sum_{i=1}^{k} n_i$.

To illustrate the Bartlett test using our example, $\mathbf{s}_1^2 = 1.2967$, $\mathbf{s}_2^2 = 1.1062$, $\mathbf{s}_3^2 = 1.0048$, $f_1 = 0.32$, and $f_2 = f_3 = 0.34$, so that $\mathbf{b} = 0.9945136$. Using the Dyer–Keating approximation, we find that the critical value of b is

$$\frac{17 \times 0.8796 + 18 \times 0.8865 + 18 \times 0.8865}{53} = 0.8843,$$

and so we conclude that the populations have equal variances.

To use the Bartlett approximation we calculate $c = 0.0205338$, and so $[-\log \mathbf{b}]/c = 0.2679$, which is to be compared with the 5% point of the $\chi^2(2)$ distribution, i.e., 0.1026. Thus the approximation leads us to conclude that the homoscedasticity hypothesis is to be accepted.

Theoretical Background

Since this test is unbiased, in particular it provides an unbiased procedure for testing the two sample homoscedasticity hypothesis $\sigma_1^2 = \sigma_2^2$ against the al-

[1] Dixon and Massey [1969] give a variant of this statistic, which they dub the "Bartlett Test". In their variant, let

$$a = \left[\frac{1}{3(k-1)}\right]\left[\sum_{i=1}^{k} \frac{1}{(n_i - 1)} - \frac{1}{\displaystyle\sum_{i=1}^{k} (n_i - 1)}\right],$$

$$l = 3(k-1)/a^2,$$

$$g = l/(1 - a + 2/l).$$

Then $l[-\log \mathbf{b}]/(g + \log \mathbf{b})$ has approximately an F distribution with $(k-1, l)$ degrees of freedom.

Table 4. Exact Critical Values of the Bartlett Test.

	Number of Populations, k								
n	2	3	4	5	6	7	8	9	10
					5% points				
3	.3123	.3058	.3173	.3299	*	*	*	*	*
4	.4780	.4699	.4803	.4921	.5028	.5122	.5204	.5277	.5341
5	.5845	.5762	.5850	.5952	.6045	.6126	.6197	.6260	.6315
6	.6563	.6483	.6559	.6646	.6727	.6798	.6860	.6914	.6961
7	.7075	.7000	.7065	.7142	.7213	.7275	.7329	.7376	.7418
8	.7456	.7387	.7444	.7512	.7574	.7629	.7677	.7719	.7757
9	.7751	.7686	.7737	.7798	.7854	.7903	.7946	.7984	.8017
10	.7984	.7924	.7970	.8025	.8076	.8121	.8160	.8194	.8224
11	.8175	.8118	.8160	.8210	.8257	.8298	.8333	.8365	.8392
12	.8332	.8280	.8317	.8364	.8407	.8444	.8477	.8506	.8531
13	.8465	.8415	.8450	.8493	.8533	.8568	.8598	.8625	.8648
14	.8578	.8532	.8564	.8604	.8641	.8673	.8701	.8726	.8748
15	.8676	.8632	.8662	.8699	.8734	.8764	.8790	.8814	.8834
16	.8761	.8719	.8747	.8782	.8815	.8843	.8868	.8890	.8909
17	.8836	.8796	.8823	.8856	.8886	.8913	.8936	.8957	.8975
18	.8902	.8865	.8890	.8921	.8949	.8975	.8997	.9016	.9033
19	.8961	.8926	.8949	.8979	.9006	.9030	.9051	.9069	.9086
20	.9015	.8980	.9003	.9031	.9057	.9080	.9100	.9117	.9132
21	.9063	.9030	.9051	.9078	.9103	.9124	.9143	.9160	.9175
22	.9106	.9075	.9095	.9120	.9144	.9165	.9183	.9199	.9213
23	.9146	.9116	.9135	.9159	.9182	.9202	.9219	.9235	.9248
24	.9182	.9153	.9172	.9195	.9217	.9236	.9253	.9267	.9280
25	.9216	.9187	.9205	.9228	.9249	.9267	.9283	.9297	.9309
26	.9246	.9219	.9236	.9258	.9278	.9296	.9311	.9325	.9336
27	.9275	.9249	.9265	.9286	.9305	.9322	.9337	.9350	.9361
28	.9301	.9276	.9292	.9312	.9330	.9347	.9361	.9374	.9385
29	.9326	.9301	.9316	.9336	.9354	.9370	.9383	.9396	.9406
30	.9348	.9325	.9340	.9358	.9376	.9391	.9404	.9416	.9426
40	.9513	.9495	.9506	.9520	.9533	.9545	.9555	.9564	.9572
50	.9612	.9597	.9606	.9617	.9628	.9637	.9645	.9652	.9658
60	.9677	.9665	.9672	.9681	.9690	.9698	.9705	.9710	.9716
80	.9758	.9749	.9754	.9761	.9768	.9774	.9779	.9783	.9787
100	.9807	.9799	.9804	.9809	.9815	.9819	.9823	.9827	.9830

SOURCE: Dyer and Keating [1980]. Reprinted from the *Journal of the American Statistical Association*, Vol. 75, with the permission of the ASA.

ternative $\sigma_1^2 \neq \sigma_2^2$. In this case

$$\mathbf{b} = \frac{(\mathbf{s}_1^2)^{f_1}(\mathbf{s}_2^2)^{f_2}}{f_1\mathbf{s}_1^2 + f_2\mathbf{s}_2^2}.$$

Since $\mathbf{v} = \mathbf{s}_1^2/\mathbf{s}_2^2$ and $f_1 + f_2 = 1$,

$$\mathbf{b} = \frac{\mathbf{v}^{f_1}(\mathbf{s}_2^2)^{f_1+f_2}}{(f_1\mathbf{v} + f_2)(\mathbf{s}_2^2)} = \frac{\mathbf{v}^{f_1}}{f_1\mathbf{v} + f_2}$$

is a monotonically increasing function of $\mathbf{v}$ for $\mathbf{v} < 1$ and decreasing function of $\mathbf{v}$ for $\mathbf{v} > 1$. Thus this procedure, using the Dyer–Keating table, should be equivalent to the Ramachandran two-tailed procedure based on $\mathbf{v}$.

To illustrate the equivalence, first let $n_1 = 21$, $n_2 = 17$. The Ramachandran tables give $F_1' = 2.65$ and $F_2' = 1/2.56 = 0.391$ as the critical values for $\mathbf{v}$. The value of $\mathbf{b}$ corresponding to each of these values of $\mathbf{v}$ is 0.897. Now the Dyer–Keating table gives $b(0.05, 21) = 0.9063$ and $b(0.05, 17) = 0.8836$, so that the approximate critical value of b is $(21 \times 0.9063 + 17 \times 0.8836)/38 = 0.896$. As another illustration, let $n_1 = n_2 = 21$, so that $F_1' = 2.46$, $F_2' = 1/2.46 = 0.4056$, and the corresponding value of $\mathbf{b}$ is 0.907. The Dyer–Keating table gives 0.9063 as the critical value of $\mathbf{b}$.

Bartlett's test is known to be highly sensitive to departures from normality (see Box [1953]). One way of adjusting this test is to modify the statistic to reflect the kurtosis of the distribution of the populations from which the data were sampled. The correction factor multiplies c by $(1 + \mathbf{g}/2)$, where $\mathbf{g}$ is an overall measure of the kurtosis of the k populations. The estimator $\mathbf{g}$ which produces a robust and most powerful Bartlett test is

$$\mathbf{g} = \frac{n \sum_{i=1}^{k} \sum_{j=1}^{n_i} (\mathbf{x}_{ij} - \tilde{\mathbf{x}}_i)^4}{\left[\sum_{i=1}^{k} (n_i - 1)\mathbf{s}_i^2\right]^2} - 3,$$

Table 5. Absolute Deviations from the Median.

Ford	Chevrolet	Plymouth
1.610	1.355	1.400
1.430	1.175	0.740
1.200	1.075	0.626
1.040	0.973	0.580
1.000	0.655	0.550
0.810	0.615	0.494
0.490	0.415	0.486
0.300	0.335	0.440
0.000	0.095	0.030
0.140	0.095	0.030
0.764	0.137	0.050
0.980	0.139	0.090
1.120	0.315	0.140
1.380	0.571	0.330
1.516	0.725	0.470
1.519	0.895	0.520
1.530	2.035	1.930
	2.545	2.930

where $\tilde{\mathbf{x}}_i$ is the median of the n_i observations from population i (see Conover, Johnson, and Johnson [1981]). The reader should note that this is not the usual estimator of the kurtosis.

To apply this procedure to our data we first determine that $\tilde{\mathbf{x}}_1 = 3.170$, $\tilde{\mathbf{x}}_2 = 3.355$, and $\tilde{\mathbf{x}}_3 = 3.010$. Next we determine the variables $|\mathbf{x}_{ij} - \tilde{\mathbf{x}}_i|$ for the three data sets. These are given in Table 5. From these data we calculate $\mathbf{g}$ to be 0.28124, so that the correction factor for c is 1.1406, and so our statistic is $[-\log \mathbf{b}]/c(1 + \mathbf{g}/2) = 0.2769/1.1406 = 0.2428$, still not significant when compared with $\chi^2_{0.05}(2)$.

(b) Normal Score Procedure (Fligner–Killeen)

Prescription

The Fligner–Killeen [1976] procedure, as modified by Conover, Johnson, and Johnson [1981], for testing homogeneity of variances consists of the following.

Step 1. Rank the variables $|\mathbf{x}_{ij} - \tilde{\mathbf{x}}_i|$ from low to high, where $\tilde{\mathbf{x}}_i$ is the median of the n_i observations from population i.

Step 2. Define

$$a_{n,i} = \Phi^{-1}\left[\frac{1}{2} + \frac{i}{2(n+1)}\right],$$

where $\Phi(z)$ is the cumulative $N(0, 1)$ distribution from $-\infty$ to z and so $\Phi^{-1}(p)$ is the 100pth percentile of the $N(0, 1)$ distribution. (A FORTRAN program for producing the $a_{n,i}$'s is given in Appendix I.)

Step 3. Let $\bar{\mathbf{a}}_i = \sum_{j \in G_i} \mathbf{a}_{n,j}/n_i$, where G_i denotes the sample from population i, $i = 1, \ldots, k$, and $\bar{\mathbf{a}} = \sum_{j=1}^{n} \mathbf{a}_{n,j}/n$. Then

$$\mathbf{x} = \frac{\sum_{i=1}^{k} n_i(\bar{\mathbf{a}}_i - \bar{\mathbf{a}})^2}{\sum_{j=1}^{n} (\mathbf{a}_{n,j} - \bar{\mathbf{a}})^2/(n-1)}$$

has an approximate chi-squared distribution with $k - 1$ degrees of freedom. Also

$$\mathbf{y} = \frac{\mathbf{x}/(k-1)}{(n-1-\mathbf{x})/(n-k)} = \frac{\sum_{i=1}^{k} n_i(\bar{\mathbf{a}}_i - \bar{\mathbf{a}})^2/(k-1)}{\left[\sum_{j=1}^{n} (\mathbf{a}_{n,j} - \bar{\mathbf{a}})^2 - \sum_{i=1}^{k} n_i(\bar{\mathbf{a}}_i - \bar{\mathbf{a}})^2\right] \Big/ (n-k)}$$

has an approximate F distribution with $(k - 1, n - k)$ degrees of freedom.

To apply this procedure to our data, we array our 53 ranked $|x_{ij} - \tilde{x}_i|$ variables, identified as to population, and with the corresponding associated value of $a_{53,i}$, in Table 6. We see that $\bar{\mathbf{a}}_1 = 0.950828$, $\bar{\mathbf{a}}_2 = 0.725826$, $\bar{\mathbf{a}}_3 =$

Table 6. Computations for the Fligner–Killeen Procedure.

$\lvert x_{ij} - \tilde{x}_i \rvert$	Auto	$a_{53,i}$	$\lvert x_{ij} - \tilde{x}_i \rvert$	Auto	$a_{53,i}$
0.000	1	0.02314	0.655	2	0.69547
0.030	3	0.04628	0.725	2	0.72468
0.030	3	0.06944	0.740	3	0.75448
0.050	3	0.09263	0.764	1	0.78492
0.090	3	0.11586	0.810	1	0.81607
0.095	2	0.13915	0.895	2	0.84797
0.095	2	0.16250	0.973	2	0.88072
0.137	2	0.18593	0.980	1	0.91438
0.139	2	0.20945	1.000	1	0.94904
0.140	1	0.23307	1.040	1	0.98483
0.140	3	0.25682	1.075	2	1.02184
0.300	1	0.28069	1.120	1	1.06024
0.315	2	0.30472	1.175	2	1.10018
0.330	3	0.32891	1.200	1	1.14188
0.335	2	0.35327	1.355	2	1.18556
0.415	2	0.37784	1.380	1	1.23152
0.440	3	0.40261	1.400	3	1.28014
0.470	3	0.42762	1.430	1	1.33189
0.486	3	0.45288	1.516	1	1.38736
0.490	1	0.47841	1.519	1	1.44739
0.494	3	0.50423	1.530	1	1.51310
0.520	3	0.53037	1.610	1	1.58615
0.550	3	0.55686	1.930	3	1.66913
0.571	2	0.58371	2.035	2	1.76643
0.580	3	0.61097	2.545	2	1.88679
0.615	2	0.63866	2.930	3	2.05319
0.626	3	0.66681			

0.601069, $\bar{\mathbf{a}} = 0.755626$, and $\sum_{j=1}^{53} (\mathbf{a}_{53,j} - \bar{\mathbf{a}})^2/52 = 0.275708$, so that $\mathbf{x} = 3.779$ and $\mathbf{y} = 1.959$. We compare $\mathbf{x}$ with $\chi^2_{0.95}(2) = 5.99$ and/or $\mathbf{y}$ with $F_{0.95}(2, 50) = 3.18$, and conclude that the hypothesis of homoscedasticity is accepted.

Theoretical Background

One approach toward developing a robust procedure for homoscedasticity testing is to look at the realm of nonparametric procedures. It has been shown (Hájek and Sĭdák [1967]) that only linear rank tests, i.e., tests based on replacing each raw datum by some "score" which is only a function of the relative rank of the datum, are locally most powerful. Let $\mathbf{a}_{n,j}$ denote the score for the jth largest observation in the combined sample of size $n = n_1 + n_2 + \cdots + n_k$. Then the statistics $\mathbf{x}$ and $\mathbf{y}$ as defined above have $\chi^2(k-1)$ and $F(k-1, n-k)$ distributions, respectively, for *all* such choices of score. The particular score noted above in Step 2 was found to be both the most robust and most powerful by Conover, Johnson, and Johnson.

(c) Use of Analysis of Variance (Levene)

Prescription

Let $\mathbf{y}_{ij} = |\mathbf{x}_{ij} - \bar{\mathbf{x}}_i|$. To test for homoscedasticity, perform an analysis of variance to test for equality of the means of the $\mathbf{y}_{ij}$.

The analysis of variance statistic is

$$\mathbf{z} = \frac{\sum_{i=1}^{k} n_i(\bar{\mathbf{y}}_{i\cdot} - \bar{\mathbf{y}}_{\cdot\cdot})^2/(k-1)}{\sum_{i=1}^{k}\sum_{j=1}^{n_i} (\mathbf{y}_{ij} - \bar{\mathbf{y}}_{i\cdot})^2/(n-k)},$$

where

$$\bar{\mathbf{y}}_{i\cdot} = \sum_{j=1}^{n_i} \mathbf{y}_{ij}/n_i \quad \text{and} \quad \bar{\mathbf{y}}_{\cdot\cdot} = \sum_{i=1}^{k}\sum_{j=1}^{n_i} \mathbf{y}_{ij}\Big/\sum_{i=1}^{k} n_i,$$

and is compared with the 95% point of the $F(k-1, n-k)$ distribution to determine the significance of the null hypothesis.

In our data set (see Table 1), we find that $\bar{\mathbf{y}}_{1\cdot} = 0.989941$, $\bar{\mathbf{y}}_{2\cdot} = 0.786111$, $\bar{\mathbf{y}}_{3\cdot} = 0.657555$, $\bar{\mathbf{y}}_{\cdot\cdot} = 0.807830$, and the analysis of variance table is given by:

Source of variation	Sum of squares	df	Mean square
Between	0.978770	2	0.4894
Within	21.238999	50	0.4248
Total	22.217769	52	

Therefore $\mathbf{z} = 1.152$, which, when compared with $F_{0.95}(2, 50) = 3.18$, leads us to accept the hypothesis of homoscedasticity.

Theoretical Background

It is well known that the analysis of variance is quite insensitive to deviations from normality of the underlying populations in testing for equality of population means (see Box [1953], Box and Andersen [1955]). Now consider the variables $\mathbf{y}_{ij} = (\mathbf{x}_{ij} - \bar{\mathbf{x}}_i)^2$, $i = 1, \ldots, k$, $j = 1, \ldots, n_i$. In general, $\mathscr{E}\mathbf{y}_{ij} = \sigma_i^2$. Even when $\mathbf{x}_{ij}$ is distributed as $N(\mu_i, \sigma_i^2)$, the distribution of $\mathbf{y}_{ij}$ is σ_i^2 times that of a $\chi^2(1)$ variable, i.e., nonnormal. But relying on the robustness of the analysis of variance, Levene [1960] suggests that one test the homoscedasticity hypothesis by applying the analysis of variance to the $\mathbf{y}_{ij}$.

Conover, Johnson, and Johnson [1981] investigated variants of this idea and found that the most powerful robust variant is that based on $\mathbf{y}_{ij} = |\mathbf{x}_{ij} - \mathbf{x}_i|$. The $\mathbf{y}_{ij}$ are not independent (nor were the $\mathbf{y}_{ij}$'s defined earlier), but the analysis of variance is even robust against that deviation from basic assumptions. Moreover, the correlation between $\mathbf{y}_{ij}$ and $\mathbf{y}_{ik}$ is of order n_i^{-2} (see

Fisher [1920]), and so should not affect the F distribution of the analysis of variance statistic.

(d) Evaluation

The major evaluative study of homoscedasticity testing procedures is that of Conover, Johnson, and Johnson [1981], who show through simulation studies that the most powerful robust procedures are the variants of the Fligner–Killeen and the Levene tests given herein. They also show that the modification of the Bartlett test to take into account kurtosis is also robust, i.e., the maximum type I error is less than 0.10 when the 0.05 level of significance is used, but is not as powerful as the other two procedures.

With respect to our automobile example, it was clearly homoscedastic and found so by all procedures. The interesting comparisons are between the actual "P values", i.e., levels at which the procedures would have dubbed the data to be heteroscedastic. They are:

Bartlett	0.115
Bartlett adjusted	0.114
Fligner–Killeen	0.151
Levene	0.324

We see then that the Levene procedure yielded a P value twice as large as that of the other procedures. This is consistent with the small shred of evidence in the CJJ study, where additional unreported simulations bore out that for the Levene procedure "the type I error rate sometimes becomes inflated to an unsatisfactory level".

3. Regression Residuals

(a) Standard Estimates

Consider the regression model

$$\mathbf{y}_i = \beta_0 + \beta_1 x_{1i} + \cdots + \beta_p x_{pi} + \mathbf{u}_i, \qquad i = 1, \ldots, n.$$

The variables $\mathbf{u}_1, \ldots, \mathbf{u}_n$ are called the "true residuals" of the regression, and in the standard model are assumed to be independent $N(0, \sigma^2)$ random variables. Unfortunately, these true residuals are not observable, but only estimable from the observations on the $\mathbf{y}_i$ and $\mathbf{x}_{ji}$'s. Thus, we will have to base any test of the hypothesis of homoscedasticity of the true residuals on some estimates $\hat{\mathbf{u}}_i$ of the $\mathbf{u}_i$. Again unfortunately, as will be shown later, the estimated residuals are not independent, nor are they homoscedastic, even though the true residuals satisfy these conditions.

Theoretical Background

It is simplest to express our estimates of the u_i using matrix notation. Let $\mathbf{Y}$ be the $n \times 1$ vector of $\mathbf{y}_i$'s, X be the $n \times p + 1$ matrix

$$\begin{bmatrix} 1 & x_{11} & \cdots & x_{p1} \\ \vdots & & & \\ 1 & x_{1n} & \cdots & x_{pn} \end{bmatrix},$$

B be the $p + 1 \times 1$ vector of regression coefficients

$$B' = [\beta_0 \beta_1 \dots \beta_p],$$

and $\mathbf{U}$ be the $n \times 1$ vector of u_i's. Then the model is expressible as

$$\mathbf{Y} = XB + \mathbf{U}$$

and the usual estimator of B is given by

$$(X'X)^{-1}X'\mathbf{Y}.$$

The usual estimator of the true residual vector is given by

$$\hat{\mathbf{U}} = \mathbf{Y} - X\hat{\mathbf{B}} = [I - X(X'X)^{-1}X']\mathbf{Y}.$$

Let $M = I - X(X'X)^{-1}X'$. Then $\hat{\mathbf{U}} = M\mathbf{Y}$, so that

$$\mathscr{E}\hat{\mathbf{U}} = \mathscr{E}M\mathbf{Y} = 0,$$

and

$$\mathscr{V}\hat{\mathbf{U}} = \mathscr{V}[M\mathbf{Y}] = \sigma^2 M^2 = \sigma^2 M,$$

since M is idempotent. Thus we see that the usual estimated residuals are themselves heteroscedastic and also, since M is not a diagonal matrix, not independent.

We also note that, since $\hat{\mathbf{U}}$ is a linear transformation of $\mathbf{Y}$, it has the multivariate normal distribution $N(0, \sigma^2 M)$. Now

$$\begin{aligned} \mathscr{E}\hat{\mathbf{U}}'\hat{\mathbf{U}} &= \mathscr{E}\mathbf{Y}'M\mathbf{Y} = \operatorname{tr} \mathscr{E}\mathbf{Y}'M\mathbf{Y} = \mathscr{E} \operatorname{tr} \mathbf{Y}'M\mathbf{Y} = \operatorname{tr} M\mathscr{E}\mathbf{Y}\mathbf{Y}' \\ &= \operatorname{tr} M\sigma^2 I = \sigma^2 \operatorname{tr} M = \sigma^2 \operatorname{tr}[I - X(X'X)^{-1}X'] \\ &= \sigma^2[n - \operatorname{tr}(X'X)^{-1}(X'X)] \\ &= \sigma^2[n - (p + 1)]. \end{aligned}$$

Therefore the rank of M is $n - p - 1$, and so one can find an orthogonal matrix Q such that $QMQ' = I_{n-p-1}$, where I_{n-p-1} is a diagonal matrix whose first $n - p - 1$ diagonal elements are equal to 1 and all others are equal to 0. Then, letting $\mathbf{V} = Q\hat{\mathbf{U}}$, we find that the covariance matrix of $\mathbf{V}$ is $Q(\sigma^2 M)Q' =$

$\sigma^2 I_{n-p-1}$. Therefore

$$\mathbf{V}'\mathbf{V} = \hat{\mathbf{U}}'Q'Q\hat{\mathbf{U}} = \hat{\mathbf{U}}'\hat{\mathbf{U}}$$

is distributed as a $\sigma^2\chi^2(n - p - 1)$ variable.

The matrix $P = I - M$ is sometimes referred to as the "hat matrix". Since $\mathscr{V}\hat{\mathbf{u}}_i = \sigma^2(1 - p_{ii})$, the set of quantities defined by

$$\mathbf{w}_i = \frac{\hat{\mathbf{u}}_i}{\hat{\boldsymbol{\sigma}}\sqrt{1 - p_{ii}}}, \qquad i = 1, \ldots, n,$$

has been found useful as a way of rescaling residuals to make them comparable to each other. These quantities are called "studentized residuals" (or sometimes "internally studentized residuals", to contrast them from quantities wherein the $\hat{\mathbf{u}}_i$ are divided by an estimate of σ which is independent of the $\hat{\mathbf{u}}_i$).[2] The studentized residual has the property that $\mathscr{E}\mathbf{w}_i = 0$ and $\mathscr{V}\mathbf{w}_i = 1$, but its distribution is neither normal nor Student's t. The distribution of $\mathbf{w}_i$ is tabulated in Lund [1975]. These residuals are useful in the analysis of potential outliers in the regression data set.

(b) Homoscedastic Estimated Residuals

Prescription

As noted earlier, the estimated residuals are neither independent nor homoscedastic. Two procedures for creating homoscedastic estimated residuals are the Theil BLUS procedure and the Hedayat–Robson method of recursive residuals. Each of these methods involves considerable matrix calcuation, described in the subsequent theoretical background sections. Results of the use of each of these procedures, along with the least squares estimation of the residuals for our example data set of Table 1, are given in Table 7.

These recursive and BLUS residuals can be used in the Goldfeld–Quandt procedure (Section 4(a) of this chapter) for testing for homoscedasticity of residuals. Harvey and Phillips [1974] point out that such residuals are most useful when applying the Goldfeld–Quandt procedure many times with different orderings of the observations.

Theoretical Background

(i) BLUS

Theil [1965] introduced the concept of the BLUS (Best Linear Unbiased Scalar covariance matrix) residual vector, and found a procedure for determining a set of estimated residuals $\mathbf{U}^*$ whose covariance matrix is $\sigma^2 I$, which is calculated as a linear transformation of $\mathbf{Y}$, whose expected value is 0, and

[2] MINITAB produces the p_{ii} when the subcommand HI is given. It automatically produces the $\mathbf{w}_i$, which are called "standardized residuals" in this program. MINITAB also produces two other "residuals", the $\hat{\mathbf{u}}_i$ (with the subcommand RESIDS) and a quantity called a "studentized" or "studentized deleted" residual (with the subcommand TRESIDS). This latter residual will be studied in Chapter 6. It is unfortunate that the MINITAB nomenclature does not conform with the statistical literature.

Table 7. Comparison of Residuals. BLUS and Recursive Residuals Excluding Observations 1, 8, and 14.

Row	RESID	BLUS	RECUR
1	0.01420		
2	0.00297	−0.00171	−0.00569
3	−0.00255	−0.00691	−0.00773
4	−0.01418	−0.01716	−0.01410
5	0.00856	0.00586	0.00697
6	−0.01632	−0.01853	−0.01141
7	−0.00586	−0.00794	0.00001
8	0.00546		
9	0.01351	0.01274	0.01291
10	0.00534	0.00437	0.00155
11	0.01295	0.01179	0.00490
12	−0.00037	−0.36394	−0.00805
13	−0.01307	−0.01533	−0.01595
14	0.00356		
15	−0.01164	−0.01253	−0.01218
16	−0.02413	0.03514	−0.02132
17	0.02157	−0.26262	0.02157

which is "best" in the sense that $\mathscr{E}\mathbf{U}^*\mathbf{U}^{*\prime} \leq \mathscr{E}\tilde{\mathbf{U}}\tilde{\mathbf{U}}'$ for any $\tilde{\mathbf{U}}$ linear in $\mathbf{Y}$ with scalar covariance matrix. To calculate $\mathbf{U}^*$ one must first partition X into submatrices

$$X = \begin{bmatrix} X_0 \\ X_1 \end{bmatrix},$$

where X_0 is $(p + 1) \times (p + 1)$ and X_1 is $(n - p - 1) \times (p + 1)$ and where X_0 is nonsingular. Now consider the $(p + 1) \times (p + 1)$ matrix $X_0(X'X)^{-1}X_0'$. It is of full rank, and has $p + 1 - k$ eigenvalues equal to 1, where k is the rank of X_1, and h eigenvalues less than 1. Let these h eigenvalues be $d_1^2, \ldots, d_h^2$ with corresponding $(p + 1) \times 1$ eigenvectors $Q_1, \ldots, Q_h$. Let $\hat{\mathbf{U}}_0 = \mathbf{Y}_0 - X_0\hat{\mathbf{B}}$, $\hat{\mathbf{U}}_1 = \mathbf{Y}_1 - X_1\hat{\mathbf{B}}$ denote the usual estimated residuals corresponding to the partition of X. Then

$$\mathbf{U}^* = \hat{\mathbf{U}}_1 - X_1 X_0'^{-1}\left[\sum_{i=1}^{h} \frac{d_i}{1 + d_i} Q_i Q_i'\right]\hat{\mathbf{U}}_0.$$

In order to develop a most stable set of BLUS residuals, we should select for the matrix X_0 the most uncollinear set of three independent variables. For our example, we find that observations 1, 8, and 14 produce the X_0 matrix with the largest determinant, and so we will have

$$X_0 = \begin{bmatrix} 1 & 1.98543 & 2.00432 \\ 1 & 1.05038 & 1.91803 \\ 1 & 1.98945 & 1.72099 \end{bmatrix}.$$

We record that

$$X'X = \begin{bmatrix} 17 & & \\ 34.2078 & 68.8418 & \\ 31.8339 & 64.0646 & 59.7595 \end{bmatrix},$$

so that

$$(X'X)^{-1} = \begin{bmatrix} 518.890 & & \\ -254.252 & 132.704 & \\ 0.416672 & -6.82419 & 7.11057 \end{bmatrix},$$

and

$$X_0(X'X)^{-1}X_0' = \begin{bmatrix} 0.325685 & & \\ -0.0602665 & 0.243032 & \\ -0.00943665 & -0.0589487 & 0.243939 \end{bmatrix}.$$

We find that $d_1^2 = 0.361245$, $d_2^2 = 0.283368$, $d_3^2 = 0.168043$, and the matrix of eigenvectors is given by

$$Q = \begin{bmatrix} -0.831080 & 0.520884 & -0.194899 \\ 0.458076 & 0.442378 & -0.771017 \\ -0.315391 & -0.730056 & -0.606256 \end{bmatrix}.$$

Then

$$R = \sum_{i=1}^{h} \frac{d_i}{1 + d_i} Q_i Q_i' = \begin{bmatrix} 0.361106 & & \\ -0.0251688 & 0.324802 & \\ -0.00629532 & -0.027917 & 0.327638 \end{bmatrix},$$

and

$$X_1 X_0^{-1} R = \begin{bmatrix} 0.32943 & 0.00884 & -0.01368 \\ 0.29303 & 0.05466 & -0.02928 \\ 0.15059 & 0.16529 & -0.01779 \\ 0.12574 & 0.16352 & 0.00646 \\ 0.06501 & 0.26746 & -0.04861 \\ 0.04829 & 0.29609 & 0.06377 \\ -0.06388 & 0.25371 & 0.08205 \\ -0.03142 & 0.16194 & 0.14973 \\ -0.00110 & 0.07629 & 0.21289 \\ 0.12887 & -0.07584 & 0.25647 \\ 0.11873 & -0.05024 & 0.23857 \\ -0.02813 & 0.09169 & 0.22098 \\ -0.01524 & 0.07279 & 0.22933 \\ -0.03934 & 0.12518 & 0.19570 \end{bmatrix}.$$

Finally, the BLUS residuals are given as $\hat{U}_1 - X_1 X_0^{-1} R \hat{U}_0$ and are recorded in Table 8, along with the least squares residuals.

It has been suggested that these residuals be augmented by $p + 1$ additional residuals so that one might have available a full set of n uncorrelated residuals to use in further analysis. Dent and Styan [1973] have shown that such augmentation produces residuals with lower variance if and only if $\sum_{i=1}^{p+1} d_i/(p+1) < \frac{1}{2}$. In our case, this sum is 0.51443, and so we see that no augmentation procedure will improve on the BLUS set of residuals.

(ii) Recursive Residuals

Another procedure for constructing estimated residuals is the method of recursive residuals described by Hedayat and Robson [1970] and studied by Harvey and Phillips [1974]. This procedure is based on the following observation. Let $Y_{(k)}$ be the first k rows of Y and

$$X_{(k)} = \begin{bmatrix} X_1' \\ \vdots \\ X_k' \end{bmatrix}$$

be the first k rows of the $n \times p + 1$ matrix X. Let $\hat{\mathbf{B}}_k$ be the least squares estimator of B based on the first k rows of X, i.e., $\hat{\mathbf{B}}_k = (X_{(k)}' X_{(k)})^{-1} X_{(k)}' \mathbf{Y}_{(k)}$, $k \geq p + 1$. Let $\tilde{\mathbf{u}}_k = \mathbf{y}_k - X_k' \hat{\mathbf{B}}_k$. Then the sequence of residuals $\tilde{\mathbf{u}}_j$, $j = p + 2, \ldots, n$, are independent.

Thus we can begin by taking $k = p + 2$ and estimating $\mathbf{u}_{p+2}$ by $\tilde{\mathbf{u}}_{p+2}$. Next we take $k = p + 3$ and determine $\tilde{\mathbf{u}}_{p+3}$, and so on until $\tilde{\mathbf{u}}_n$ has been calculated. In all, we can estimate $n - p - 1$ residuals using this successive procedure.

This procedure can be made recursive, i.e., there is no need to actually rerun a regression program $n - p - 1$ times, by virtue of the relationship

$$\hat{\mathbf{B}}_j = \hat{\mathbf{B}}_{j-1} + \frac{(\mathbf{y}_j - X_j' \hat{\mathbf{B}}_{j-1})(X_{(j-1)}' X_{(j-1)})^{-1} X_j}{1 + X_j'(X_{(j-1)}' X_{(j-1)})^{-1} X_j}.$$

The recursive residuals are given by

$$\tilde{\mathbf{u}}_j = \frac{\mathbf{y}_j - X_j' \hat{\mathbf{B}}_j}{1 - X_j'(X_{(j)}' X_{(j)})^{-1} X_j}.$$

Moreover,

$$(X_{(j)}' X_{(j)})^{-1} = (X_{(j-1)}' X_{(j-1)})^{-1} - \frac{(X_{(j-1)}' X_{(j-1)})^{-1} X_j X_j' (X_{(j-1)}' X_{(j-1)})^{-1}}{1 + X_j'(X_{(j-1)}' X_{(j-1)})^{-1} X_j}.$$

When applied to our data set (reorganized so that the first three observations are observations 1, 8, and 14, in order to provide comparability with the BLUS residuals), we found that recursive residuals procedure produced residuals with even lower variance than that of BLUS. The recursive residuals are given in Table 7. The respective variances of the BLUS and recursive residuals (using 13 as denominator) are 0.0134764 and 0.0001447. Note that the recursive

Table 8. *C*-Matrix for Anscombe's Test.

0.326	0.294	0.256	0.115	0.093	0.028	0.010	−0.060	−0.089	−0.050	−0.015	0.119	0.107	−0.009	−0.041	−0.027	−0.054
0.294	0.267	0.237	0.117	0.096	0.045	0.031	−0.031	−0.064	−0.038	−0.013	0.096	0.088	−0.018	−0.036	−0.025	−0.045
0.256	0.237	0.215	0.121	0.101	0.070	0.062	0.009	−0.034	−0.023	−0.013	0.064	0.061	−0.034	−0.032	−0.025	−0.035
0.115	0.117	0.121	0.109	0.099	0.118	0.123	0.105	0.059	0.034	0.010	−0.001	0.006	−0.036	0.005	0.003	0.015
0.093	0.096	0.101	0.099	0.091	0.110	0.115	0.104	0.066	0.043	0.021	0.005	0.012	−0.018	0.019	0.016	0.028
0.028	0.045	0.070	0.118	0.110	0.173	0.190	0.193	0.129	0.068	0.011	−0.071	−0.054	−0.069	0.016	0.005	0.038
0.010	0.031	0.062	0.123	0.115	0.190	0.211	0.218	0.147	0.075	0.008	−0.092	−0.072	−0.083	0.015	0.002	0.041
−0.060	−0.031	0.009	0.105	0.104	0.193	0.218	0.243	0.183	0.104	0.030	−0.101	−0.079	−0.059	0.044	0.027	0.072
−0.089	−0.064	−0.034	0.059	0.066	0.129	0.147	0.183	0.166	0.118	0.072	−0.033	−0.019	0.031	0.088	0.076	0.105
−0.050	−0.038	−0.023	0.034	0.043	0.068	0.075	0.104	0.118	0.104	0.091	0.038	0.043	0.091	0.101	0.096	0.106
−0.015	−0.013	−0.013	0.010	0.021	0.011	0.008	0.030	0.072	0.091	0.108	0.105	0.100	0.147	0.114	0.115	0.107
0.119	0.096	0.064	−0.001	0.005	−0.071	−0.092	−0.101	−0.033	0.038	0.105	0.208	0.188	0.195	0.097	0.111	0.071
0.107	0.088	0.061	0.006	0.012	−0.054	−0.072	−0.079	−0.019	0.043	0.100	0.188	0.171	0.179	0.094	0.106	0.071
−0.009	−0.018	−0.034	−0.036	−0.018	−0.069	−0.083	−0.059	0.031	0.091	0.147	0.195	0.179	0.244	0.152	0.160	0.129
−0.041	−0.036	−0.032	0.005	0.019	0.016	0.015	0.044	0.088	0.101	0.114	0.097	0.094	0.152	0.123	0.123	0.118
−0.027	−0.025	−0.025	0.003	0.016	0.005	0.002	0.027	0.076	0.096	0.115	0.111	0.106	0.160	0.123	0.124	0.115
−0.054	−0.045	−0.035	0.015	0.028	0.038	0.041	0.072	0.105	0.106	0.107	0.071	0.071	0.129	0.118	0.115	0.117

residuals do not satisfy one of the BLUS requirements, the scalar covariance matrix condition. Though they are independent, they are not homoscedastic. To convert recursive residuals to homoscedasticity, one must calculate appropriate scale factors with which to deflate them. Examples of such scale factors are given in Hedayat and Robson [1970], but no general formula is derived.

4. Testing Homoscedasticity of Regression Residuals

Before discussing formal procedures for testing for homoscedasticity of residuals, let me direct the reader to an informal graphic procedure which gives some insight into this issue, the residuals versus fitted (RVSF) plot (see Anscombe and Tukey [1963], Section 9]. Such a graph plots the fitted as the abscissa and the residual as the ordinate, and is illustrated by Figure 1 based on the Koerts–Abrahamse data. Absence of heteroscedasticity is visually indicated by a scatter of points uniformly distributed across the graph. Inspection of Figure 1 would lead one to conclude that the residuals were homoscedastic. A cornucopia shaped scatter, by contrast, indicates residuals increasing with magnitude of $\hat{\mathbf{y}}$. Since $\hat{\mathbf{y}}$ is a monotonic function of each of the

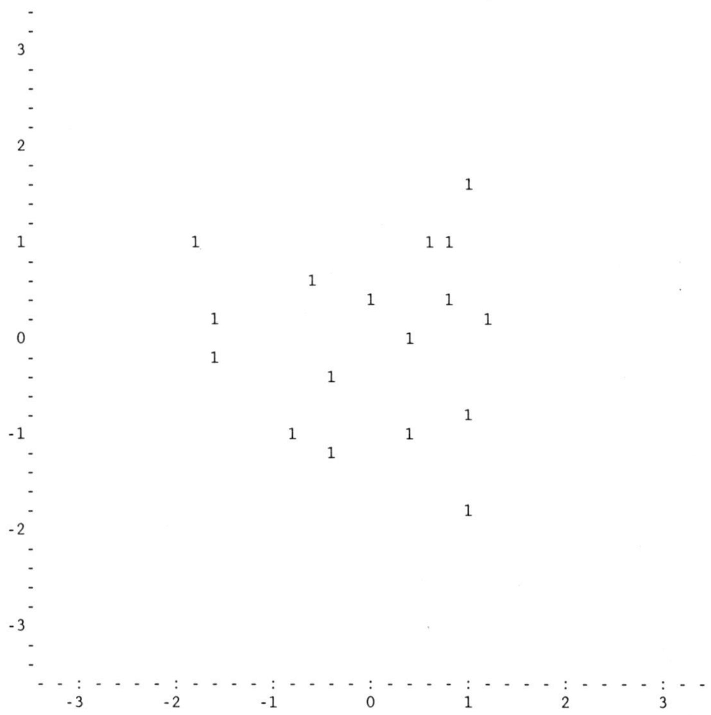

Figure 1. Scatter Plot of 17 Standardized Values of RESID versus FITTED.

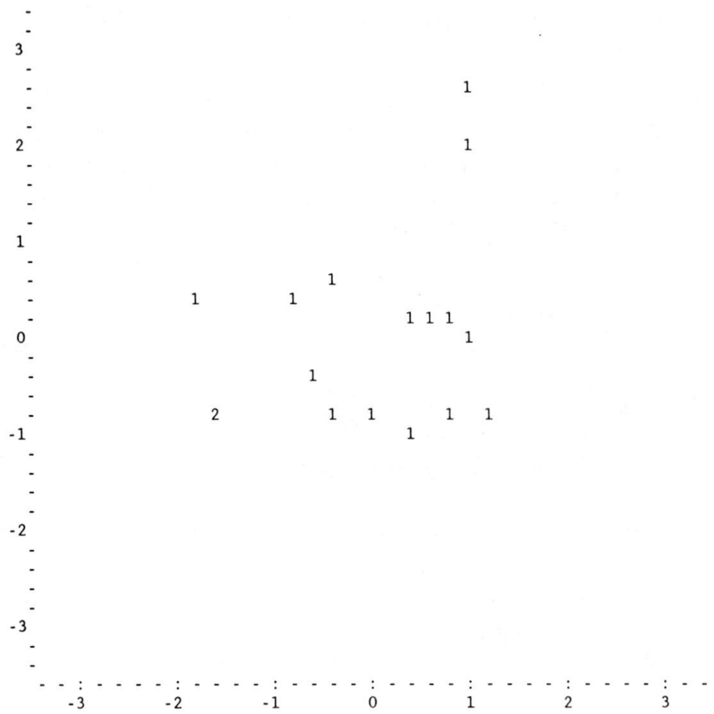

Figure 2. Scatter Plot of 17 Standardized Values of Squared RESID versus FITTED.

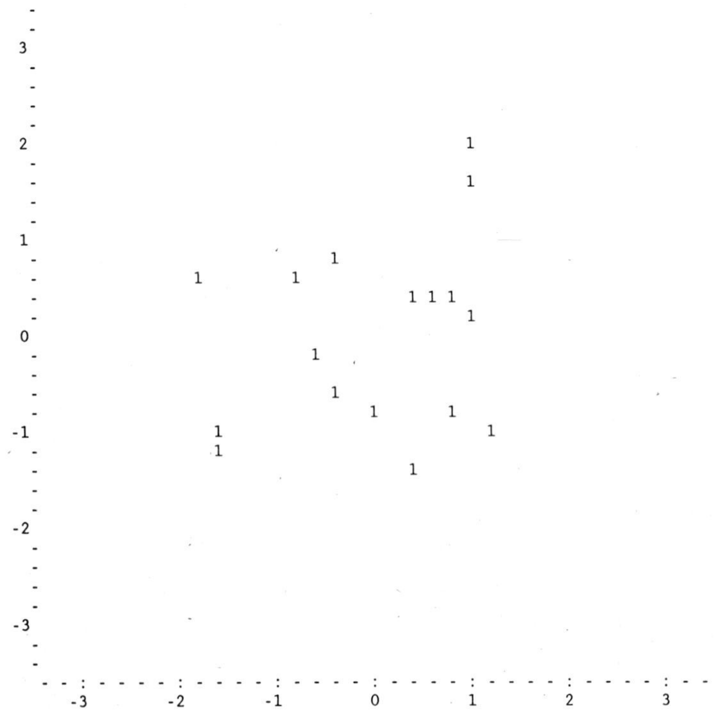

Figure 3. Scatter Plot of 17 Standardized Values of Absolute Value of RESID versus FITTED.

independent variables, it may be possible, by subsequently plotting $\hat{y}$ against each of the independent variables, to ascertain which independent variable(s) cause the heteroscedasticity.

The correlation between the fitted $\hat{\mathbf{y}}_i$ and the residuals $\hat{\mathbf{u}}_i$ is always equal to zero. However, if there is heteroscedasticity it can be detected by a comparison of a suitable function of the $\hat{\mathbf{u}}_i$ with the fitted $\hat{\mathbf{y}}_i$. One function, suggested by Anscombe [1961] and Anscombe and Tukey [1963], is $\hat{\mathbf{u}}_i^2$. A computation of the correlation between the $\hat{\mathbf{u}}_i^2$ and the $\hat{\mathbf{y}}_i$ is the basis for the Anscombe procedure for testing for homoscedasticity given in Section 3(d). This idea has been generalized by Bickel [1978] to allow the data analyst to select two functions, one a function $\alpha(\hat{\mathbf{y}}_i)$ of the fitted and the other a function $b(\hat{\mathbf{u}}_i)$ of the residual, and calculate the correlation between the $\alpha(\hat{\mathbf{y}}_i)$ and $b(\hat{\mathbf{u}}_i)$. A simple choice might be $\alpha(\hat{y}_i) = \hat{y}_i$ and $b(\hat{u}_i) = |u_i|$. Figure 2 depicts the plot of $\hat{u}_i^2$ versus $\hat{y}_i$, and Figure 3 depicts the plot of $|u_i|$ versus $\hat{y}_i$ for our example data. Both these figures show much greater scatter at the largest value of $\hat{y}_i$, thus indicating potential heteroscedasticity.

Another procedure which has its origins in a plotting diagnostic is that of the Lagrange multiplier test described in Section 4(b). Here $\hat{u}_i^2$ is plotted versus each of the x_{ji}'s. A significant regression slope in any of these plots not only indicates heteroscedasticity but pinpoints its source. Figures 4 and 5 plot $\hat{u}_i^2$

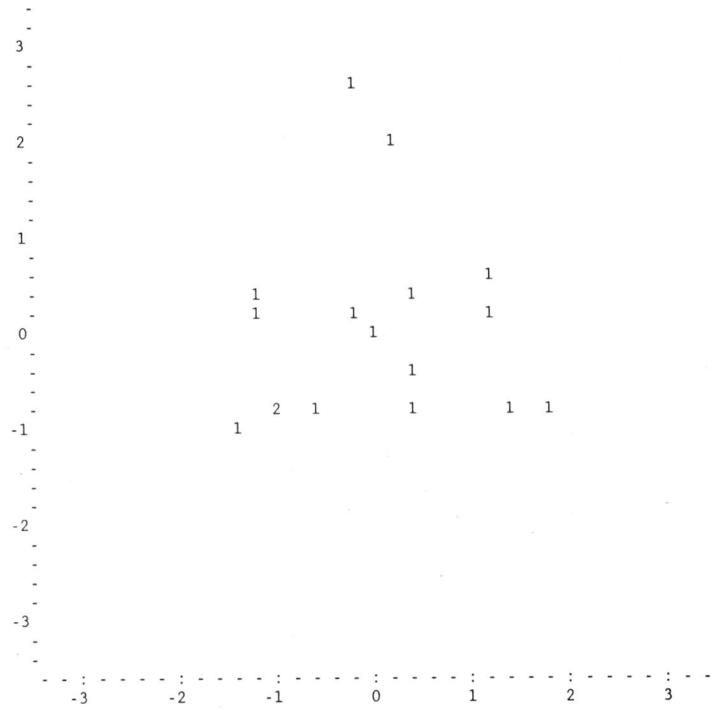

Figure 4. Scatter Plot of 17 Standardized Values of Squared RESID versus LOGY/C.

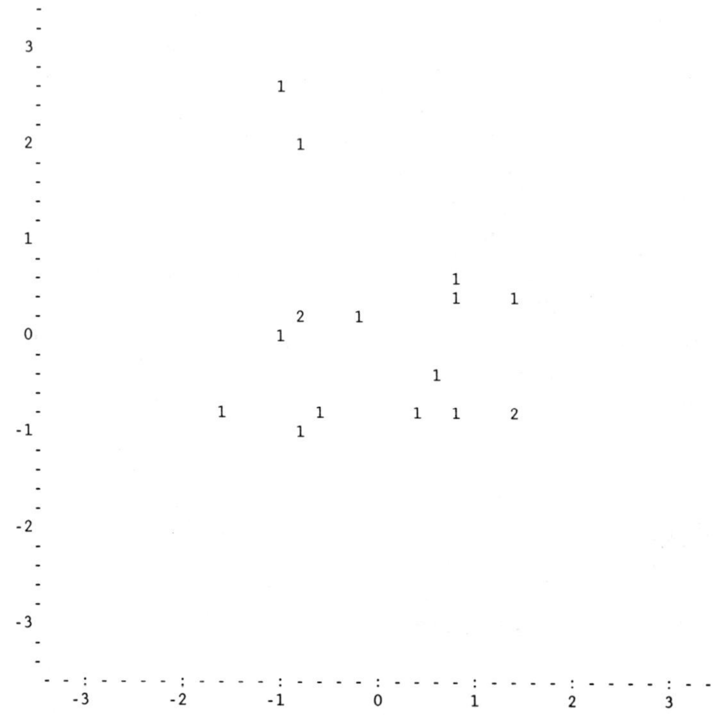

Figure 5. Scatter Plot of 17 Standardized Values of Squared RESID versus LOGDPI.

against x_{1i} and x_{2i}, respectively. These figures indicate potential heteroscedasticity induced by each of their variables. We will see in succeeding sections of this chapter how well the various techniques fared in detecting heteroscedasticity of residuals for this example.

(a) Goldfeld–Quandt Procedure

Prescription

Suppose we split the data into three disjoint sets of k_1, k_2, and k_3 observations, and disregard the set of k_2 observations. If we separately estimate B using the first set of k_1 observations and the last set of k_3 observations, then the vectors of estimated residuals from these two regressions, $\hat{\mathbf{U}}_1$ and $\hat{\mathbf{U}}_3$, are independent vectors under the null hypothesis of homoscedasticity, and the ratio

$$\mathbf{q} = \frac{(k_3 - p - 1)\mathbf{U}'\mathbf{U}_1}{(k_1 - p - 1)\mathbf{U}_3'\mathbf{U}_3}$$

has an $F(k_1 - p - 1, k_3 - p - 1)$ distribution.

Goldfeld and Quandt [1965] introduced this idea for testing for homoscedasticity of regression residuals. Their paper discusses the procedure in the context of determining whether an independent variable x_l, suspected of being the source of heteroscedasticity in the true residuals [i.e., $\mathscr{V}\mathbf{u}_i = \phi(x_{li})$], is indeed culpable. They therefore suggest sorting the n observations on the values of the x_l, so that the data ordering conforms to $x_{l1} \leq x_{l2} \leq \cdots \leq x_{ln}$. If the variance of the true residual increases with the size of x_l, then small values of q will be significant. Likewise, if the variances of the $\mathbf{u}_i$ decreases with the size of the x_l, then large values of $\mathbf{q}$ will be significant.

With regard to the choice of k_1, k_2, and k_3, it is clear that as k_2 goes to zero $\mathbf{q}$ will be based on more and more observations. However, as k_2 goes to zero the effect of the lower values of x_l in the first k_1 observations on the numerator will tend to be less sharply delineated from that of the higher values of x_l in the last k_3 observations in the denominator. A rough rule of thumb based on simulation studies is to take $k_1 = k_3$ and $k_2 = n/4$, i.e., use three-quarters of the observations, for maximum power.

To apply the Goldfeld–Quandt procedure to the data of Table 2, we sort the data on x_1, eliminate the middle five observations, separately regress $\mathbf{y}$ on x_1 and x_2 for the first six observations and the last six observations, and record the mean square of the residuals for each of these two expressions. We find them to be 1.0498×10^{-4} and 3.37913×10^{-5}, respectively, yielding an $F(3, 3)$ statistic of 3.1067. When this is compared with $F_{0.95}(3, 3) = 9.28$, we see that x_1 does not induce heteroscedasticity in the residuals.

If we sort the data on x_2 and follow the same procedure, we find the mean square of the residuals to be 4.45945×10^{-4} and 5.53103×10^{-5}, respectively, for an $F(3, 3) = 8.0626$. Again we conclude that x_2 does not contribute to heteroscedasticity in the residuals. From this we conclude that the residuals are homoscedastic.

Theoretical Background

Suppose we partitioned the vector $\mathbf{Y}$ into three subvectors, $\mathbf{Y}' = [\mathbf{Y}_1' \ \mathbf{Y}_2' \ \mathbf{Y}_3']$, where $\mathbf{Y}_1$ is a k_1 vector, $\mathbf{Y}_2$ is a k_2 vector, and $\mathbf{Y}_3$ is a k_3 vector, with $k_1 + k_2 + k_3 = n$. Similarly, let us partition X into

$$X = \begin{bmatrix} X_1' \\ X_2' \\ X_3' \end{bmatrix},$$

where X_i is a $p + 1 \times k_i$ matrix, and $\mathbf{U}'$ into $\mathbf{U}' = [\mathbf{U}_1' \ \mathbf{U}_2' \ \mathbf{U}_3']$. Then our model is expressible as

$$\mathbf{Y}_1 = X_1'B + \mathbf{U}_1,$$

$$\mathbf{Y}_2 = X_2'B + \mathbf{U}_2,$$

$$\mathbf{Y}_3 = X_3'B + \mathbf{U}_3.$$

If we separately estimated B using the first set of k_1 observations and the last set of k_3 observations, and estimated the residuals $\mathbf{U}_1$ and $\mathbf{U}_3$, respectively, by

$$\hat{\mathbf{U}}_1 = \mathbf{Y}_1[I - X_1(X_1'X_1)^{-1}X_1'] = \mathbf{Y}_1 M_1,$$
$$\hat{\mathbf{U}}_3 = \mathbf{Y}_3[I - X_3(X_3'X_3)^{-1}X_3'] = \mathbf{Y}_3 M_3,$$

then $\hat{\mathbf{U}}_1$ and $\hat{\mathbf{U}}_3$ are independent vectors. Under the null hypothesis of homoscedasticity, $\mathscr{E}\hat{\mathbf{U}}_i'\hat{\mathbf{U}}_i = \sigma^2(k_i - p - 1)$, and $\hat{\mathbf{U}}_i'\hat{\mathbf{U}}_i$ is distributed as a $\sigma^2\chi^2(k_i - p - 1)$ variable. Therefore the ratio

$$\mathbf{q} = \frac{(k_3 - p - 1)\mathbf{U}_1'\mathbf{U}_1}{(k_1 - p - 1)\mathbf{U}_3'\mathbf{U}_3}$$

will have an $F(k_1 - p - 1, k_3 - p - 1)$ distribution under the null hypothesis.

Harrison and McCabe [1979] point out that one might generalize the approach taken by Goldfeld and Quandt and consider using a ratio of the form

$$\mathbf{q} = \frac{\hat{\mathbf{U}}'A\hat{\mathbf{U}}}{\hat{\mathbf{U}}'\hat{\mathbf{U}}}$$

to test for heteroscedasticity, where A is a diagonal "selector" matrix with 1's in positions on the diagonal selected to highlight the particular suspected deviation from heteroscedasticity. For example, if the heteroscedasticity is related to increasing values of x_l, then m diagonal elements of A corresponding to the smallest values of x_l will be set to 1. They found a beta distribution approximation to the distribution of q, but the approximation depends on the eigenvalues of MA. In the most common case, where $m > p + 1$ and $n - m > p + 1$, there will be $m - p - 1$ eigenvalues equal to 1, $n - m - p - 1$ eigenvalues equal to 0, and $p + 1$ eigenvalues λ_i between 0 and 1.

Harrison and McCabe show that

$$\mathscr{E}\mathbf{q} = \frac{m - p - 1 + \sum_{i=1}^{p+1} \lambda_i}{n - k - 1} = \bar{\lambda},$$

$$\mathscr{V}\mathbf{q} = \frac{2[(m - p - 1)(1 - \bar{\lambda})^2 + (n - m - p - 1)\bar{\lambda}^2 + \sum_{i=1}^{p+1} (\lambda_i - \bar{\lambda})^2}{(n - p - 1)(n - p + 1)}.$$

The moment-matching equations

$$1 + \alpha + \beta = \frac{\mathscr{E}\mathbf{q}[1 - \mathscr{E}\mathbf{q}]}{\mathscr{V}\mathbf{q}},$$

$$\frac{\alpha}{\alpha + \beta} = \mathscr{E}\mathbf{q},$$

determine the parameters (α, β) of the beta approximation to the distribution of $\mathbf{q}$. It appears from limited simulation studies that this statistic has higher power than that of Goldfeld–Quandt and that based on recursive or BLUS residuals (see Section 3(e)).

Szroeter [1978] describes another generalization of the Goldfeld–Quandt procedure. Let A be some nonempty subset of $\{1, 2, \ldots, n\}$, and let $\{h_t^A\}$ be a set of scalars indexed on the elements of A such that $h_t^A \leq h_s^A$ if $t < s$. Let

$$\mathbf{w}_i^A = \hat{\mathbf{u}}_i^2 \Big/ \left(\sum_{s \in A} \hat{\mathbf{u}}_s^2 \right), \qquad i \in A,$$

$$\tilde{\mathbf{h}}_A = \sum_{i \in A} \mathbf{w}_i^A h_i^A,$$

$$\bar{\mathbf{h}}_A = \sum_{i \in A} h_i^A / n_A.$$

Then homoscedasticity is rejected if $\tilde{\mathbf{h}}_A$ is larger than $\bar{\mathbf{h}}_A$. The Goldfeld–Quandt procedure is of this form if we take

$$A = \{1, 2, \ldots, k_1, k_2 + 1, \ldots, n\}$$

$$\mathbf{h}_i^A = \begin{cases} -1, & i \leq k_1, \\ 1, & i \geq k_2 + 1. \end{cases}$$

(b) Lagrange Multiplier Test

Prescription

Breusch and Pagan [1979] (see also Cook and Weisberg [1983] for an alternate development of the Breusch–Pagan test) have considered a class of alternatives to homoscedasticity of the form

$$\mathscr{V}\mathbf{u}_i = h\left(\alpha_1 + \sum_{j=2}^{k} \alpha_j z_{ji} \right),$$

where h is an arbitrary function not indexed by i, the α's are functionally unrelated to the β's, and the z_j's are some exogenous variables. They derived the Lagrange multiplier test for the hypothesis that $\alpha_2 = \cdots = \alpha_k = 0$, i.e., that $\mathscr{V}\mathbf{u}_i = h(\alpha_1) \equiv \sigma^2$, and found that it had an extremely simple form.

Let $\hat{\boldsymbol{\sigma}}^2 = \sum_{i=1}^n \hat{\mathbf{u}}_i^2 / n$, and define $\mathbf{g}_i = \hat{\mathbf{u}}_i^2 / \hat{\boldsymbol{\sigma}}^2$. Suppose we regress the $\mathbf{g}_i$ onto the z_{ji}. The explained sum of squares of regression of $\mathbf{g}_i$ onto the z_{ji} is given by

$$\begin{aligned} \mathbf{s} &= \sum_{i=1}^{n} (\mathbf{g}_i - \hat{\mathbf{g}}_i)^2 \\ &= \mathbf{G}' Z (Z'Z)^{-1} Z' \mathbf{G} - \left(\sum_{i=1}^{n} \mathbf{g}_i \right)^2 \Big/ n, \end{aligned}$$

where $\hat{\mathbf{g}}_i$ is the estimate of $\mathbf{g}_i$ from the regression on the $\mathbf{z}_{ji}$'s, $\mathbf{G}$ is the vector of the $\mathbf{g}_i$'s, and Z is the $n \times k$ matrix

$$Z = \begin{bmatrix} 1 & z_{11} & \cdots & z_{k1} \\ \vdots & \vdots & & \vdots \\ 1 & z_{1n} & \cdots & z_{kn} \end{bmatrix}.$$

Breusch and Pagan have shown that $\mathbf{s}/2$ has a $\chi^2(k-1)$ distribution. Cook and Weisberg recommend plotting the $\mathbf{g}_i$'s against each of the z_{ji}'s, $j = 2, \ldots, k$, as a visual check for homoscedasticity.

In particular, let $k = 2$ and $z_{ji} = x_{li}$. Then our hypothesis is that the heteroscedasticity is due to the lth independent variable. In our example, $\hat{\boldsymbol{\sigma}}^2 = 0.000150987$. When we create the $\mathbf{g}_i$ and regress them on the x_{1i}, we obtain $\mathbf{s} = 19.0925$; $\mathbf{s}/2 = 9.546$ is to be compared with $\chi^2_{0.05}(1) = 3.48$, and we see that there is heteroscedasticity. When we regress the g_i on the x_{2i}, we obtain $\mathbf{s} = 17.7823$; $\mathbf{s}/2 = 8.891$, and so we conclude that x_2 also contributes to heteroscedasticity using this test.

(c) White Procedure

Prescription

White [1980] has considered the problem of testing for heteroscedasticity in an even more general context, wherein the independent variables are also allowed to be random. In our exposition, though, we will retain the non-random nature of the x's. White proposes the following procedure. First, define

$$\hat{\mathbf{V}} = \frac{1}{n} \sum_{i=1}^{n} \hat{\mathbf{u}}_i^2 X_i X_i'.$$

Also, let $\hat{\boldsymbol{\sigma}}^2 = \sum_{i=1}^{n} (\mathbf{y}_i - X_i'\hat{\mathbf{B}})^2/n$, the maximum likelihood estimator of σ^2 if the $\mathbf{u}_i$ are homoscedastic. Let

$$\mathbf{D} = \hat{\mathbf{V}} - \hat{\boldsymbol{\sigma}}^2 X'X/n$$

$$= \frac{1}{n} \sum_{i=1}^{n} (\hat{\mathbf{u}}_i^2 - \hat{\boldsymbol{\sigma}}^2) X_i X_i'.$$

Under the homoscedasticity hypothesis, $\mathscr{E}\hat{\mathbf{u}}_i^2 = \sigma^2$, so $\text{plim}_{n\to\infty} \mathbf{D} = 0$.

Now $\mathbf{D}$ is a $p \times p$ matrix but, being symmetric, only has $s = p(p+1)/2$ distinct elements. Let us spell these out by defining W_i to be the s vector containing the lower triangular submatrix of $X_i X_i'$, i.e.,

$$W_i' = [x_{i1}^2 \; x_{i1}x_{i2} \quad x_{i2}^2 \; x_{i1}x_{i3} \; x_{i2}x_{i3} \quad x_{i3}^2 \quad \ldots \quad x_{ip}^2],$$

and let $\overline{W} = \sum_{i=1}^{n} W_i/n$. Now define

$$\mathbf{D}^* = \sum_{i=1}^{n} (\hat{\mathbf{u}}_i^2 - \hat{\boldsymbol{\sigma}}^2) W_i/n$$

and

$$\mathbf{E} = \frac{1}{n} \sum_{i=1}^{n} (\hat{\mathbf{u}}_i^2 - \hat{\boldsymbol{\sigma}}^2)^2 (W_i - \overline{W})(W_i - \overline{W})'.$$

White has shown that under homoscedasticity the statistic $\mathbf{v} = n\mathbf{D}^{*\prime}\mathbf{E}^{-1}\mathbf{D}^*$ has a $\chi^2(s)$ distribution.

Based on the Koerts–Abrahamse data of Table 2, we find that

$$E = \begin{bmatrix} 0.388205E-21 & & & & & \\ 0.273136E-16 & 0.103202E-09 & & & & \\ 0.112919E-15 & 0.516026E-09 & 0.167717E-08 & & & \\ 0.690082E-15 & 0.180581E-09 & 0.738123E-09 & 0.410093E-08 & & \\ 0.144019E-14 & 0.557622E-09 & 0.226846E-08 & 0.857475E-08 & 0.182702E-07 & \\ 0.257561E-14 & 0.658522E-09 & 0.269266E-09 & 0.153547E-07 & 0.320724E-07 & \end{bmatrix},$$

and

$$D^* = \begin{bmatrix} -0.434284E-10 \\ 0.225364E-06 \\ 0.880630E-06 \\ -0.402681E-05 \\ -0.771058E-04 \\ -0.152012E-04 \end{bmatrix},$$

so that $\mathbf{v} = 127.9$, which, when compared with the 95% point of the $\chi^2(6)$ distribution indicates that there is heteroscedasticity.

(d) Robust Procedures

Prescription

Anscombe [1961] proposed the use of the statistic

$$a = \frac{\dfrac{\sqrt{n-p+1}}{\sqrt{2(n-p-1)}} \sum_{i=1}^{n} \hat{\mathbf{u}}_i^2(\hat{\mathbf{y}}_i - \bar{\mathbf{y}}^*)}{\mathbf{s}^2 \sqrt{\sum_{i=1}^{n} \sum_{j=1}^{n} c_{ij}^2(\hat{\mathbf{y}}_i - \bar{\mathbf{y}}^*)(\hat{\mathbf{y}}_j - \bar{\mathbf{y}}^*)}}$$

as the test for homoscedasticity of residuals, where

$$\mathbf{s}^2 = \sum_{i=1}^{n} \hat{\mathbf{u}}_i^2/(n-p-1), \qquad \bar{\mathbf{y}}^* = \sum_{i=1}^{n} c_{ii}\hat{\mathbf{y}}_i/(n-p),$$

and the matrix C is given by

$$C = X(X'X)^{-1}X'.$$

This statistic is approximately normally distributed, and is approximately the locally most powerful test of the homoscedasticity hypothesis.

Applying Anscombe's procedure to our data, we find that the C matrix is as given in Table 8, that $\bar{\mathbf{y}}^* = 0.451562$, and that $\mathbf{h} = 2.61316$, thus rejecting the hypothesis of homoscedasticity.

Theoretical Background

Anscombe [1961] proposes as a model of the heteroscedasticity that $\mathbf{y}_i = X_i'B + \sigma(X_i'B)\mathbf{v}_i$, i.e., $\mathbf{u}_i$ is a product of a homoscedastic variable $\mathbf{v}_i$ and a positive function σ of the true value $X_i'B$. Let $\eta = X_i'B$. If $\sigma(\eta)$ is of the form

$$\sigma(\eta_i) = 1 + \theta\alpha(\eta_i) + o(\theta),$$

then homoscedasticity is equivalent to $\theta = 0$.

Bickel [1978] generalizes the Anscombe statistic in the following way. Suppose the function $\alpha(\eta)$ is known. Then

$$\mathbf{a}_b = \frac{\sqrt{n-p-1}\sum_{i=1}^{n}(\alpha(\hat{\mathbf{y}}_i) - \bar{\boldsymbol{\alpha}})(b(\hat{\mathbf{u}}_i) - \bar{b}(\hat{\mathbf{u}}))}{\sqrt{\sum_{i=1}^{n}(\alpha(\hat{\mathbf{y}}_i) - \bar{\boldsymbol{\alpha}})^2 \sum_{i=1}^{n}(b(\hat{\mathbf{u}}_i) - \bar{b}(\hat{\mathbf{u}}))^2}}$$

is approximately normally distributed where

$$\bar{\boldsymbol{\alpha}} = \sum_{i=1}^{n} \alpha(\hat{\mathbf{y}}_i)/n,$$

$$\bar{b}(\hat{\mathbf{u}}) = \sum_{i=1}^{n} b(\hat{\mathbf{u}}_i)/n,$$

and $b(\hat{u})$ is a function selected to produce a reasonable robust test. Some choices of b are:

(1) $b(\hat{u}) = |\hat{u}|$;

(2) $b(\hat{u}) = \begin{cases} \hat{u}^2, & |u| \leq k, \\ k^2, & |u| > k. \end{cases}$

One example is $\alpha(\hat{y}_i) = \hat{y}_i$. Then, if $b(\hat{u}) = |\hat{u}|$ the statistic is

$$\mathbf{a}_b = \frac{\sqrt{n-p-1}\sum_{i=1}^{n}(\hat{\mathbf{y}}_i - \bar{\mathbf{y}})(|\hat{\mathbf{u}}_i| - |\bar{\mathbf{u}}|)}{\sqrt{\sum_{i=1}^{n}(\hat{\mathbf{y}}_i - \bar{\mathbf{y}})^2 \sum_{i=1}^{n}(|\hat{\mathbf{u}}_i| - |\bar{\mathbf{u}}|)^2}},$$

where $\bar{\mathbf{u}} = \sum_{i=1}^{n} |\mathbf{u}_i|/n$. More generally, the best choice of $b(\hat{u})$ is

$$b(\hat{u}) = -\hat{u}f_{\mathbf{u}}'(u)/f_{\mathbf{u}}(u),$$

where $f_{\mathbf{u}}(u)$ is the density of $\mathbf{u}$ and $f_{\mathbf{u}}'(u)$ is its derivative. If $f_{\mathbf{u}}(u)$ is normal, $b(\hat{u}) = \hat{u}^2$; $b(\hat{u}) = |\hat{u}|$ if $f_{\mathbf{u}}(u)$ is double exponential.

In general, $\mathbf{a}_b$ is essentially the sample correlation between $b(\hat{\mathbf{u}}_i)$ and $\alpha(\hat{\mathbf{y}}_i)$, i.e., between the part of the variance of the residual that produces the heteroscedasticity ($\alpha(\mathbf{y}_i)$) and an even function of the residual ($b(\hat{\mathbf{u}}_i)$). If this correlation is small, $b(\hat{\mathbf{u}}_i)$ is deemed not to co-vary with $\alpha(\hat{\mathbf{y}}_i)$. In this example, we look at the correlation between the absolute value of the residual and the fitted as a test statistic. This corresponds with our aforementioned diagnostic procedure of looking at RVSF plots to check for heteroscedasticity.

(e) Nonparametric Tests

As noted earlier, a robust procedure for testing for heteroscedasticity is to correlate the absolute estimated residual with the fitted dependent variable. If the estimated residuals are uncorrelated, one can test for heteroscedasticity by calculating a rank correlation statistic, or other measure of concordance such as Kendall's tau, between these two variables, and test for the significance of that statistic. The use of Kendall's tau in this way was advocated by Brown, Durbin, and Evans [1975] as a nonparametric procedure for testing for heteroscedasticity. As applied to the Koerts–Abrahamse data, excluding observations 1, 8, and 14, and using the BLUS procedure, we record the BLUS residuals also with the fitted dependent variables in Table 9. From this we compute Kendall's tau to be 0.36. The 95% point of the distribution of tau for these data is 0.39 (see Siegel [1956]), and so we do not reject the hypothesis of homoscedasticity. Similarly, a computation of the rank correlation coefficient yields $r = 0.49$, also not significant. As we have seen from parametric

Table 9. Computations for the Rank Correlation Calculation.

			Ranks	
	BLUS Residuals	BLUS Fitted dependent variable	Absolute residuals	Fitted dependent variable
1	—	—	—	—
2	−0.00171	1.99735	14	14
3	−0.00691	2.00691	11	13
4	−0.01716	2.06482	5	12
5	−0.00886	2.08121	12	11
6	−0.01853	2.08894	4	10
7	−0.00794	2.09108	10	9
8	—	—		
9	−0.01274	2.17534	7	6
10	0.00437	2.18202	13	5
11	0.01179	2.18824	9	4
12	−0.36394	2.51193	1	1
13	−0.01533	2.14951	6	7
14	—	—		
15	−0.01253	2.20190	8	3
16	0.03514	2.13805	3	8
17	−0.26262	2.48142	2	2

analyses, the residuals are heteroscedastic. Thus this nonparametric procedure is not quite sensitive to the existence of heteroscedasticity in this example.

Goldfeld and Quandt [1965] also developed a nonparametric test for heteroscedasticity. Unfortunately (see Goldfeld–Quandt errata [1967]), their presentation of this procedure was erroneously couched in terms of application directly to the estimated residuals. If their procedure were instead applied to uncorrelated residuals such as those developed in Section 3(e) above, it would be applied validly. (See also Hedayat, Raktoe, and Talwar [1977], who in a Monte Carlo study compare the use of BLUS residuals with the use of the usual estimated residuals in this test. They find not only that the use of BLUS residuals yields a test of higher power, but that its power is even greater than that of the Goldfeld–Quandt F test.)

Their procedure in brief consists of the following steps:

Step 1. Regress the dependent variable on the independent variable suspected of causing the heteroscedasticity, and create a variable equal to the absolute value of the BLUS or other uncorrelated residual.

Step 2. Sort the absolute residuals based on the value of the independent variable.

Step 3. Define a *peak* as a value of the sorted absolute uncorrelated residual $|\hat{u}_i|$ such that $|\hat{u}_i| > |\hat{u}_j|$ for all $j < i$. Count the number of peaks in the list of sorted absolute uncorrelated residuals.

Step 4. Compare the count of the number of peaks with value of x from Table 10 associated with either a 0.025 or 0.975 cumulative probability for the appropriate value of n. If the count is either too small or too large, then reject the hypothesis of homoscedasticity.

A significantly large number of peaks means that the absolute residuals rise with an increase in the independent variable; a significantly small number of

Table 10. Cumulative Probabilities for the Distribution of Peaks.

	P (number of peaks $\leq x$)										
n	$x=0$	$x=1$	$x=2$	$x=3$	$x=4$	$x=5$	$x=6$	$x=7$	$x=8$	$x=9$	$x=10$
5	.2000	.6167	.9083	.9917	1.0000						
10	.1000	.3829	.7061	.9055	.9797	.9971	.9997	1.0000			
15	.0667	.2834	.5833	.8211	.9433	.9866	.9976	.9997	1.0000		
20	.0500	.2274	.5022	.7530	.9056	.9720	.9935	.9988	.9998	1.0000	
25	.0400	.1910	.4441	.6979	.8705	.9559	.9879	.9973	.9995	.9999	1.0000
30	.0333	.1654	.4001	.6525	.8386	.9395	.9815	.9953	.9990	.9998	1.0000
35	.0286	.1462	.3654	.6144	.8098	.9234	.9745	.9929	.9984	.9997	.9999
40	.0250	.1313	.3373	.5818	.7837	.9078	.9674	.9903	.9975	.9995	.9999
45	.0222	.1194	.3138	.5536	.7600	.8930	.9601	.9874	.9966	.9992	.9998
50	.0200	.1096	.2940	.5288	.7383	.8788	.9530	.9844	.9956	.9989	.9998
55	.0182	.1014	.2769	.5068	.7184	.8653	.9456	.9813	.9944	.9986	.9997
60	.0167	.0944	.2620	.4871	.7001	.8524	.9384	.9780	.9932	.9982	.9996

SOURCE: Goldfeld and Quandt [1965]. Reprinted from the *Journal of the American Statistical Association*, Vol. 60, with the permission of the ASA.

Table 11. BLUS Residuals Excluding Observations 1 and 2.

i	Residual ($\hat{u}_i$)
1	—
2	—
3	−0.0144623
4	−0.00171812
5	0.0213131
6	−0.0174972
7	0.311669
8	0.0523166
9	−0.124710
10	0.0217079
11	0.0163528
12	−0.0327195
13	−0.0463203
14	−0.0227008
15	−0.146338
16	−0.244608
17	0.0311140

Table 12. Absolute BLUS Residuals Sorted on Value of x_1 and x_2.

i	x_{1i}	$\|\hat{u}_i\|$	i	x_{2i}	$\|\hat{u}_i\|$
12	1.97955	0.03272*	14	1.72099	0.02270
13	1.98408	0.04632*	16	1.77452	0.02446
14	1.98945	0.02270	15	1.77597	0.01463
3	2.00000	0.01446	17	1.78746	0.03111*
16	2.00689	0.02446	11	1.78746	0.01635
11	2.00732	0.01635	12	1.79588	0.03272*
15	2.01030	0.01463	13	1.80346	0.04632*
17	2.01620	0.03111	10	1.81558	0.02171
5	2.02078	0.02131	9	1.84572	0.12471*
4	2.02078	0.00172	8	1.91803	0.05232
10	2.02243	0.02171	5	1.93702	0.02131
9	2.03862	0.12471	6	1.95279	0.01750
6	2.03941	0.01750	4	1.95713	0.00172
7	2.04454	0.31167*	7	1.95713	0.31167*
8	2.05038	0.05232	3	2.00000	0.01446

peaks conversely means that the absolute residuals fall with an increase in the independent variable.

As an example of this procedure let us consider over again the Koerts–Abrahamse data, excluding observations 1 and 2. Upon using the BLUS procedure, we obtain BLUS residuals as given in Table 11. Table 12 shows these residuals sorted based on the value of x_1 and also of x_2, and peaks are designated by an $*$. We see that based on x_2 there are seven peaks in 15 observations. Since the probability of obtaining seven or more peaks is 0.0003 (from Table 10), we reject the hypothesis of homoscedasticity. On the other hand, based on x_1 there are three peaks, and, since the probability of obtaining three or more peaks is 0.1789, we have determined that x_1 does not contribute to the heteroscedasticity.

Horn [1981] has proposed an alternative nonparametric test statistic, namely $\mathbf{d} = \sum_{i=1}^{n} (\mathbf{r}_i - i)^2$, where $\mathbf{r}_i$ is the rank of the ith absolute residual. For large samples, $\mathbf{d}$ is approximately normally distributed with mean $(n^3 - n)/6$ and variance $[n^2(n + 1)^2(n - 1)]/36$. The results of a Monte Carlo comparison with the peak test show that this test is more powerful, even when based on least squares residuals instead of uncorrelated residuals.

(f) Adapting Tests of Homogeneity of Variances

In addition to the procedures described in this section, one can adapt the procedures of Section 2 to the problem of testing regression residuals for homoscedasticity. One can, as seen above, avoid the problems associated with the lack of independence of the $\hat{\mathbf{u}}_i$'s by using BLUS or recursive residuals. One cannot automatically use the procedures of Section 2 on the residuals unless one makes some simplifying assumptions about the structure of the alternative hypothesis of heteroscedasticity. If, for example, each residual is assumed (in the alternative hypothesis) to have a different variance, then we are faced with the problem of having only one observation from each of the alternative populations.

Ramsey [1969] makes the simplifying assumption that the residuals are drawn from k populations and that one can identify each residual with its appropriate population, and then suggests using the Bartlett test on these grouped residuals to test for homoscedasticity of the residuals. He suggests $k = 3$ as a reasonable choice. Moreover, the study of Ramsey and Gilbert [1972] indicates that there is a no need to restrict the application of this procedure to uncorrelated residuals.

(g) Evaluation

There have been a few evaluative studies of the power of the various procedures for testing regression residuals for homoscedasticity. Ali and Giaccotto

[1984] studied all the tests described above and found that they all tend to have higher power when least squares residuals are used than when uncorrelated residuals are used. Griffiths and Surekha [1986] compared the power of the Goldfeld–Quandt test, Szroeter's generalization of that test, the Ramsey adaptation of the Bartlett test (dubbed BAMSET by Ramsey), and the Lagrange multiplier test. They found that if it is possible to order the observations according to increasing residual variances, then the Szroeter test is most powerful; otherwise BAMSET is.

Two of the procedures cited here are also diagnostic, i.e., not only test for homoscedasticity but, if heteroscedastic, isolate the independent variable(s) which contribute to the heteroscedasticity. These, the Goldfeld–Quandt and the Lagrange multiplier procedure, led to quite opposite conclusions in our example, the former concluding homoscedasticity and the latter concluding heteroscedasticity due to both x_1 and x_2. Finally, the Goldfeld–Quandt nonparametric procedure concluded that heteroscedasticity is due only to x_2. Our visual inspection of Figures 2–5 indicated heteroscedasticity due to both x_1 and x_2. Thus, at least from this example, one could infer that the Lagrange multiplier procedure reached the appropriate conclusion.

Appendix I

```
      read(5,*)n
      xn=n
      do 10 i=1,n
      p1=i/xn
      p=p1
      if(p1.gt..5)p=1.-p1
      t=sqrt(-2.*alog(p))
      x=t-(2.515517+.802853*t+.010328*t*t)/(1.+1.432788*t
     +     +.189269*t*t+.001308*t**3)
      if(p1.lt..5)x=-x
      write(5,600)i,p1,x
600   format(1h i3,2(x,e12.6))
10    continue
      end
```

References

Ali, M. M. and Giaccotto, C. 1984. A study of several new and existing tests for heteroscedasticity in the general linear model. *Journal of Econometrics* **26** (December): 355–73.

Anscombe, F. J. 1961. Examination of residuals. In *Proceedings of the Fourth Berkeley Symposium on Mathematical Statistics and Probability*, Vol. 1, ed. J. Neyman. Berkeley: University of California Press, pp. 1–36.

Anscombe, F. J. and Tukey, J. W. 1963. The examination and analysis of residuals. *Technometrics* **5** (May): 141–60.

Bartlett, M. S. 1937. Properties of sufficiency and statistical tests. *Proceedings of the Royal Statistical Society, Series A* **160**: 268–82.

Bickel, P. J. 1978. Using residuals robustly I: Tests for heteroscedasticity, nonlinearity. *Annals of Statistics* **6** (March): 266–91.

Box, G. E. P. 1953. Non-normality and tests on variances. *Biometrika* **40** (December): 318–35.

Box, G. E. P. 1954a. Some theorems on quadratic forms applied in the study of analysis of variance problems. I. Effects of inequality of variance in the one-way classification. *Annals of Mathematical Statistics* **25** (June): 290–302.

Box, G. E. P. 1954b. Some theorems on quadratic forms applied in the study of analysis of variance problems. II. Effects of inequality of variance and of correlation between errors in the two-way classification. *Annals of Mathematical Statistics* **25** (September): 484–98.

Box, G. E. P. and Andersen, S. L. 1955. Permutation theory in the derivation of robust criteria and the study of departures from assumption. *Journal of the Royal Statistical Society, Series B* **17** (March): 1–26.

Breusch, T. S. and Pagan, A. R. 1979. A simple test for heteroscedasticity and random coefficient variation. *Econometrica* **47** (September): 1287–94.

Brown, R. L., Durbin, J., and Evans, J. 1975. Techniques for testing the constancy of the regression relationships over time. *Journal of the Royal Statistical Society, Series B* **37**: 149–63.

Carroll, R. J. 1982. Adapting for heteroscedasticity in linear models. *Annals of Statistics* **10** (December): 1224–33.

Carroll, R. J. and Ruppert, D. 1982. Robust estimation in heteroscedastic linear models. *Annals of Statistics* **10** (June): 429–41.

Chao, Min-te and Glaser, R. E. 1978. The exact distribution of Bartlett's test statistic for homogeneity of variances with unequal sample sizes. *Journal of the American Statistical Association* **73** (June): 422–26.

Conover, W. J., Johnson, M. E., and Johnson, M. M. 1981. A comparative study of tests for homogeneity of variances, with applications to the outer continental shelf bidding data. *Technometrics* **23** (November): 351–61.

Cook, R. D. and Weisberg, S. 1983. Diagnostics for heteroscedasticity in regression. *Biometrika* **70** (April): 1–10.

Dent, W. T. and Styan, G. P. H. 1973. Uncorrelated residuals from linear models. Technical Report No. 88 (January): The Economics Series, Institute for Mathematical Studies in the Social Sciences. Stanford, Ca. Stanford Uhiversity.

Dixon, W. J. and Massey, F. J., Jr. 1969. *Introduction to Statistical Analysis*. New York: McGraw-Hill.

Dyer, D. D. and Keating, J. P. 1980. On the determination of critical values for Bartlett's test. *Journal of the American Statistical Association* **75** (June): 313–19.

Fisher, R. A. 1920. A mathematical examination of the methods of determining the accuracy of an observation by the mean error, and by the mean square error. *Monthly Notices of the Royal Astronomical Society* **80**: 758–70.

Fligner, M. A. and Killeen, T. J. 1976. Distribution-free two-sample tests for scale. *Journal of the American Statistical Association* **71** (March): 210–13.

Goldfeld, S. M. and Quandt, R. E. 1965. Some tests for homoscedasticity. *Journal of the American Statistical Association* **60** (June): 539–47. (see also, corrigenda 1967, *Journal of the American Statistical Association* **62** (December): 1518).

Griffiths, W. E. and Surekha, K. 1986. A Monte Carlo evaluation of the power of some tests for heteroscedasticity. *Journal of Econometrics* **31** (March): 219–331.

Hájek, J. and Sĭdák, Z. 1967. *Theory of Rank Tests*. New York: Academic Press.

Harrison, M. J. and McCabe, B. P. M. 1979. A test for heteroscedasticity based on ordinary least squares residuals. *Journal of the American Statistical Association* **74** (June): 494–99.

Harvey, A. C. and Phillips, G. D. A. 1974. A comparison of the power of some tests for heteroskedasticity in the general linear model. *Journal of Econometrics* **2** (December): 307–16.

Hedayat, A. and Robson, D. S. 1970. Independent stepwise residuals for testing homoscedasticity. *Journal of the American Statistical Association* **65** (December): 1573–81.

Hedayat, A., Raktoe, B. L., and Talwar, P. P. 1977. Examination and analysis of residuals: a test for detecting a monotonic relation between mean and variance in regression through the origin. *Communications in Statistics—Theory and Methods* **A6**(6): 497–506.

Horn, P. 1981. Heteroscedasticity of residuals: a non-parametric alternative to the Goldfeld-Quandt peak test. *Communications in Statistics—Theory and Methods* **A10**(8): 795–808.

Koerts, J. and Abrahamse, A. P. J. 1969. *On the Theory and Application of the General Linear Model*. Rotterdam: Universitaire Pers Rotterdam.

Levene, H. 1960. Robust tests for the equality of variances, in *Contributions to Probability and Statistics*, ed. I. Olkin. Palo Alto, Ca: Stanford University Press, pp. 278–92.

Lund, R. E. 1975. Tables for an approximate test for outliers in linear models. *Technometrics* **17** (November): 473–76.

Neyman, J. and Pearson, E. S. 1931. On the problem of *k* samples. *Bulletin Academie Polonaise Sciences et Lettres, Series A* **3**: 460–81.

Ramachandran, K. V. 1958. A test of variances. *Journal of the American Statistical Association* **53** (September): 741–47.

Ramsey, J. B. 1969. Tests for specification error in the general linear model. *Journal of the Royal Statistical Society B* **31**: 250–71.

Ramsey, J. B. and Gilbert, R. 1972. Some small sample properties of tests for specification error. *Journal of the American Statistical Association* **67** (March): 180–86.

Rao, C. R. 1970. Estimation of heteroscedastic variances in linear models. *Journal of the American Statistical Association* **65** (March): 161–72.

Siegel, S. 1956. *Nonparametric Statistics for the Behavioral Sciences*. New York: McGraw-Hill.

Szroeter, J. 1978. A class of parametric tests for heteroscedasticity in linear econometric models. *Econometrica* **46** (November): 1311–27.

Theil, H. 1965. The analysis of disturbances in regression analysis. *Journal of the American Statistical Association* **60** (December): 1067–79.

Theil, H. 1971. *Principles of Econometrics*. New York: Wiley.

White, H. 1980. A heteroskedasticity-consistent covariance matrix estimator and a direct test for heteroskedasticity. *Econometrica* **48** (May): 817–38.

CHAPTER 3

Testing for Independence of Observations

"The ideal of independence requires resistance to the herd spirit."
Mason [1927]

"We hold these truths to be self-evident . . . ."
Declaration of Independence [1776]

0. Introduction

The problem addressed in this chapter is one of testing whether a sequence of random variables $\mathbf{x}_1, \ldots, \mathbf{x}_n$ are independent based on a set of observations $x_1, \ldots, x_n$ of these random variables. One approach to this problem is to make an assumption about the distribution of the $\mathbf{x}_i$, both when they are independent and under the alternative that they are not independent, and test one hypothesis against the other. The simplest such pair of assumptions is that the $\mathbf{x}_i$ are identically distributed as $N(0, \sigma^2)$ random variables and, in the case of dependence, that the dependence can be modeled as a first-order autoregressive series wherein

$$\mathbf{x}_1 = \mathbf{u}_1,$$

$$\mathbf{x}_i = \boldsymbol{\rho}\mathbf{x}_{i-1} + \mathbf{u}_i, \qquad i = 2, \ldots, n,$$

and the $\mathbf{u}_i$ are independent $N(0, \sigma^2)$ random variables. The case $\rho = 0$ corresponds to the case in which the $\mathbf{x}_i$ are independent $N(0, \sigma^2)$ random variables. Under the alternative hypothesis, $\mathscr{C}(\mathbf{x}_i, \mathbf{x}_j) = \rho^{|i-j|}$ and $\mathscr{V}\mathbf{x}_i = \sigma^2(1 + \rho + \cdots + \rho^{i-1}) = \sigma^2(1 - \rho^i)/(1 - \rho)$, so that the $\mathbf{x}_i$ are not even identically distributed. Section 1 studies tests of this and more complex autoregressive models as alternatives to independence, both for a sequence of observations and for regression residuals.

An alternative approach to the problem of testing for independence of the $\mathbf{x}_i$'s is nonparametric in nature and capitalizes on the properties that a sequence of independent identically distributed random variables should exhibit. The nonparametric procedures considered in Section 2 include the runs tests, both runs above and below the mean (or median) and runs up and down, some rank correlation tests, and a rank analog of the von Neumann ratio given in Section 1.

1. Parametric Procedures

(a) Independence of Observations

Prescription

Let us consider the most general case, that $\mathbf{X} = (\mathbf{x}_1, \ldots, \mathbf{x}_n)'$ is distributed as the multivariate normal distribution $N(\mu, \Sigma)$, where Σ is a known nondiagonal covariance matrix. The independence hypothesis we shall consider is that $\mathbf{X}$ is distributed as $N(\mu, \Delta)$, where Δ is a known diagonal covariance matrix. Let $\psi = \Delta^{1/2}\Sigma^{-1}\Delta^{1/2}$, where $\Delta^{1/2}$ denotes the diagonal matrix whose elements are the square roots of those of Δ. Then the likelihood ratio test of the null hypothesis that the true covariance matrix is Σ versus the alternative that the true covariance matrix is Δ consists of calculating the statistic

$$\mathbf{v} = \frac{(\mathbf{X} - \bar{\mathbf{x}}E)'\psi(\mathbf{X} - \bar{\mathbf{x}}E)}{(\mathbf{X} - \bar{\mathbf{x}}E)'(\mathbf{X} - \bar{\mathbf{x}}E)},$$

where $E' = (1, \ldots, 1)$ and $\bar{\mathbf{x}} = \sum_{i=1}^{n} \mathbf{x}_i/n$, and rejecting the null hypothesis if $\mathbf{v}$ is either too large or too small.

When $\Delta = \sigma^2 I$ and the alternative is that the $\mathbf{x}_i$ are an autoregressive AR(1) series

$$\mathbf{v} = \frac{\sum_{i=2}^{n} (\mathbf{x}_i - \mathbf{x}_{i-1})^2}{\sum_{i=1}^{n} (\mathbf{x}_i - \bar{\mathbf{x}})^2}.$$

Under the null hypothesis of independence, $\mathscr{E}\mathbf{v} = 2$ and $\mathscr{V}\mathbf{v} = 4(n-2)/(n^2-1)$ (see Williams [1941] for a simple derivation of this result, except for a typographical error in his final result) and $\mathbf{v}$ is asymptotically normally distributed (von Neumann [1941]). The exact distribution of $[n/(n-1)]\mathbf{v}$ for $n = 4(1)12, 15(5)30(10)60$ is tabulated in Hart and von Neumann [1942] and reproduced herein as Table 1.

We illustrate this test on data in column 1 of Table 2 taken from Appendix A of Box and Jenkins [1972]. Here $\sum_{i=2}^{197}(\mathbf{x}_i - \mathbf{x}_{i-1})^2 = 26.5923$ and $\sum_{i=1}^{197}(\mathbf{x}_i - \bar{\mathbf{x}})^2 = 31.0999$, so that $\mathbf{v} = 0.855$. Then $(\mathbf{v} - 2)/0.142 = -8.056$ and so $\mathbf{v}$ is over eight standard deviations from its expected value under independence of 2. We thus reject the hypothesis of independence.

Table 1. Distribution of the von Neumann Ratio.

k \ n	4	5	6	7	8	9	10	11	12
.25				.00001	.00001	.00001	.00001		
.30				.00007	.00007	.00005	.00004	.00002	.00001
.35			.00006	.00027	.00021	.00014	.00009	.00005	.00003
.40			.00047	.00065	.00047	.00031	.00019	.00012	.00007
.45			.00126	.00126	.00088	.00059	.00038	.00025	.00016
.50		.00038	.00246	.00214	.00150	.00103	.00069	.00046	.00031
.55		.00223	.00409	.00333	.00237	.00168	.00116	.00080	.00055
.60		.00493	.00615	.00486	.00355	.00259	.00185	.00132	.00094
.65		.00830	.00865	.00678	.00511	.00382	.00282	.00208	.00152
.70		.01225	.01161	.00913	.00710	.00544	.00414	.00313	.00235
.75		.01673	.01505	.01197	.00958	.00753	.00587	.00455	.00351
.80	.00356	.02171	.01900	.01534	.01263	.01015	.00809	.00642	.00508
.85	.01302	.02717	.02348	.01932	.01631	.01338	.01089	.00883	.00714
.90	.02257	.03310	.02851	.02403	.02068	.01729	.01436	.01188	.00980
.95	.03223	.03949	.03412	.02957	.02579	.02196	.01858	.01565	.01316
1.00	.04199	.04634	.04035	.03598	.03171	.02745	.02363	.02025	.01733
1.05	.05186	.05364	.04728	.04325	.03849	.03384	.02959	.02578	.02241
1.10	.06184	.06140	.05500	.05137	.04618	.04120	.03655	.03232	.02852
1.15	.07194	.06963	.06361	.06036	.05482	.04957	.04458	.03997	.03577
1.20			.07323	.07020	.06445	.05901	.05375	.04882	.04425
1.25						.06956	.06412	.05894	.05407
1.30								.07040	.06531

k \ n	15	20	25	30	40	50	60
.35	.00001						
.40	.00002						
.45	.00004						
.50	.00009	.00001					
.55	.00018	.00002					
.60	.00033	.00005	.00001				
.65	.00059	.00012	.00002				
.70	.00100	.00024	.00005	.00001			
.75	.00161	.00044	.00011	.00003			
.80	.00250	.00076	.00023	.00007	.00001		
.85	.00375	.00127	.00044	.00015	.00002		
.90	.00547	.00206	.00079	.00030	.00004	.00001	
.95	.00778	.00323	.00135	.00057	.00010	.00002	
1.00	.01079	.00489	.00222	.00102	.00022	.00005	.00001
1.05	.01465	.00720	.00355	.00176	.00044	.00012	.00003
1.10	.01950	.01033	.00550	.00294	.00085	.00026	.00008
1.15	.02550	.01448	.00826	.00474	.00158	.00054	.00019
1.20	.03280	.01986	.01208	.00738	.00280	.00108	.00043
1.25	.04155	.02670	.01723	.01117	.00476	.00206	.00092
1.30	.05189	.03524	.02402	.01644	.00780	.00376	.00185
1.35	.06396	.04571	.03276	.02357	.01235	.00656	.00355
1.40	.07787	.05834	.04379	.03298	.01892	.01098	.00649
1.45		.07333	.05743	.04511	.02810	.01769	.01133
1.50			.07398	.06038	.04055	.02750	.01893
1.55				.07920	.05696	.04131	.03034
1.60					.07797	.06006	.04675
1.65						.08465	.06942
1.70							.09949

SOURCE: Hart and von Neumann [1942]. Reprinted from *Annals of Mathematical Statistics*, Vol. **13**, with the permission of the IMS.

Table 2. Analysis of Chemical Process Concentration Readings.

Row	Data	Residuals	Above or below Median	Above or below Mean	Increase or Decrease	Rank
1	17.01701	-0.04291	0	0		90
2	16.64279	-0.41741	0	0	0	29
3	16.34549	-0.71499	0	0	0	6
4	16.13574	-0.92501	0	0	0	1
5	17.14196	0.08093	1	1	1	115
6	16.92342	-0.13789	0	0	0	68
7	16.82243	-0.23916	0	0	0	51
8	17.44727	0.38540	1	1	1	167
9	17.10715	0.04500	1	1	0	102
10	17.00802	-0.05441	0	0	0	86
11	16.72704	-0.33567	0	0	0	35
12	17.43577	0.37278	1	1	1	163
13	17.20577	0.14250	1	1	0	119
14	17.41582	0.35227	1	1	1	156
15	17.40504	0.34122	1	1	0	149
16	17.02139	-0.04271	0	0	0	92
17	17.30917	0.24479	1	1	1	139
18	17.21732	0.15266	1	1	0	124
19	17.43508	0.37014	1	1	1	162
20	16.81662	-0.24860	0	0	0	49
21	17.11937	0.05387	1	1	1	108
22	17.43622	0.37044	1	1	1	164
23	17.40515	0.33909	1	1	0	150
24	17.52443	0.45809	1	1	1	171
25	17.42751	0.36090	1	1	0	160
26	17.62973	0.56284	1	1	1	181
27	17.42850	0.36133	1	1	0	161
28	17.30947	0.24202	1	1	0	141
29	17.02417	-0.04356	0	0	0	93
30	17.84613	0.77812	1	1	1	191
31	17.53345	0.46516	1	1	0	175
32	18.12911	1.06054	1	1	1	196
33	17.53133	0.46248	1	1	0	174
34	17.40421	0.33508	1	1	0	148
35	17.41231	0.34291	1	1	1	154
36	17.13702	0.06734	1	1	0	112
37	17.61644	0.54648	1	1	1	179
38	17.74935	0.67911	1	1	1	186
39	17.40680	0.33628	1	1	0	151
40	17.82466	0.75386	1	1	1	189
41	17.63522	0.56414	1	1	0	183
42	17.51122	0.43986	1	1	0	170
43	16.54536	-0.52628	0	0	0	19
44	17.81725	0.74533	1	1	1	188
45	17.30575	0.23356	1	1	0	138
46	17.34136	0.26889	1	1	1	144
47	17.14145	0.06870	1	1	0	114
48	17.44136	0.36833	1	1	1	166
49	16.91203	-0.16128	0	0	0	64

(continued)

Table 2 (*continued*)

Row	Data	Residuals	Above or below: Median	Above or below: Mean	Increase or Decrease	Rank
50	17.30944	0.23585	1	1	1	140
51	17.60290	0.52903	1	1	1	177
52	16.91756	-0.15659	0	0	0	66
53	16.72419	-0.35024	0	0	0	34
54	16.83642	-0.23829	0	0	1	55
55	16.84212	-0.23286	0	0	1	57
56	17.21486	0.13960	1	1	1	122
57	16.80581	-0.26973	0	0	0	46
58	17.62023	0.54441	1	1	1	180
59	17.23957	0.16347	1	1	0	132
60	16.63141	-0.44497	0	0	0	27
61	17.11224	0.03558	1	1	1	105
62	16.94915	-0.12779	0	0	0	78
63	16.60564	-0.47158	0	0	0	23
64	18.00471	0.92721	1	1	1	194
65	17.24429	0.16652	1	1	0	134
66	17.34941	0.27136	1	1	1	147
67	17.00077	-0.07756	0	0	0	80
68	16.93112	-0.14749	0	0	0	70
69	17.30539	0.22650	1	1	1	137
70	16.80600	-0.27317	0	0	0	47
71	17.34616	0.26671	1	1	1	146
72	17.41789	0.33816	1	1	1	157
73	17.74585	0.66584	1	1	1	185
74	16.83519	-0.24510	0	0	0	53
75	16.94512	-0.13545	0	0	1	77
76	17.01128	-0.06956	0	0	1	89
77	16.94023	-0.14089	0	0	0	74
78	17.00231	-0.07909	0	0	1	83
79	16.60195	-0.47973	0	0	0	22
80	16.73071	-0.35125	0	0	1	37
81	16.80083	-0.28141	0	0	1	45
82	16.70289	-0.37963	0	0	0	31
83	16.44897	-0.63383	0	0	0	13
84	16.44326	-0.63982	0	0	0	12
85	16.51188	-0.57148	0	0	1	14
86	16.60155	-0.48208	0	0	1	21
87	16.54200	-0.54191	0	0	0	17
88	16.71919	-0.36500	0	0	1	33
89	16.40486	-0.67961	0	0	0	8
90	16.42993	-0.65482	0	0	1	10
91	16.20718	-0.87785	0	0	0	2
92	16.40511	-0.68020	0	0	1	9
93	16.32073	-0.76486	0	0	0	4
94	16.40444	-0.68143	0	0	1	7
95	17.04371	-0.04244	1	0	1	99
96	16.90725	-0.17917	0	0	0	61
97	17.11638	0.02968	1	1	1	107
98	17.13196	0.04498	1	1	1	111

(*continued*)

Table 2 (*continued*)

Row	Data	Residuals	Above or below Median	Above or below Mean	Increase or Decrease	Rank
99	16.74435	-0.34291	0	0	0	40
100	16.92096	-0.16658	0	0	1	67
101	16.54355	-0.54427	0	0	0	18
102	17.22097	0.13287	1	1	1	125
103	16.43859	-0.64979	0	0	0	11
104	17.00160	-0.08706	0	0	1	82
105	17.03201	-0.05693	0	0	1	96
106	16.70810	-0.38111	0	0	0	32
107	16.21747	-0.87202	0	0	0	3
108	16.63017	-0.45960	0	0	1	26
109	16.92418	-0.16587	0	0	1	69
110	16.52448	-0.56585	0	0	0	15
111	16.61878	-0.47183	0	0	1	25
112	16.61731	-0.47358	0	0	0	24
113	17.04021	-0.05096	0	0	1	97
114	17.12720	0.03575	1	1	1	109
115	17.14703	0.05530	1	1	1	116
116	16.73656	-0.35544	0	0	0	38
117	16.82746	-0.26482	0	0	1	52
118	16.33188	-0.76068	0	0	0	5
119	16.63698	-0.45586	0	0	1	28
120	16.84890	-0.24422	0	0	1	58
121	16.90669	-0.18671	0	0	1	60
122	17.10601	0.01233	1	1	1	101
123	16.83639	-0.25757	0	0	0	54
124	17.00564	-0.08860	0	0	1	85
125	17.22547	0.13095	1	1	1	127
126	17.31650	0.22171	1	1	1	142
127	17.20568	0.11061	1	1	0	118
128	17.30042	0.20507	1	1	1	136
129	17.24962	0.15399	1	1	0	135
130	17.23628	0.14037	1	1	0	131
131	17.52850	0.43231	1	1	1	173
132	16.93604	-0.16043	0	0	0	73
133	16.94480	-0.15195	0	0	1	76
134	16.90932	-0.18771	0	0	0	62
135	17.01803	-0.07928	0	0	1	91
136	16.54779	-0.54979	0	0	0	20
137	16.74236	-0.35550	0	0	1	39
138	16.81500	-0.28314	0	0	1	48
139	16.74825	-0.35017	0	0	0	42
140	16.73015	-0.36855	0	0	0	36
141	16.64815	-0.45083	0	0	0	30
142	16.53733	-0.56193	0	0	0	16
143	17.00865	-0.09089	0	0	1	87
144	16.74964	-0.35018	0	0	0	44
145	16.74866	-0.35144	0	0	0	43
146	16.90432	-0.19606	0	0	1	59
147	17.44922	0.34857	1	1	1	168

(*continued*)

Table 2 (*continued*).

Row	Data	Residuals	Above or below: Median	Above or below: Mean	Increase or Decrease	Rank
148	17.10908	0.00815	1	1	0	104
149	17.00122	-0.09999	0	0	0	81
150	16.83817	-0.26332	0	0	0	56
151	17.21231	0.11054	1	1	1	121
152	17.20437	0.10232	1	1	0	117
153	17.41511	0.31278	1	1	1	155
154	17.24219	0.13958	1	1	0	133
155	16.93169	-0.17120	0	0	0	72
156	16.81695	-0.28622	0	0	0	50
157	17.02788	-0.07556	0	0	1	94
158	17.41997	0.31625	1	1	1	158
159	17.23561	0.13161	1	1	0	130
160	17.22998	0.12570	1	1	0	129
161	17.13021	0.02565	1	1	0	110
162	17.14141	0.03657	1	1	1	113
163	17.10888	0.00376	1	1	0	103
164	17.43719	0.33179	1	1	1	165
165	17.21551	0.10983	1	1	0	123
166	16.93112	-0.17484	0	0	0	71
167	16.90978	-0.19645	0	0	0	63
168	17.02873	-0.07778	0	0	1	95
169	16.74617	-0.36062	0	0	0	41
170	16.94064	-0.16643	0	0	1	75
171	17.33452	0.22717	1	1	1	143
172	17.84907	0.74144	1	1	1	192
173	17.81379	0.70588	1	1	0	187
174	17.63263	0.52444	1	1	0	182
175	17.54611	0.43764	1	1	0	176
176	17.00007	-0.10868	0	0	0	79
177	16.91617	-0.19285	0	0	0	65
178	17.11554	0.00624	1	1	1	106
179	17.22421	0.11463	1	1	1	126
180	17.44947	0.33961	1	1	1	169
181	17.52535	0.41521	1	1	1	172
182	17.90374	0.79332	1	1	1	193
183	17.04086	-0.06984	0	0	0	98
184	17.04384	-0.06714	1	0	1	100
185	17.00983	-0.10143	0	0	0	88
186	17.22613	0.11459	1	1	1	128
187	17.34321	0.23140	1	1	1	145
188	17.42100	0.30891	1	1	1	159
189	17.41083	0.29846	1	1	0	153
190	17.00513	-0.10752	0	0	0	84
191	18.01395	0.90102	1	1	1	195
192	18.21411	1.10090	1	1	1	197
193	17.60969	0.49620	1	1	0	178
194	17.83876	0.72499	1	1	1	190
195	17.71733	0.60328	1	1	0	184
196	17.20931	0.09498	1	1	0	120
197	17.40935	0.29475	1	1	0	152

SOURCE: Box and Jenkins [1976]. Reprinted from *Times Series Analysis: Forecasting and Control*, 2nd edn. with permission from Holden-Day.

Theoretical Background

We can for ease of notation define $\mathbf{Y} = \mathbf{X} - \bar{\mathbf{x}}E$ and, since ψ is a symmetric matrix, express $\psi = \Gamma\Theta\Gamma'$, where Γ is an orthogonal and Θ is a diagonal matrix. Then $\mathbf{v}$ can be rewritten as

$$\mathbf{v} = \frac{\sum_{i=1}^{n} \theta_i \mathbf{z}_i^2}{\sum_{i=1}^{n} \mathbf{z}_i^2},$$

where $\mathbf{Z} = \psi\Gamma$ and the θ_i are the diagonal elements of Θ, i.e., the eigenvalues of ψ.

Since Δ is known in this formulation of the testing problem, one can just as well take $\Delta = I$ by redefining $\mathbf{X}$ as $\mathbf{X}' = (\mathbf{x}_1/\sqrt{\delta_{11}}, \ldots, \mathbf{x}_n/\sqrt{\delta_{nn}})$. In this case the test procedure hinges on $\psi = \Sigma^{-1}$, and it is an easy exercise to show that

$$\mathscr{E}\mathbf{v} = \bar{\theta} = \sum_{i=1}^{n} \theta_i/n,$$

$$\mathscr{V}\mathbf{v} = 2\sum_{i=1}^{n} (\theta_i - \bar{\theta})^2/n(n+2),$$

and its third and fourth moments are given by

$$\mu_3 = \frac{8\sum_{i=1}^{n} (\theta_i - \bar{\theta})^3}{n(n+2)(n+4)},$$

$$\mu_4 = \frac{48\sum_{i=1}^{n} (\theta_i - \bar{\theta})^4 + 12\left[\sum_{i=1}^{n} (\theta_i - \bar{\theta})^2\right]^2}{n(n+2)(n+4)(n+6)}$$

(see, for example, Madansky [1976], p. 233 and Durbin and Watson [1950]).

When, in particular, $\Delta = \sigma^2 I$ and Σ^{-1} is of the form

$$\Sigma^{-1} = \frac{1}{\sigma^2}\begin{bmatrix} 1 & -1 & 0 & \ldots & 0 & 0 & 0 \\ -1 & 2 & -1 & \ldots & 0 & 0 & 0 \\ 0 & -1 & 2 & \ldots & 0 & 0 & 0 \\ & & & \ldots & & & \\ \vdots & \vdots & \vdots & \ldots & \vdots & \vdots & \vdots \\ & & & \ldots & & & \\ 0 & 0 & 0 & \ldots & -1 & 2 & -1 \\ 0 & 0 & 0 & \ldots & 0 & -1 & 1 \end{bmatrix}$$

the numerator of $\mathbf{v}$ reduces to

$$2\sum_{i=2}^{n-1} (\mathbf{x}_i - \bar{\mathbf{x}})^2 - 2\sum_{i=2}^{n} (\mathbf{x}_i - \bar{\mathbf{x}})(\mathbf{x}_{i-1} - \bar{\mathbf{x}}) + (\mathbf{x}_1 - \bar{\mathbf{x}})^2 + (\mathbf{x}_n - \bar{\mathbf{x}})^2$$

$$= \sum_{i=2}^{n} (\mathbf{x}_i - \mathbf{x}_{i-1})^2,$$

and so $\mathbf{v}$ reduces to the von Neumann ratio given above.

More generally, suppose Σ^{-1} is of the form $\sigma^{-2}(I + \lambda\Phi)$. Then its eigenvalues would be $\theta_i = \sigma^{-2}(1 + \lambda v_i)$, where the v_i are the eigenvalues of Φ. Suppose further that $\Delta = \sigma^2 I$. Then

$$\mathbf{v} = \frac{\sum_{i=1}^{n} (1 + \lambda v_i)\mathbf{z}_i^2}{\sum_{i=1}^{n} \mathbf{z}_i^2} = 1 + \lambda \frac{\sum_{i=1}^{n} v_i \mathbf{z}_i^2}{\sum_{i=1}^{n} \mathbf{z}_i^2}.$$

And if Φ is of the form

$$\Phi = \begin{bmatrix} 1 & -1 & 0 & \cdots & 0 & 0 & 0 \\ -1 & 2 & -1 & \cdots & 0 & 0 & 0 \\ \vdots & \vdots & \vdots & & \vdots & \vdots & \vdots \\ 0 & 0 & 0 & & -1 & 2 & -1 \\ 0 & 0 & 0 & & 0 & -1 & 1 \end{bmatrix},$$

then $\mathbf{v}$ reduces to a monotonic function of the von Neumann ratio.

Why did we introduce this apparently arbitrary general form of Σ^{-1}? Because it recurs (approximately) in practice. Suppose the $\mathbf{x}_i$ are an autoregressive AR(1) series

$$\mathbf{x}_1 = \mathbf{u}_1,$$
$$\mathbf{x}_i = \rho \mathbf{x}_{i-1} + \mathbf{u}_i, \qquad i = 2, \ldots, n,$$

where the $\mathbf{u}_i$ are independent $N(0, \sigma^2)$ random variables. Then

$$\mathscr{C}(\mathbf{x}_i, \mathbf{x}_j) = \sigma^2 \rho^{|i-j|}(1 - \rho)^2/(1 - \rho^2),$$

so that

$$\Sigma = \sigma^2 \frac{(1 - \rho)^2}{1 - \rho^2} \begin{bmatrix} 1 & \rho & \rho^2 & \cdots & \rho^{n-1} \\ \rho & 1 & \rho & \cdots & \rho^{n-2} \\ \vdots & \vdots & \vdots & & \vdots \\ \rho^{n-1} & \rho^{n-2} & \rho^{n-3} & \cdots & 1 \end{bmatrix},$$

and

$$\Sigma^{-1} = \frac{1}{\sigma^2(1 - \rho)^2} \begin{bmatrix} 1 & -\rho & 0 & 0 & \cdots & 0 & 0 & 0 \\ -\rho & 1 + \rho^2 & -\rho & 0 & \cdots & 0 & 0 & 0 \\ 0 & -\rho & 1 + \rho^2 & 0 & \cdots & 0 & 0 & 0 \\ \vdots & \vdots & \vdots & \vdots & & \vdots & \vdots & \vdots \\ 0 & 0 & 0 & 0 & & -\rho & 1 + \rho^2 & -\rho \\ 0 & 0 & 0 & 0 & & 0 & & 1 \end{bmatrix}.$$

If Σ^{-1} had $1 + \rho^2 - \rho$ instead of 1 in its upper-left and lower-right corners, it would be the form $\sigma^{-2}(I + \lambda\Phi)$, where $\lambda = \rho/(1 - \rho)^2$ and Φ is of the tridiagonal form given above. Thus if the $\mathbf{x}_i$ are an autoregressive AR(1) series, the von Neumann ratio is approximately equivalent to the likelihood ratio procedure for testing that hypothesis against the hypothesis of independence.

(b) Autocorrelation

The standard attack on the question of whether the $\mathbf{x}_i$ are independent is to check for serial correlation by correlating the series $\{\mathbf{x}_i\}$ with the series $\{\mathbf{x}_{i-l}\}$, where l is the size of the gap (or lag) between observations being correlated. With a gap of l one has a set of $n - l$ pairs of observations $(\mathbf{x}_i, \mathbf{x}_{i-l})$, $i = l + 1, \ldots, n$, to be used in calculating the correlation coefficient.

There is a definitional ambiguity in what constitutes the serial correlation coefficient. For example, what estimator of the population mean should be used, $\bar{x}$, the mean of the entire set of n observations or the mean for each series separately, in calculating the "deviation from the mean"? Similarly, what estimator of the population variance should be used, one based on all the n observations or a separate estimate based on each series of $n - l$ observations? The usual resolution is to define the lth autocorrelation (or serial correlation) as

$$\mathbf{r}_l = \frac{\sum_{i=l+1}^{n} (\mathbf{x}_i - \bar{\mathbf{x}})(\mathbf{x}_{i-l} - \bar{\mathbf{x}})/(n - l)}{\sum_{i=1}^{n} (\mathbf{x}_i - \bar{\mathbf{x}})^2/n},$$

where $\bar{\mathbf{x}} = \sum_{i=1}^{n} \mathbf{x}_i/n$.

```
            S.E.
      AUTO- RANDOM
ORDER CORR. MODEL   -1  -.75 -.50 -.25  0   .25  .50  .75  +1   ADJ.B-P
                    :----:----:----:----:----:----:----:----:
  1   0.571 0.071                    +  :  +       *            65.15
  2   0.496 0.071                    +  :  +      *             114.6
  3   0.398 0.070                    +  :  +    *               146.6
  4   0.356 0.070                    +  :  +   *                172.3
  5   0.324 0.070                    +  :  +  *                 193.7
  6   0.349 0.070                    +  :  +   *                218.7
  7   0.398 0.070                    +  :  +    *               251.4
  8   0.324 0.069                    +  :  +  *                 273.1
  9   0.303 0.069                    +  :  +  *                 292.3
 10   0.263 0.069                    +  :  + *                  306.8
 11   0.194 0.069                    +  :  +*                   314.7
 12   0.168 0.069                    +  :  *                    320.7
 13   0.197 0.069                    +  :  +*                   328.9
 14   0.242 0.068                    +  :  + *                  341.4
 15   0.142 0.068                    +  :  *                    345.8
 16   0.180 0.068                    +  :  +*                   352.8
 17   0.197 0.068                    +  :  +*                   361.3
 18   0.201 0.068                    +  :  +*                   370.1
 19   0.140 0.067                    +  :  *                    374.4
 20   0.180 0.067                    +  :  +*                   381.6
                    :----:----:----:----:----:----:----:----:
                    -1  -.75 -.50 -.25  0   .25  .50  .75  +1

                    * : AUTOCORRELATIONS
                    + : 2 STANDARD ERROR LIMITS (APPROX.)
```

Figure 1. Autocorrelation Plot of Data of Table 2.

Under the hypothesis of independence, the lth serial correlation has for large values of n a normal distribution with mean $-1/(n-1)$ and variance $1/n$ (see Kendall and Stuart [1966], pp. 434 and 432, respectively).

The plot of the autocorrelations as a function of l, the *order* of the autocorrelation, is called the correlogram of the data. An example based on the data of Table 2 is given in Figure 1. This output, produced by IDA, approximates the expected value of $\mathbf{r}_l$ with 0 and the variance with $1/(n-l)$. Thus, for example, $0.071 = 1/\sqrt{197-1}$ is used as the standard error of $\mathbf{r}_1$ (S. E. RANDOM MODEL). (The adjusted Box–Pierce statistic given in the right-hand column will be described in Section 2(d) of this chapter.) The +'s in Figure 1 are a plot of the approximate two standard error limits for $\mathbf{r}_l$, i.e., $0 \pm 2/\sqrt{n-l}$; the *'s are a plot of the $\mathbf{r}_l$. We see from this output that the data are highly dependent.

(c) Independence of Regression Residuals

In the case of regression residuals, we are as always confronted with the problem that, although the $\mathbf{u}_i = \mathbf{y}_i - X_i'B$ are independent $N(0, \sigma^2)$ variables, the estimated residuals $\hat{\mathbf{u}}_i = \mathbf{y}_i - X_i'\hat{\mathbf{B}}$ are correlated, so that if $\hat{\mathbf{U}}' = (\hat{\mathbf{u}}_1, \ldots, \hat{\mathbf{u}}_n)$, then $\hat{\mathbf{U}}$ has a $N(0, \sigma^2(I - X(X'X)^{-1}X'))$ distribution. One can work with $\hat{\mathbf{U}}$ in the framework of section (a) above, as has been done by Durbin and Watson ([1950, 1951, 1971]). They conclude that the von Neumann ratio based on the $\hat{\mathbf{u}}_i$ is an appropriate test statistic for testing the independence of the $\mathbf{u}_i$. They did not obtain the exact distribution of the statistic, but did find bounds d_L and d_U such that, when testing for positive correlation, if the statistic $\mathbf{d}$ is below d_L it is significant and if it is above d_U it is not significant. The decision for values of $\mathbf{d}$ between d_L and d_U is indeterminate. When testing for negative correlation, the statistic $\mathbf{d}' = 4 - \mathbf{d}$ is to be compared with the values of d_L and d_U, in essence reflecting $\mathbf{d}$ around its approximate expected value of 2. A two-tailed test is based on a comparison of $\mathbf{d}$ and $4 - \mathbf{d}$ with d_L and d_U. A table of these bounds at the 5% level is given as Table 3.

To illustrate this, we consider once again the data of Table 2, and take as an independent variable the sequence number of the observation. Here $p = 2$ (since we assume a nonzero intercept) and column 2 of Table 2 contains the set of $\hat{\mathbf{u}}_i$ from this regression. The regression equation is $x_t = 17.06 + 0.00027901t$, with an insignificant slope. Since

$$\sum_{i=2}^{197} (\hat{u}_i - \hat{u}_{i-1})^2 = 26.592143 \quad \text{and} \quad \sum_{i=1}^{197} \hat{u}_i^2 = 31.050304,$$

we see that $v = 0.856$, as in the case when we considered the $\mathbf{x}_i$ themselves. We thus conclude that the residuals from this regression are not independent.

Table 3. Bounds for the Durbin–Watson Statistic.

n	$k' = 1$		$k' = 2$		$k' = 3$		$k' = 4$		$k' = 5$	
	d_L	d_U	d_L	d_U	d_L	d_U	d_L	d_U	d_L	d_U
15	1·08	1·36	0·95	1·54	0·82	1·75	0·69	1·97	0·56	2·21
16	1·10	1·37	0·98	1·54	0·86	1·73	0·74	1·93	0·62	2·15
17	1·13	1·38	1·02	1·54	0·90	1·71	0·78	1·90	0·67	2·10
18	1·16	1·39	1·05	1·53	0·93	1·69	0·82	1·87	0·71	2·06
19	1·18	1·40	1·08	1·53	0·97	1·68	0·86	1·85	0·75	2·02
20	1·20	1·41	1·10	1·54	1·00	1·68	0·90	1·83	0·79	1·99
21	1·22	1·42	1·13	1·54	1·03	1·67	0·93	1·81	0·83	1·96
22	1·24	1·43	1·15	1·54	1·05	1·66	0·96	1·80	0·86	1·94
23	1·26	1·44	1·17	1·54	1·08	1·66	0·99	1·79	0·90	1·92
24	1·27	1·45	1·19	1·55	1·10	1·66	1·01	1·78	0·93	1·90
25	1·29	1·45	1·21	1·55	1·12	1·66	1·04	1·77	0·95	1·89
26	1·30	1·46	1·22	1·55	1·14	1·65	1·06	1·76	0·98	1·88
27	1·32	1·47	1·24	1·56	1·16	1·65	1·08	1·76	1·01	1·86
28	1·33	1·48	1·26	1·56	1·18	1·65	1·10	1·75	1·03	1·85
29	1·34	1·48	1·27	1·56	1·20	1·65	1·12	1·74	1·05	1·84
30	1·35	1·49	1·28	1·57	1·21	1·65	1·14	1·74	1·07	1·83
31	1·36	1·50	1·30	1·57	1·23	1·65	1·16	1·74	1·09	1·83
32	1·37	1·50	1·31	1·57	1·24	1·65	1·18	1·73	1·11	1·82
33	1·38	1·51	1·32	1·58	1·26	1·65	1·19	1·73	1·13	1·81
34	1·39	1·51	1·33	1·58	1·27	1·65	1·21	1·73	1·15	1·81
35	1·40	1·52	1·34	1·58	1·28	1·65	1·22	1·73	1·16	1·80
36	1·41	1·52	1·35	1·59	1·29	1·65	1·24	1·73	1·18	1·80
37	1·42	1·53	1·36	1·59	1·31	1·66	1·25	1·72	1·19	1·80
38	1·43	1·54	1·37	1·59	1·32	1·66	1·26	1·72	1·21	1·79
39	1·43	1·54	1·38	1·60	1·33	1·66	1·27	1·72	1·22	1·79
40	1·44	1·54	1·39	1·60	1·34	1·66	1·29	1·72	1·23	1·79
45	1·48	1·57	1·43	1·62	1·38	1·67	1·34	1·72	1·29	1·78
50	1·50	1·59	1·46	1·63	1·42	1·67	1·38	1·72	1·34	1·77
55	1·53	1·60	1·49	1·64	1·45	1·68	1·41	1·72	1·38	1·77
60	1·55	1·62	1·51	1·65	1·48	1·69	1·44	1·73	1·41	1·77
65	1·57	1·63	1·54	1·66	1·50	1·70	1·47	1·73	1·44	1·77
70	1·58	1·64	1·55	1·67	1·52	1·70	1·49	1·74	1·46	1·77
75	1·60	1·65	1·57	1·68	1·54	1·71	1·51	1·74	1·49	1·77
80	1·61	1·66	1·59	1·69	1·56	1·72	1·53	1·74	1·51	1·77
85	1·62	1·67	1·60	1·70	1·57	1·72	1·55	1·75	1·52	1·77
90	1·63	1·68	1·61	1·70	1·59	1·73	1·57	1·75	1·54	1·78
95	1·64	1·69	1·62	1·71	1·60	1·73	1·58	1·75	1·56	1·78
100	1·65	1·69	1·63	1·72	1·61	1·74	1·59	1·76	1·57	1·78

SOURCE: Durbin and Watson [1951]. Reprinted from *Biometrika*, Vol. 38, with the permission of the Biometrika Trustees. (Their k' is our p-1)

Theoretical Background

Durbin and Watson [1951] recommend as an approximation to the distribution of $\mathbf{d}/4$ a beta distribution with parameters α and β, where α and β satisfy

$$\alpha + \beta = \frac{\mathscr{E}\mathbf{d}[4 - \mathscr{E}\mathbf{d}]}{\mathscr{V}\mathbf{d}} - 1,$$

$$\alpha = \tfrac{1}{4}(\alpha + \beta)\mathscr{E}\mathbf{d}.$$

Let A be the tridiagonal matrix

$$A = \begin{bmatrix} 1 & -1 & 0 & \cdots & 0 & 0 & 0 \\ -1 & 2 & -1 & \cdots & 0 & 0 & 0 \\ 0 & -1 & 2 & \cdots & 0 & 0 & 0 \\ \vdots & \vdots & \vdots & & \vdots & \vdots & \vdots \\ 0 & 0 & 0 & & -1 & 2 & -1 \\ 0 & 0 & 0 & & 0 & -1 & 1 \end{bmatrix}$$

and $M = I - X(X'X)^{-1}X'$. Then

$$\mathscr{E}\mathbf{d} = \frac{\text{tr MA}}{n-p},$$

$$\mathscr{V}\mathbf{d} = \frac{2\,\text{tr}(MA)^2 + (\text{tr } MA)^2}{(n-p)(n-p+2)}.$$

Other approximations have been recommended, notably that of Theil and Nagar [1961], but Durbin and Watson [1971] have shown that their approximation is superior. The problem with their approximation is that the computation of $\mathscr{E}\mathbf{d}$ and $\mathscr{V}\mathbf{d}$ requires a fair amount of matrix calculation. The Theil–Nagar approximation reduces to approximating

$$\mathbf{w} = \frac{n^2\mathbf{d} - (4p^2 - 1)}{4n^2 - 4p^2 + 2}$$

with a beta distribution with parameters

$$\alpha = (n+p)/2,$$

$$\beta = (n-p+2)/2,$$

i.e., does not involve M or A except in the computation of the moments of $\mathbf{d}$.

(d) Box–Pierce Statistic

Suppose we knew the true residuals $\mathbf{u}_1, \ldots, \mathbf{u}_n$ and calculated from them the m lag correlation coefficients $\mathbf{r}_1, \ldots, \mathbf{r}_m$, where

$$\mathbf{r}_l = \sum_{i=l+1}^{n} \mathbf{u}_i \mathbf{u}_{i-l} \Big/ \sum_{i=1}^{n} \mathbf{u}_i^2.$$

Then the statistic $\mathbf{w} = n(n+1)\sum_{l=1}^{m} \mathbf{r}_l^2/(n-l)$ is approximately distributed as a chi-square variable with m degrees of freedom (see Box and Pierce [1970]), since under the hypothesis of independence of the $\mathbf{u}_i$, $c(\mathbf{r}_l, \mathbf{r}_{l'}) = 0$, $l > l'$, and

$$\mathscr{V}\mathbf{r}_l = \frac{n-l}{n(n+2)} \approx \frac{1}{n}$$

for large n. Thus the statistic

$$\mathbf{q} = n \sum_{l=1}^{m} \mathbf{r}_l^2$$

is approximately distributed as χ^2 with m degrees of freedom as well.

Now suppose $\mathbf{r}_1, \ldots, \mathbf{r}_m$ are replaced by $\hat{\mathbf{r}}_1, \ldots, \hat{\mathbf{r}}_m$ in the definition of $\mathbf{q}$, i.e.,

$$\hat{\mathbf{q}} = n \sum_{i=1}^{m} \hat{\mathbf{r}}_l^2,$$

where

$$\hat{\mathbf{r}}_l = \sum_{i=l+1}^{n} \hat{\mathbf{u}}_i \hat{\mathbf{u}}_{i-l} \Big/ \sum_{i=1}^{n} \hat{\mathbf{u}}_i^2,$$

and the $\hat{\mathbf{u}}_i$ are the residuals estimated from the regression. Then, Box and Pierce [1970] suggest, $\hat{\mathbf{q}}$ is approximately distributed as χ^2 with $m - p$ degrees of freedom, where p is the number of parameters estimated in the regression. Box and Ljung [1978] have modified the definition of $\hat{\mathbf{q}}$ to obtain a statistic whose distribution is more closely approximated by a χ^2 distribution with $m - p$ degrees of freedom, namely

$$\hat{\mathbf{q}}^* = n(n+2) \sum_{l=1}^{m} \hat{\mathbf{r}}_l^2/(n-l).$$

This statistic is sometimes called the "adjusted Box–Pierce" statistic and is useful as an overall test of whether the residuals are independent.

One can of course test each $\hat{\mathbf{r}}_l$ to see if it is significantly different from 0, by noting that the variance of $\hat{\mathbf{r}}_l$ is approximately $(n - l)/n(n + 2)$ and that for large n the distribution of $\hat{\mathbf{r}}_l$ is approximately normal. The problem with this approach is that the probability of rejecting the null hypothesis of independence when in fact it is true has to be dealt with carefully. If, for example, one used an $\alpha = 0.05$ for testing each of the $\hat{\mathbf{r}}_l$, then even if independence prevailed one would on average reject the null hypothesis once in 20 tests, i.e., once in every 20 lags. So to be conservative one should use $1 - (1 - \alpha)^{1/m}$ as the significance level for the tests of the individual $\hat{\mathbf{r}}_l$'s (so that the overall level would be α, since the tests are approximately independent).

As can be seen from Figures 1 and 2, the autocorrelation plot of the residuals mirrors that of the raw data. The adjusted Box–Pierce statistic for 20 lags is 390.4, and since $m = 20$ and $p = 2$, we compare this with a $\chi^2(18)$ variable and find that it is highly significant. If $\alpha = 0.05$, then the individual $\hat{\mathbf{r}}_l$'s should be judged at the $1 - (0.95)^{1/20} = 0.0026$ level, i.e., critical values to be used are approximately $0 \pm 3.01/\sqrt{197} = \pm 0.229$, where 3.01 is the 0.13% point of the $N(0, 1)$ distribution. Using this critical value, we see that only the autocorrelations of order 10 or below are significant.

```
            S.E.
      AUTO- RANDOM
ORDER CORR. MODEL  -1  -.75 -.50 -.25  0   .25  .50  .75  +1   ADJ.B-P
                   :----:----:----:----:----:----:----:----:
1     0.570 0.071                   +  :  +       *              65.07
2     0.496 0.071                   +  :  +      *               114.5
3     0.399 0.070                   +  :  +    *                 146.7
4     0.358 0.070                   +  :  +   *                  172.7
5     0.327 0.070                   +  :  +   *                  194.5
6     0.353 0.070                   +  :  +   *                  220.1
7     0.403 0.070                   +  :  +    *                 253.7
8     0.328 0.069                   +  :  +   *                  276.1
9     0.308 0.069                   +  :  +  *                   295.9
10    0.268 0.069                   +  :  + *                    310.9
11    0.199 0.069                   +  :  +*                     319.3
12    0.174 0.069                   +  :  *                      325.7
13    0.203 0.069                   +  :  +*                     334.5
14    0.247 0.068                   +  :  + *                    347.5
15    0.147 0.068                   +  :  *                      352.2
16    0.186 0.068                   +  :  +*                     359.7
17    0.203 0.068                   +  :  +*                     368.7
18    0.207 0.068                   +  :  +*                     378.1
19    0.146 0.067                   +  :  *                      382.7
20    0.186 0.067                   +  :  +*                     390.4
                   :----:----:----:----:----:----:----:----:
                   -1  -.75 -.50 -.25  0   .25  .50  .75  +1

                   * : AUTOCORRELATIONS
                   + : 2 STANDARD ERROR LIMITS (APPROX.)
```

Figure 2. Autocorrelation Plot of Residuals of Table 2.

2. Nonparametric Procedures

The essence of nonparametric procedures for testing for independence of $\mathbf{x}_1, \ldots, \mathbf{x}_n$ is not so much to analyze the magnitudes of the sequence of observations on these random variables as to analyze the sequential "pattern" of these observations. For example, if the sequence is one of independent identically distributed random variables then with probability 1/2 each observation will be above (or below) the population median, and, by extension, the sample median as well. For another example, if the sequence is one of independent random variables then with probability 1/2 each observation will be above (or below) its predecessor in the sample. These implications do not depend on detailed distributional assumptions about the $\mathbf{x}$'s, and so any tests of whether our data are consonant with these implications of independence will be "nonparametric" (in the sense that this term has come to be used in statistics parlance).

(a) Runs Above and Below the Median

Let v be the median of the distribution of $\mathbf{x}$. Suppose we associate with each $\mathbf{x}_i$ a variable $\mathbf{u}_i$, where

$$\mathbf{u}_i = \begin{cases} 1 & \text{if } \mathbf{x}_i > v, \\ 0 & \text{if } \mathbf{x}_i < v, \end{cases}$$

and where $\mathbf{u}_i$ is undefined if $\mathbf{x}_i = v$ (alternatively, if $\mathbf{x}_i = v$ set $\mathbf{u}_i = 1$ with probability 1/2 and $\mathbf{u}_i = 0$ with probability 1/2). Consider now the sequence $\mathbf{u}_1, \ldots, \mathbf{u}_n$, and let $\mathbf{p}$ denote the number of $\mathbf{u}_i$ which are equal to 1. Conditional on the value of $\mathbf{p}$ there are $\binom{n}{p}$ possible sequences of p 1's and $n - p$ 0's, of which our observed set is one of these sequences.

One statistic based on the sequence of $\mathbf{u}_i$ which is a useful indicator of the independence of the $\mathbf{u}_i$ (and hence of the $\mathbf{x}_i$) is the "runs count". We define a "run" as a maximal consecutive set of $\mathbf{u}_i$'s having the same value. The sequence of $\mathbf{u}_i$'s can be subdivided into runs, and the number of runs in the sequence can be counted. For example, if the sequence is

0 1 1 1 0 0 1 0 1 0 0,

then the runs can be demarcated thus

0|111|00|1|0|1|00

and the runs count for this sequence is 7.

A low runs count is indicative of one kind of deviation from independence, namely a tendency for below-median $\mathbf{x}$'s and above-median $\mathbf{x}$'s to be observed in clusters. A high runs count is indicative of another kind of deviation from independence, namely a tendency for a below-median observation to be followed by an above-median observation.

What is the sampling distribution of the runs count? Let $\mathbf{r}$ be the runs count and let $\mathbf{p}$ be the number of 1's in the sequence of signs. Both $\mathbf{r}$ and $\mathbf{p}$ are random variables, in that another sample of $\mathbf{x}_1, \ldots, \mathbf{x}_n$ would produce other values of both $\mathbf{r}$ and $\mathbf{p}$. In the runs test, however, we disregard the random nature of $\mathbf{p}$ in determining the sampling distribution of $\mathbf{r}$. The sampling distribution of $\mathbf{r}$ is conditional on the observed value of $\mathbf{p}$. That is, suppose we had p 1's and $n - p$ 0's. We consider all the possible sequences of p 1's and $n - p$ 0's and, for each possible sequence, we record the runs count r. This collection of possible runs counts, given that p is fixed, is the basic population of the sample statistic r from which we determine the sampling distribution.

For example, if $n = 10$ and $p = 5$, there are $10!/5!\,5! = 252$ possible sequences of 1's and 0's, and associated with each sequence is a runs count. The list of these sequences, their runs counts, and a histogram of the runs counts, is given in Table 4 and Figure 3. The sampling distribution of the runs count in general is approximately normal with mean $2p(n - p)/n + 1$ and standard deviation

$$\sqrt{\frac{2p(n-p)[2p(n-p)-n]}{n^2(n-1)}}$$

(see Wald and Wolfowitz [1940]). For example, the normal approximation yields a mean of 6 and a standard deviation of 1.490712. This is in contrast to the true mean and standard deviation of the sampling distribution of $\mathbf{r}$, namely 6 and 1.4937. (Figure 4 shows a normal probability paper plot of the

Table 4. All Possible Sequences of Runs with Five 1's and Five 0's.

001	1111100000	2	064	1011010010	8	127	0111110000	3	190	0100101011	8
002	1111010000	4	065	1011010001	7	128	0111101000	5	191	0100100111	6
003	1111001000	4	066	1011001100	6	129	0111100100	5	192	0100011110	5
004	1111000100	4	067	1011001010	8	130	0111100010	5	193	0100011101	6
005	1111000010	4	068	1011001001	7	131	0111100001	4	194	0100011011	6
006	1111000001	3	069	1011000110	6	132	0111011000	5	195	0100010111	6
007	1110110000	4	070	1011000101	7	133	0111010100	7	196	0100001111	4
008	1110101000	6	071	1011000011	5	134	0111010010	7	197	0011111000	3
009	1110100100	6	072	1010111000	6	135	0111010001	6	198	0011110100	5
010	1110100010	6	073	1010110100	8	136	0111001100	5	199	0011110010	5
011	1110100001	5	074	1010110010	8	137	0111001010	7	200	0011110001	4
012	1110011000	4	075	1010110001	7	138	0111001001	6	201	0011101100	5
013	1110010100	6	076	1010101100	8	139	0111000110	5	202	0011101010	7
014	1110010010	6	077	1010101010	10	140	0111000101	6	203	0011101001	6
015	1110010001	5	078	1010101001	9	141	0111000011	4	204	0011100110	5
016	1110001100	4	079	1010100110	8	142	0110111000	5	205	0011100101	6
017	1110001010	6	080	1010100101	9	143	0110110100	7	206	0011100011	4
018	1110001001	5	081	1010100011	7	144	0110110010	7	207	0011011100	5
019	1110000110	4	082	1010011100	6	145	0110110001	6	208	0011011010	7
020	1110000101	5	083	1010011010	8	146	0110101100	7	209	0011011001	6
021	1110000011	3	084	1010011001	7	147	0110101010	9	210	0011010110	7
022	1101110000	4	085	1010010110	8	148	0110101001	8	211	0011010101	8
023	1101101000	6	086	1010010101	9	149	0110100110	7	212	0011010011	6
024	1101100100	6	087	1010010011	7	150	0110100101	8	213	0011001110	5
025	1101100010	6	088	1010001110	6	151	0110100011	6	214	0011001101	6
026	1101100001	5	089	1010001101	7	152	0110011100	5	215	0011001011	6
027	1101011000	6	090	1010001011	7	153	0110011010	7	216	0011000111	4
028	1101010100	8	091	1010000111	5	154	0110011001	6	217	0010111100	5
029	1101010010	8	092	1001111000	4	155	0110010110	7	218	0010111010	7
030	1101010001	7	093	1001110100	6	156	0110010101	8	219	0010111001	6
031	1101001100	6	094	1001110010	6	157	0110010011	6	220	0010110110	7
032	1101001010	8	095	1001110001	5	158	0110001110	5	221	0010110101	8
033	1101001001	7	096	1001101100	6	159	0110001101	6	222	0010110011	6
034	1101000110	6	097	1001101010	8	160	0110001011	6	223	0010101110	7
035	1101000101	7	098	1001101001	7	161	0110000111	4	224	0010101101	8
036	1101000011	5	099	1001100110	6	162	0101111000	5	225	0010101011	8
037	1100111000	4	100	1001100101	7	163	0101110100	7	226	0010100111	6
038	1100110100	6	101	1001100011	5	164	0101110010	7	227	0010011110	5
039	1100110010	6	102	1001011100	6	165	0101110001	6	228	0010011101	6
040	1100110001	5	103	1001011010	8	166	0101101100	7	229	0010011011	6
041	1100101100	6	104	1001011001	7	167	0101101010	9	230	0010010111	6
042	1100101010	8	105	1001010110	8	168	0101101001	8	231	0010001111	4
043	1100101001	7	106	1001010101	9	169	0101100110	7	232	0001111100	3
044	1100100110	6	107	1001010011	7	170	0101100101	8	233	0001111010	5
045	1100100101	7	108	1001001110	6	171	0101100011	6	234	0001111001	4
046	1100100011	5	109	1001001101	7	172	0101011100	7	235	0001110110	5
047	1100011100	4	110	1001001011	7	173	0101011010	9	236	0001110101	6
048	1100011010	6	111	1001000111	5	174	0101011001	8	237	0001110011	4
049	1100011001	5	112	1000111100	4	175	0101010110	9	238	0001101110	5
050	1100010110	6	113	1000111010	6	176	0101010101	10	239	0001101101	6
051	1100010101	7	114	1000111001	5	177	0101010011	8	240	0001101011	6
052	1100010011	5	115	1000110110	6	178	0101001110	7	241	0001100111	4
053	1100001110	4	116	1000110101	7	179	0101001101	8	242	0001011110	5
054	1100001101	5	117	1000110011	5	180	0101001011	8	243	0001011101	6
055	1100001011	5	118	1000101110	6	181	0101000111	6	244	0001011011	6
056	1100000111	3	119	1000101101	7	182	0100111100	5	245	0001010111	6
057	1011110000	4	120	1000101011	7	183	0100111010	7	246	0001001111	4
058	1011101000	6	121	1000100111	5	184	0100111001	6	247	0000111110	3
059	1011100100	6	122	1000011110	4	185	0100110110	7	248	0000111101	4
060	1011100010	6	123	1000011101	5	186	0100110101	8	249	0000111011	4
061	1011100001	5	124	1000011011	5	187	0100110011	6	250	0000110111	4
062	1011011000	6	125	1000010111	5	188	0100101110	7	251	0000101111	4
063	1011010100	8	126	1000001111	3	189	0100101101	8	252	0000011111	2

```
    -               *
    -               *
  70-               *
    -               *
    -               *
    -               *
    -               *
  65-               *
    -               *
    -               *
    -               *
    -               *
  60-               *
    -               *
    -               *
    -               *
    -               *
  55-               *
    -               *
    -               *
    -               *
    -               *
  50-               *
    -               *
    -              ***
    -              ***
    -              ***
  45-              ***
    -              ***
    -              ***
    -              ***
    -              ***
  40-              ***
    -              ***
    -              ***
    -              ***
    -              ***
  35-              ***
    -              ***
    -              ***
    -             *****
    -             *****
  30-             *****
    -             *****
    -             *****
    -             *****
    -             *****
  25-             *****
    -             *****
    -             *****
    -             *****
    -             *****
  20-             *****
    -             *****
    -             *****
    -             *****
    -             *****
  15-             *****
    -             *****
    -             *****
    -             *****
    -             *****
  10-             *****
    -             *****
    -            *******
    -            *******
    -            *******
   5-            *******
    -            *******
    -            *******
    -           *********
    -           *********
     ------:---------:---------:---------:---------:---------:------
                               6
```

Figure 3. Histogram of Runs Counts of Table 4.

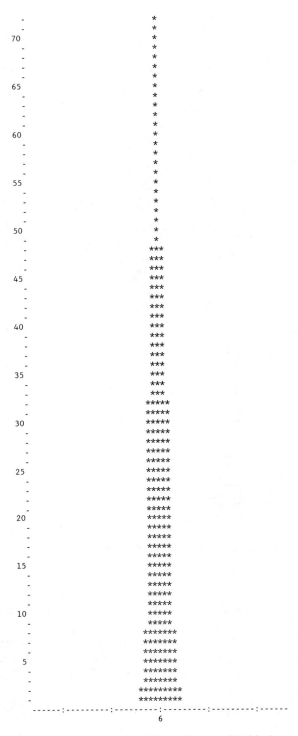

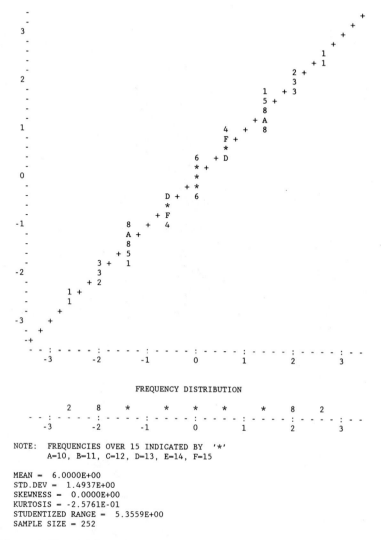

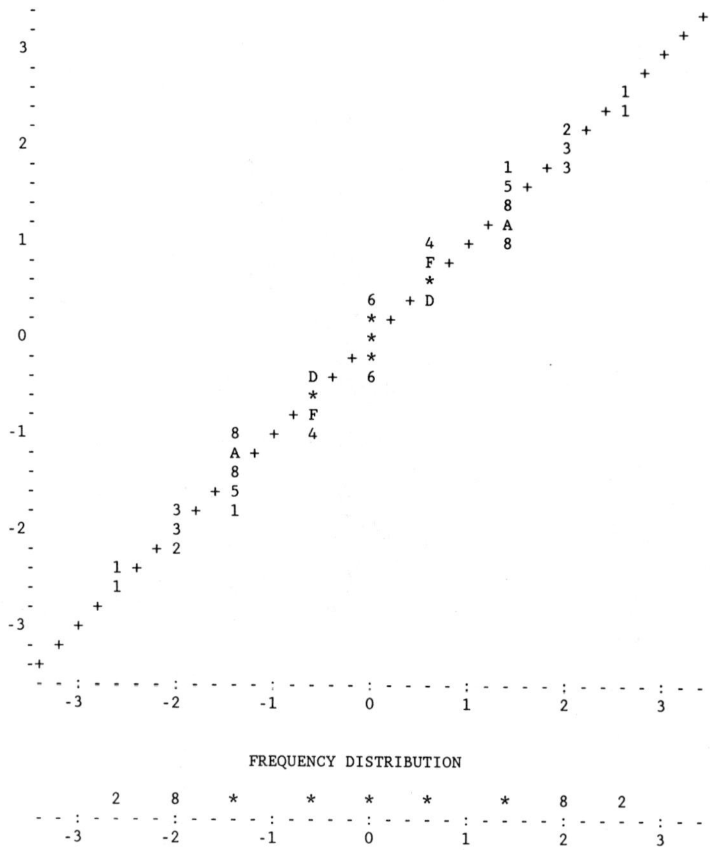


```
NOTE:   FREQUENCIES OVER 15 INDICATED BY  '*'
        A=10, B=11, C=12, D=13, E=14, F=15

MEAN =   6.0000E+00
STD.DEV =   1.4937E+00
SKEWNESS =   0.0000E+00
KURTOSIS = -2.5761E-01
STUDENTIZED RANGE =   5.3559E+00
SAMPLE SIZE = 252
```

Figure 4. Normal Cumulative Probability Plot of Runs Counts of Table 4.

population of **r**'s, to check on how good the normal approximation is for our example. Obviously, the discreteness shows up on the plot, but the distribution is perfectly symmetric and the kurtosis is quite low.)

For the data of Table 2, $\bar{x} = 17.0873$ and $\tilde{x}$ is $x_{183} = 17.04086$. Column 3 of Table 2 contains the u_i based on $\tilde{x}$. We find $p = 97$ and count 52 runs above and below the mean, so our statistic is $(52 - 99.47716)/6.998277 = -6.78$. We find $p = 99$ and count 54 runs above and below the median, so our statistic (based on 196 observations, since the median observation is excluded from

Table 5. Distribution of the Runs Count.

n_1 \ n_2	2	3	4	5	6	7	8	9	10	11	12	13	14	15	16	17	18	19	20
2											2	2	2	2	2	2	2	2	2
3					2	2	2	2	2	2	2	2	2	3	3	3	3	3	3
4				2	2	2	3	3	3	3	3	3	3	3	4	4	4	4	4
5			2	2	3	3	3	3	3	4	4	4	4	4	4	4	5	5	5
6		2	2	3	3	3	3	4	4	4	4	5	5	5	5	5	5	6	6
7		2	2	3	3	3	4	4	5	5	5	5	5	6	6	6	6	6	6
8		2	3	3	3	4	4	5	5	5	6	6	6	6	6	7	7	7	7
9		2	3	3	4	4	5	5	5	6	6	6	7	7	7	7	8	8	8
10		2	3	3	4	5	5	5	6	6	7	7	7	7	8	8	8	8	9
11		2	3	4	4	5	5	6	6	7	7	7	8	8	8	9	9	9	9
12	2	2	3	4	4	5	6	6	7	7	7	8	8	8	9	9	9	10	10
13	2	2	3	4	5	5	6	6	7	7	8	8	9	9	9	10	10	10	10
14	2	2	3	4	5	5	6	7	7	8	8	9	9	9	10	10	10	11	11
15	2	3	3	4	5	6	6	7	7	8	8	9	9	10	10	11	11	11	12
16	2	3	4	4	5	6	6	7	8	8	9	9	10	10	11	11	11	12	12
17	2	3	4	4	5	6	7	7	8	9	9	10	10	11	11	11	12	12	13
18	2	3	4	5	5	6	7	8	8	9	9	10	10	11	11	12	12	13	13
19	2	3	4	5	6	6	7	8	8	9	10	10	11	11	12	12	13	13	13
20	2	3	4	5	6	6	7	8	9	9	10	10	11	12	12	13	13	13	14

n_1 \ n_2	2	3	4	5	6	7	8	9	10	11	12	13	14	15	16	17	18	19	20
2																			
3																			
4				9	9														
5			9	10	10	11	11												
6			9	10	11	12	12	13	13	13	13								
7				11	12	13	13	14	14	14	14	15	15	15					
8				11	12	13	14	14	15	15	16	16	16	16	17	17	17	17	17
9					13	14	14	15	16	16	16	17	17	18	18	18	18	18	18
10					13	14	15	16	16	17	17	18	18	18	19	19	19	20	20
11					13	14	15	16	17	17	18	19	19	19	20	20	20	21	21
12					13	14	16	16	17	18	19	19	20	20	21	21	21	22	22
13						15	16	17	18	19	19	20	20	21	21	22	22	23	23
14						15	16	17	18	19	20	20	21	22	22	23	23	23	24
15						15	16	18	18	19	20	21	22	22	23	23	24	24	25
16							17	18	19	20	21	21	22	23	23	24	25	25	25
17							17	18	19	20	21	22	23	23	24	25	25	26	26
18							17	18	19	20	21	22	23	24	25	25	26	26	27
19							17	18	20	21	22	23	23	24	25	26	26	27	27
20							17	18	20	21	22	23	24	25	25	26	27	27	28

SOURCE: Siegel [1956]. Reprinted from *Nonparametric Statistics for the Behavioral Sciences* with permission from McGraw-Hill.

Table 6. Changes in Stock Levels for 1968–69 to 1977–78 Deflated by the Australian Gross Domestic Product Price Index.

528
348
264
−20
−167
575
410
−4
430
−122

SOURCE: Bartels [1982]. Reprinted from the *Journal of the American Statistical Association*, Vol. 77, with the permission of the ASA.

consideration) is $(54 - 98.9898)/6.981297 = -6.44$. From both these computations we conclude that the data are not independently drawn.

The exact sampling distribution of $\mathbf{r}$ has been tabulated by Swed and Eisenhart [1943] and recast in the form given in Table 5 by Siegel [1956] for p and $n - p = 2(1)20$. As an example of the use of the small sample distributions of these nonparametric statistics, consider the data set of Table 6 taken from Bartels [1982]. Here the set of v_i is 1100011010, whether the sample mean or median is used as the statistic for comparison. Since $n = 10$, $p = 4$, and $r = 6$, we see that the critical values from Table 5 are 2 and 9, and so we cannot from this test conclude that the data are dependent.

Theoretical Background

We have defined $\mathbf{u}_i$ by comparing $\mathbf{x}_i$ with v, the population median. If the median is unknown, it can be replaced by $\tilde{\mathbf{x}}$, the sample median, without invalidating the procedure given above. Intuitively, this is because the covariance between $\mathbf{x}_i - \tilde{\mathbf{x}}$ and $\mathbf{x}_j - \tilde{\mathbf{x}}$ is approximately[1] of order $1/n$, and so the $\mathbf{u}_i$ are approximately independent for large enough n.

Similarly, we can replace v by $\overline{\mathbf{x}}$, the sample mean, even if the distribution of the x's is not necessarily symmetric because

$$C(\mathbf{x}_i - \overline{\mathbf{x}}, \mathbf{x}_j - \overline{\mathbf{x}}) = -\sigma^2/n,$$

so that for large n the $\mathbf{u}_i$ are approximately independent.

[1] More precisely, let $\mathbf{x}_i$ be an estimate of the p_1th percentile ξ_{p_1} and $\mathbf{x}_j$ be an estimate of the p_2th percentile ξ_{p_2} of the distributon $\mathbf{x}$. Then

$$C(\mathbf{x}_i - \tilde{\mathbf{x}}, \mathbf{x}_j - \tilde{\mathbf{x}}) \approx \frac{1}{n}\left[\frac{p_1 p_2}{f_{\mathbf{x}}(\xi_{p_1}) f_{\mathbf{x}}(\xi_{p_2})} - \frac{p_1}{2f_{\mathbf{x}}(\xi_{p_1}) f_{\mathbf{x}}(v)} - \frac{p_2}{2f_{\mathbf{x}}(\xi_{p_2}) f(v)} + \frac{1}{4f_{\mathbf{x}}^2(v)}\right].$$

Thus whether the sequence $\mathbf{x}_1, \ldots, \mathbf{x}_n$ is compared with ν, $\tilde{\mathbf{x}}$, or $\bar{\mathbf{x}}$ and the resulting deviations used as the basis for this runs test, the asymptotic distribution of the runs count conditional on the number of 1's given above can be used to test the hypothesis of independence. For small samples, though, one needs not only to ascertain the exact distribution of the runs count but also has to be careful about which benchmark one uses to compare against each of the $\mathbf{x}_i$.

(b) Runs Up-and-Down

Another assessment of independence can be made based on the successive differences between the $\mathbf{x}$'s, i.e., the $\mathbf{y}_i = \mathbf{x}_{i+1} - \mathbf{x}_i$, $i = 1, \ldots, n-1$. We define

$$\mathbf{v}_i = \begin{cases} 1 & \text{if } \mathbf{y}_i > 0, \\ 0 & \text{if } \mathbf{y}_i < 0, \end{cases}$$

and disregard $\mathbf{v}_i$ if $\mathbf{y} = 0$ (or assign a value of 1 to $\mathbf{v}_i$ with probability 1/2 and a value of 0 with probability 1/2 if $\mathbf{y}_i = 0$). Once more we compute the runs count $\mathbf{r}$ from the sequence $\mathbf{v}_1, \ldots, \mathbf{v}_{n-1}$. If the runs count is 1, the $\mathbf{x}_i$'s are a monotonic sequence (either increasing or decreasing) and thus the $\mathbf{x}_i$'s are not independent. If the runs count is $n - 1$ the sequence of $\mathbf{x}_i$'s oscillates and so are not independent.

One can show (see Levene and Wolfowitz [1944]) that

$$\mathscr{E}\mathbf{r} = \frac{2n-1}{3},$$

$$\mathscr{V}\mathbf{r} = \frac{16n-29}{90},$$

and that for large n

$$\mathbf{z} = \frac{\mathbf{r} - (2n-1)/3}{\sqrt{\dfrac{16n-29}{90}}}$$

has a $N(0, 1)$ distribution. This is the test statistic usually derived from the $\mathbf{v}_i$ to test for independence.

The small sample distribution of the runs up-and-down statistic is tabulated in Edgington [1961] for $n = 2(1)25$, and reproduced herein as Table 7. In our small sample given in Table 6, we see that the sequence of v_i is 0 0 0 0 1 0 0 1 0, so that $r = 5$. The probability of five or fewer runs in ten observations is 0.2427, so again we do not conclude that the data are dependent.

As between the two nonparametric procedures considered so far, the general conclusion based on power studies is that the runs up-and-down test is

Table 7. Distribution of Runs Up-and-Down.

Number of Runs (r)	Number of Observations (o)																								
	1	2	3	4	5	6	7	8	9	10	11	12	13	14	15	16	17	18	19	20	21	22	23	24	25
1	......	1.0000	.3333	.0833	.0167	.0028	.0004	.0000	.0000	.0000	.0000	.0000	.0000	.0000	.0000	.0000	.0000	.0000	.0000	.0000	.0000	.0000	.0000	.0000	.0000
2	......	......	1.0000	.5833	.2500	.0861	.0250	.0063	.0014	.0003	.0001	.0000	.0000	.0000	.0000	.0000	.0000	.0000	.0000	.0000	.0000	.0000	.0000	.0000	.0000
3	......	......	......	1.0000	.7333	.4139	.1909	.0749	.0257	.0079	.0022	.0005	.0001	.0000	.0000	.0000	.0000	.0000	.0000	.0000	.0000	.0000	.0000	.0000	.0000
4	......	......	......	......	1.0000	.8306	.5583	.3124	.1500	.0633	.0239	.0082	.0026	.0007	.0002	.0001	.0000	.0000	.0000	.0000	.0000	.0000	.0000	.0000	.0000
5	......	......	......	......	......	1.0000	.8921	.6750	.4347	.2427	.1196	.0529	.0213	.0079	.0027	.0009	.0003	.0001	.0000	.0000	.0000	.0000	.0000	.0000	.0000
6	......	......	......	......	......	......	1.0000	.9313	.7653	.5476	.3438	.1918	.0964	.0441	.0186	.0072	.0026	.0009	.0003	.0001	.0000	.0000	.0000	.0000	.0000
7	......	......	......	......	......	......	......	1.0000	.9563	.8329	.6460	.4453	.2749	.1534	.0782	.0367	.0160	.0065	.0025	.0009	.0003	.0001	.0000	.0000	.0000
8	......	......	......	......	......	......	......	......	1.0000	.9722	.8823	.7280	.5413	.3633	.2216	.1238	.0638	.0306	.0137	.0058	.0023	.0009	.0003	.0001	.0000
9	......	......	......	......	......	......	......	......	......	1.0000	.9823	.9179	.7942	.6278	.4520	.2975	.1799	.1006	.0523	.0255	.0117	.0050	.0021	.0008	.0003
10	......	......	......	......	......	......	......	......	......	......	1.0000	.9887	.9432	.8464	.7030	.5369	.3770	.2443	.1467	.0821	.0431	.0213	.0099	.0044	.0018
11	......	......	......	......	......	......	......	......	......	......	......	1.0000	.9928	.9609	.8866	.7665	.6150	.4568	.3144	.2012	.1202	.0674	.0356	.0177	.0084
12	......	......	......	......	......	......	......	......	......	......	......	......	1.0000	.9954	.9733	.9172	.8188	.6848	.5337	.3873	.2622	.1661	.0988	.0554	.0294
13	......	......	......	......	......	......	......	......	......	......	......	......	......	1.0000	.9971	.9818	.9400	.8611	.7454	.6055	.4603	.3276	.2188	.1374	.0815
14	......	......	......	......	......	......	......	......	......	......	......	......	......	......	1.0000	.9981	.9877	.9569	.8945	.7969	.6707	.5312	.3953	.2768	.1827
15	......	......	......	......	......	......	......	......	......	......	......	......	......	......	......	1.0000	.9988	.9917	.9692	.9207	.8398	.7286	.5980	.4631	.3384
16	......	......	......	......	......	......	......	......	......	......	......	......	......	......	......	......	1.0000	.9992	.9944	.9782	.9409	.8749	.7789	.6595	.5292
17	......	......	......	......	......	......	......	......	......	......	......	......	......	......	......	......	......	1.0000	.9995	.9962	.9846	.9563	.9032	.8217	.7148
18	......	......	......	......	......	......	......	......	......	......	......	......	......	......	......	......	......	......	1.0000	.9997	.9975	.9892	.9679	.9258	.8577
19	......	......	......	......	......	......	......	......	......	......	......	......	......	......	......	......	......	......	......	1.0000	.9998	.9983	.9924	.9765	.9436
20	......	......	......	......	......	......	......	......	......	......	......	......	......	......	......	......	......	......	......	......	1.0000	.9999	.9989	.9947	.9830
21	......	......	......	......	......	......	......	......	......	......	......	......	......	......	......	......	......	......	......	......	......	1.0000	.9999	.9993	.9963
22	......	......	......	......	......	......	......	......	......	......	......	......	......	......	......	......	......	......	......	......	......	......	1.0000	1.0000	.9995
23	......	......	......	......	......	......	......	......	......	......	......	......	......	......	......	......	......	......	......	......	......	......	......	1.0000	1.0000
24	......	......	......	......	......	......	......	......	......	......	......	......	......	......	......	......	......	......	......	......	......	......	......	......	1.0000

SOURCE: Edgington [1961]. Reprinted from the *Journal of the American Statistical Association*, Vol. 56, with the permission of the ASA.

inferior to the runs above and below the mean or median for detecting trend, but superior for detecting autoregression as the alternative to independence (see Levene [1952] for a statement of this finding; unfortunately, the subsequent publication referred to for details never appeared).

Theoretical Background

One need not restrict oneself to this statistic, however. For example, it can be shown (see Levene and Wolfowitz [1944]) that the expected number of runs of length exactly p is given by

$$\mathscr{E}\mathbf{r}_p = \frac{2n(p^2 + 3p + 1) - 2(p^3 + 3p^2 - p - 4)}{(p + 3)!}$$

for $p \leq n - 2$ and that the expected number of runs of length at least p is given by

$$\mathscr{E}\mathbf{r}'_p = \frac{2n(p + 1) - 2(p^2 + p - 1)}{(p + 2)!}$$

for $p \leq n - 1$. (Thus, with $p = 1$ in the above formula one obtains $\mathscr{E}\mathbf{r}_1 = (2n - 1)/3$.)

Levene and Wolfowitz also derive the variances of a few values of $\mathbf{r}_p$ and $\mathbf{r}'_p$, namely

$$\mathscr{V}\mathbf{r}_1 = (305n - 347)/720,$$

$$\mathscr{V}\mathbf{r}_2 = (51106n - 73859)/453600,$$

$$\mathscr{V}\mathbf{r}'_1 = (16n - 29)/90,$$

$$\mathscr{V}\mathbf{r}'_2 = (57n - 43)/720,$$

$$\mathscr{V}\mathbf{r}'_3 = (21496n - 51269)/453600.$$

Since $\mathbf{r}_p$ and $\mathbf{r}'_p$ are asymptotically normally distributed, one can construct tests of independence using runs statistics other than $\mathbf{r}_1$, the statistic usually used.

One might also, as Wallis and Moore [1941] suggest, compare the number of runs of length p with its expected value for each value of p, treating

$$\frac{(\mathbf{r}_p - \mathscr{E}\mathbf{r}_p)^2}{\mathscr{V}\mathbf{r}_p}$$

as a $\chi^2(1)$ variable and combining them all by adding them and treating the sum as a χ^2 distributed variable with some number of degrees of freedom. It is unclear, however (as the authors point out), that the distribution of this statistic is well approximated by such a χ^2 distribution. A correct variant of this idea is given in Levene and Wolfowitz [1944] (see also Wolfowitz [1944] on this matter).

To illustrate these various statistics, consider once again the data of Table 2. The values of v_i are given in column 5 of this table. From these data we find the following:

Statistic	Value	Expected value	Variance	Standard score
r_1	81	131.000	82.969	−5.49
r_2	26	35.883	22.033	−2.11
r_1'	126	131.000	34.700	−0.85
r_2'	45	48.833	15.536	−0.97
r_3'	19	12.950	9.223	1.99

Our conclusion is mixed, since r_1, r_2, and r_3' lead us to rejecting the hypothesis of independence but r_1' and r_2' do not.

(c) Rank von Neumann Ratio

One can consider replacing each observation with its rank and calculating the von Neumann ratio from these ranks. Bartels [1982] has studied the effectiveness of this statistic vis-à-vis the runs up-and-down test under a variety of distributional assumptions about $\mathbf{x}$ and found it to have greater power in all cases, and in many cases have power over twice that of the runs test for all values of the first-order autocorrelation coefficient ρ. Moreover, when the underlying distribution is normal, the power efficiency of this test relative to the von Neumann ratio using the $\mathbf{x}$'s is no less than 0.89 (see also Knoke [1977], who shows that the asymptotic relative efficiency is 0.91.).

Let $\mathbf{r}_1, \ldots, \mathbf{r}_n$ denote the ranks associated with the $\mathbf{x}_i$'s. The rank von Neumann ratio is given by

$$\mathbf{v} = \frac{\sum_{i=2}^{n} (\mathbf{r}_i - \mathbf{r}_{i-1})^2}{n(n^2 - 1)/12}.$$

Critical values of $\mathbf{c} = [n(n^2 - 1)/12]\,\mathbf{v}$ are given in Table 8 for $n = 4(1)10$, and approximate critical values of $\mathbf{v}$ are given in Table 9. For large n $\mathbf{v}$ is approximately distributed as $N(2, 4/n)$, though Bartels recommends $20/(5n + 7)$ as a better approximation to the variance of $\mathbf{v}$.

The ranks of each observation in our sample of 197 observations are given in column 6 of Table 2. From this we find that $c\sum_{i=2}^{197}(r_i - r_{i-1})^2 = 549.047$, so that $v = 0.862$. Since $(v - 2)/0.1425$ has a $N(0, 1)$ distribution, we see that 0.866 is -7.99 standard deviations from its expected value of 2, and thus we conclude that the data are not independent. (If the better approximation for the variance were used, 0.1425 would be replaced by 0.1420, leading to the same conclusion.)

For the small sample of Table 6 we find $\mathbf{c} = 169$. Since the critical values at the 0.025 level for $n = 10$ are 72 and 259, we accept the hypothesis of independence.

Table 8. Exact Critical Values for the Rank von Neumann Ratio Test.

	Left Tail $Pr(NM \le c_0)$		Right Tail $Pr(NM \ge c_0)$	
T = 4	c_0 = 3	.0833	c_0 = 17	.0833
	6	.2500	14	.2500
T = 5	c_0 = 4	.0167	c_0 = 35	.0333
	7	.0500	33	.0667
	10	.1333	30	.1333
T = 6	c_0 = 5	.0028	c_0 = 65	.0028
	8	.0083	63	.0083
	11	.0250	62	.0139
	14	.0472	60	.0194
	16	.0750	59	.0306
	17	.0806	56	.0361
	19	.1306	55	.0694
			52	.0972
			51	.1139
T = 7	c_0 = 14	.0048	c_0 = 101	.0040
	15	.0079	100	.0056
	17	.0119	98	.0087
	18	.0151	97	.0103
	20	.0262	93	.0206
	24	.0444	92	.0254
	25	.0563	88	.0464
	31	.0988	87	.0536
	32	.1155	81	.0988
			80	.1115
T = 8	c_0 = 23	.0049	c_0 = 149	.0043
	24	.0073	148	.0052
	26	.0095	144	.0084
	27	.0111	143	.0105
	32	.0221	136	.0249
	33	.0264	135	.0286
	39	.0481	129	.0481
	40	.0529	128	.0530
	48	.0978	120	.0997
	49	.1049	119	.1074
T = 9	c_0 = 34	.0045	c_0 = 208	.0046
	35	.0055	207	.0053
	40	.0096	202	.0091
	41	.0109	201	.0104
	49	.0236	191	.0245
	50	.0255	190	.0262
	59	.0486	181	.0499
	60	.0516	180	.0528
	71	.0961	169	.0978
	72	.1010	168	.1030
T = 10	c_0 = 51	.0050	c_0 = 282	.0046
	59	.0100	281	.0051
	72	.0242	273	.0097
	73	.0260	272	.0103
	85	.0493	259	.0240
	86	.0517	258	.0252
	101	.0985	246	.0475
	102	.1017	245	.0504
			229	.0990
			228	.1023

SOURCE: Bartels [1982]. Reprinted from the *Journal of the American Statistical Association*, Vol. 77, with the permission of the ASA. (His T is our c; his NM is our v).

Table 9. Approximate Critical Values for the Rank von Neumann Ratio Test.

T	*Alpha* .005	.010	.025	.050	.100
10	.62	.72	.89	1.04	1.23
11	.67	.77	.93	1.08	1.26
12	.71	.81	.96	1.11	1.29
13	.74	.84	1.00	1.14	1.32
14	.78	.87	1.03	1.17	1.34
15	.81	.90	1.05	1.19	1.36
16	.84	.93	1.08	1.21	1.38
17	.87	.96	1.10	1.24	1.40
18	.89	.98	1.13	1.26	1.41
19	.92	1.01	1.15	1.27	1.43
20	.94	1.03	1.17	1.29	1.44
21	.96	1.05	1.18	1.31	1.45
22	.98	1.07	1.20	1.32	1.46
23	1.00	1.09	1.22	1.33	1.48
24	1.02	1.10	1.23	1.35	1.49
25	1.04	1.12	1.25	1.36	1.50
26	1.05	1.13	1.26	1.37	1.51
27	1.07	1.15	1.27	1.38	1.51
28	1.08	1.16	1.28	1.39	1.52
29	1.10	1.18	1.30	1.40	1.53
30	1.11	1.19	1.31	1.41	1.54
32	1.13	1.21	1.33	1.43	1.55
34	1.16	1.23	1.35	1.45	1.57
36	1.18	1.25	1.36	1.46	1.58
38	1.20	1.27	1.38	1.48	1.59
40	1.22	1.29	1.39	1.49	1.60
42	1.24	1.30	1.41	1.50	1.61
44	1.25	1.32	1.42	1.51	1.62
46	1.27	1.33	1.43	1.52	1.63
48	1.28	1.35	1.45	1.53	1.63
50	1.29	1.36	1.46	1.54	1.64
55	1.33	1.39	1.48	1.56	1.66
60	1.35	1.41	1.50	1.58	1.67
65	1.38	1.43	1.52	1.60	1.68
70	1.40	1.45	1.54	1.61	1.70
75	1.42	1.47	1.55	1.62	1.71
80	1.44	1.49	1.57	1.64	1.71
85	1.45	1.50	1.58	1.65	1.72
90	1.47	1.52	1.59	1.66	1.73
95	1.48	1.53	1.60	1.66	1.74
100	1.49	1.54	1.61	1.67	1.74

SOURCE: Bartels [1982]. Reprinted from the *Journal of the American Statistical Association* with the permission of the ASA. (His T is our c.)

References

Bartels, R. 1982. The rank version of von Neumann's ratio test for randomness. *Journal of the American Statistical Association* **77** (March): 40–46.

Box, G. E. P. and Jenkins, G. M. 1976. *Time Series Analysis: Forecasting and Control*, 2nd edn. San Francisco: Holden-Day.

Box, G. E. P. and Ljung, G. M. 1978. On a measure of lack of fit in time series models. *Biometrika* **65** (August): 297–303.

Box, G. E. P. and Pierce, D. A. 1970. Distribution of residual autocorrelations in autoregressive-integrated moving average time series models. *Journal of the American Statistical Association* **65** (December): 1509–26.

Durbin, J. and Watson, G. S. 1950. Testing for serial correlation in least squares regression. I. *Biometrika* **37** (December): 409–28.

Durbin, J. and Watson, G. S. 1951. Testing for serial correlation in least squares regression. II. *Biometrika* **38** (June): 159–78.

Durbin, J. and Watson, G. S. 1971. Testing for serial correlation in least squares regression. III. *Biometrika* **58** (April): 1–19.

Edgington, E. S. 1961. Probability table for number of runs of signs of first differences in ordered series. *Journal of the American Statistical Association* **56** (March): 156–59.

Hart, B. I. and von Neumann, J. 1942. Tabulation of the probabilities for the ratio of the mean square successive difference to the variance. *Annals of Mathematical Statistics* **13** (June): 207–14.

Kendall, M. G. and Stuart, A. 1966. *The Advanced Theory of Statistics*, Vol. 3. London: Griffin.

Knoke, J. D. 1977. Testing for randomness against autocorrelation: Alternative tests. *Biometrika* **64** (December): 523–29.

Levene, H. 1952. On the power function of tests of randomness based on runs up and down. *Annals of Mathematical Statistics* **23** (March): 34–56.

Levene, H. and Wolfowitz, J. 1944. The covariance matrix of runs up and down. *Annals of Mathematical Statistics* **15** (March): 58–69.

Madansky, A. 1976. *Foundations of Econometrics*. Amsterdam: North-Holland.

Mason, D. G. 1927. *Artistic Ideals*. New York: Norton.

Siegel, S. 1956. *Nonparametric Statistics for the Behavioral Sciences*. New York: McGraw-Hill.

Swed, F. S. and Eisenhart, C. 1943. Tables for testing randomness of grouping in a sequence of alternatives. *Annals of Mathematical Statistics* **14** (March): 83–86.

Theil, H. and Nagar, A. L. 1961. Testing the independence of regression disturbances. *Journal of the American Statistical Association* **56** (December): 793–806.

von Neumann, J. 1941. Distribution of the ratio of the mean square successive difference to the variance. *Annals of Mathematical Statistics* **12** (September): 367–95.

Wald, A. and Wolfowitz, J. 1940. On a test whether two samples are from the same population. *Annals of Mathematical Statistics* **11** (June): 147–62.

Wallis, W. A. and Moore, G. H. 1941. A significance test for time series analysis. *Journal of the American Statistical Association* **36** (September): 401–9.

Williams, J. D. 1941. Moments of the ratio of the mean square successive difference tothe mean square difference in samples from a normal universe. *Annals of Mathematical Statistics* **12** (June): 239–41.

Wolfowitz, J. 1944. On the theory of runs with some applications to quality control. *Annals of Mathematical Statistics* **15** (September): 280–88.

CHAPTER 4

Identification of Outliers

"Turn the rascals out!"
Charles A. Dana [1871] *New York Sun* (referring to Tweed ring)
Horace A. Greeley [1872] (slogan in campaign against Grant)

0. Introduction

One of the most vexing of problems in data analysis is the determination of whether or not to discard some observations because they are inconsistent with the rest of the observations and/or the probability distribution assumed to be the underlying distribution of the data. One direction of research activity related to this problem is that of the study of robust statistical procedures (cf. Huber [1981]), primarily procedures for estimating population parameters which are insensitive to the effect of "outliers", i.e., observations inconsistent with the assumed model of the random process generating the observations. Typically, the robust procedure involves some "trimming" or down-weighting procedure, wherein some fraction of the extreme observations are automatically eliminated or given less weight to guard against the potential effect of outliers.

We will not consider such procedures in this chapter.[1] Instead we take the more traditional view that every observation is costly, contains additional "information" relating to the inferential problem of interest, and so should, if possible, be used in the inferential procedure. We thus describe procedures for culling suspected outliers out from the data, leaving us with a data set onto which to apply standard statistical procedures. We will concentrate on only one parametric model, where the underlying data distribution is assumed to

[1] Neither will we consider the so-called "slippage" problem. The context of that problem is one in which the data can be divided into distinct subsamples based on some external criterion (e.g., in the one-way analysis of variance we sample from separate populations). In the slippage problem we are concerned that one or a few of the subsamples have associated population parameters which are different ("have slipped") from those of the remaining subsamples.

be $N(\mu, \sigma^2)$, in Section 1. Section 2 describes a few nonparametric procedures for isolating outliers. For a more detailed, thorough, and complete survey of outlier identification techniques, see Barnett and Lewis [1978] and Hawkins [1980]. Tietjen and Moore [1972] and Ferguson [1961b] also present brief but useful surveys. Finally, the interested reader should refer to Beckman and Cook [1983] and the concomitant discussion. In Section 3 we describe criteria and procedures for identifying outliers in regression data. Our discussion is highly influenced by the material in Cook and Weisberg [1982].

The approach taken in this chapter is that one should clean the data by rejecting outliers and then use classical statistical procedures on the remaining data, perferably parametric procedures. The reason for highlighting parametric procedures is that usually the outlier is judged to be one relative to a parametric model of the data. Typically, if one wanted to gloss over the "outlying" nature of a data point one could do so by using nonparametric procedures, e.g., substitute ranks for the data points, on all the data. The loss of efficiency of such procedures given "outlying" observations makes the choice between nonparametric procedures using all the data and parametric procedures following a rejection of outliers favor the latter.

The advocates of robustness, though, argue as follows:

> "(1) It is rarely possible to separate the two steps cleanly; for instance, in multiparameter regression problems outliers are difficult to recognize unless we have reliable, robust estimates for the parameters.
> (2) Even if the original batch of observations consists of normal observations interspersed with some gross errors, the cleaned data will not be normal (there will be statistical errors of both kinds, false rejections and false retentions), and the situation is even worse when the original batch derives from a genuine nonnormal distribution, instead of from a gross-error framework. Therefore the classical normal theory is not applicable to cleaned samples, and the actual performance of such a two-step procedure may be more difficult to work out than that of a straight robust procedure.
> (3) It is an empirical fact that the best rejection procedures do not quite reach the performance of the best robust procedures. The latter apparently are superior because they can make a smooth transition between full acceptance and full rejection of an observation."
>
> Huber [1981]

The reader should also be apprised of a nascent middle ground between the robust approach which includes all the data but modifies the statistical procedure and the classical approach of this chapter, which tests for and rejects outliers and then uses standard statistical procedures. This middle ground is the "diagnostic" approach, typified in the regression context by the work of Belsley, Kuh, and Welsch [1980]. This approach is more informal and interactive than either of the other two, and is just in its formative developmental stage.

1. Normal Distribution

(a) One Outlier

Prescription

Let $\mathbf{x}_1, \ldots, \mathbf{x}_n$ be independent observations from a $N(\mu, \sigma^2)$ distribution, where both μ and σ^2 are unknown. Let $\mathbf{u}_i = (\mathbf{x}_i - \overline{\mathbf{x}})/\mathbf{s}$, where $\overline{\mathbf{x}} = \sum_{i=1}^n \mathbf{x}_i/n$ and

$$\mathbf{s}^2 = \sum_{i=1}^n \frac{(\mathbf{x}_i - \overline{\mathbf{x}})^2}{n-1}.$$

Intuitively, it seems that the appropriate outlier detection procedure should be to discard observations with large values of $\mathbf{u}_i$ or $|\mathbf{u}_i|$. That this is indeed optimal, assuming that there is only one outlier, is a consequence of a theorem of Karlin and Truax [1960].

Let $\mathbf{u} = \max_i \mathbf{u}_i$. Then it follows from a result of Pearson and Chandra Sekar [1936] that

$$\begin{aligned}\Pr\{\mathbf{u} > c\} &= \Pr\{\text{at least one } \mathbf{u}_i > c\}\\ &= n \Pr\{\mathbf{u}_i > c\}\end{aligned}$$

if $c > \sqrt{(n-2)/2n}$. Also, let $\mathbf{u}^* = \max_i |\mathbf{u}_i|$. Then it follows from another result of Pearson and Chandra Sekar that

$$\Pr\{\mathbf{u}^* > c\} = 2n \Pr\{|\mathbf{u}_i| > c\}$$

if $c > \sqrt{0.5}$. Thus one can use the density function of the $\mathbf{u}_i$ to derive fractiles of the distribution of $\mathbf{u}$ and $\mathbf{u}^*$. A table of critical values of $\sqrt{n-1}\mathbf{u}$ is given as Table 1. Table 2 presents critical values of $\sqrt{n-1}\mathbf{u}^*$.

To illustrate these procedures, suppose these are our data:

i	x_i	u_i
1	0.55170	−0.21436
2	1.04666	0.03223
3	−0.77651	−0.84945
4	1.30871	0.14760
5	0.64365	−0.17040
6	−1.41342	−1.15399
7	0.00963	−0.47356
8	1.72238	0.34540
9	5.90736	2.34646

Then $\mathbf{u} = \mathbf{u}^* = 2.34646$. If our alternative hypothesis is the one-sided hypothesis that $\mathscr{E}\mathbf{x}_9 > \mathscr{E}\mathbf{x}_i, i = 1, \ldots, 8$, then we compare $\mathbf{u}$ with $0.7458 \times \sqrt{8} = 2.109$, reject the null hypothesis and dub $\mathbf{x}_9$ as an outlier. If our alternative

Table 1. Critical Values of $\sqrt{n-1}\,\max_i \mathbf{u}_i$ for $\alpha = 0.10$.

n	
5	0.8357
6	0.8149
7	0.7912
8	0.7679
9	0.7458
10	0.7254
11	0.7064
12	0.6889
13	0.6728
14	0.6578
15	0.6438
16	0.6308
17	0.6187
18	0.6073
19	0.5966
20	0.5865
21	0.5770
22	0.5680
23	0.5594
24	0.5513
25	0.5435
26	0.5361
27	0.5291
28	0.5224
29	0.5159
30	0.5097

SOURCE: Hawkins [1980]. Reprinted from *Identification of Outliers* with permission from Chapman and Hall.

Table 2. Critical Values of $\sqrt{n-1}\,\max_i |\mathbf{u}_i|$ for $\alpha = 0.05$.

n	
5	0.8575
6	0.8440
7	0.8246
8	0.8038
9	0.7831
10	0.7633
11	0.7445
12	0.7271
13	0.7107
14	0.6954
15	0.6811
16	0.6676
17	0.6550
18	0.6431
19	0.6319
20	0.6213
21	0.6113
22	0.6018
23	0.5927
24	0.5841
25	0.5760
26	0.5681
27	0.5607
28	0.5535
29	0.5466
30	0.5400

SOURCE: Hawkins [1980]. Reprinted from *Identification of Outliers* with permission from Chapman and Hall.

hypothesis is the two-sided hypothesis that $\mathbf{x}_9 \neq \mathbf{x}_i$, $i = 1, \ldots, 8$, then we compare $\mathbf{u}^*$ with $0.7831 \times \sqrt{8} = 2.215$, and again reject the null hypothesis.

Theoretical Background

More precisely, consider the multiple decision problem

$$H_0: \quad \mathscr{E}\mathbf{x}_i = \mu,$$

$$H_i: \quad \begin{cases} \mathscr{E}\mathbf{x}_i = \mu + \delta, & \delta > 0, \\ \mathscr{E}\mathbf{x}_j = \mu & j \neq i. \end{cases}$$

If one restricts oneself to outlier selection procedures which are invariant under linear transformations of the $\mathbf{x}_i$, then the procedure which maximizes the probability of both stating that an outlier exists, if in fact it does, and identifying it correctly in that case is the procedure based on $\max_i \mathbf{u}_i$. The procedure based on $\max_i |\mathbf{u}_i|$ has been shown by Kudo [1956] to be optimal when it is not known *a priori* whether the contaminating mean is larger or smaller. If, however, we cannot assume that there is only one outlier, then this procedure, even when applied by removing outliers one-at-a-time and recalculating $\bar{x}$ and s, will not be optimal.

To illustrate the difficulties inherent in dealing with multiple outliers, suppose $\mu = 0, \sigma = 1$, that $\mathbf{x}_1 \leq \mathbf{x}_2 \leq \cdots \leq \mathbf{x}_n$, and that $\mathbf{x}_{n-1}$ and $\mathbf{x}_n$ come from a $N(\delta, 1)$ distribution. Then

$$\bar{\mathbf{x}} = 2\delta/n + \mathbf{o}_p(1)$$

and

$$\begin{aligned}(n-1)\mathbf{s}^2 &= \sum_{i=1}^{n} (\mathbf{x}_i - \bar{\mathbf{x}})^2 \\ &= (n-2)4\delta^2/n^2 + 2(\delta - 2\delta/n)^2 + \mathbf{o}_p(1) \\ &= 2\delta^2(n-2)/n + \mathbf{o}_p(1).\end{aligned}$$

Hence

$$\begin{aligned}\mathbf{u}_n &= \frac{\mathbf{x}_n - \bar{\mathbf{x}}}{\mathbf{s}} \to \frac{\delta - 2\delta/n}{\sqrt{2\delta^2(n-2)/n(n-1)}} \\ &= \sqrt{\frac{n-2}{2n(n-1)}} \\ &= a.\end{aligned}$$

If the fractile used in rejecting outliers exceeds a, then, as $\delta \to \infty$, $\Pr\{\mathbf{u}_n > a\} \to 0$. Thus the existence of two or more outliers may even lead to nonidentification of even the largest outlier.

Moreover, if more than one outlier exists, then $\mathbf{s}$ may be very large (especially if the largest outliers are on opposite sides of $\bar{\mathbf{x}}$), so that all values of $\mathbf{u}_i$ are small because of this. Thus the existence of multiple outliers may "mask" their presence by making the stepwise procedure lead to the culling of no outliers.

Meanwhile, let us look carefully at the procedures for dealing with a single outlier. Thompson [1935] showed that $\mathbf{u}_i$ has density function

$$f_{\mathbf{u}}(u) = \frac{\Gamma(\frac{1}{2}(n-1))\sqrt{n}}{\Gamma(\frac{1}{2}(n-2))\sqrt{\pi(n-2)}}[1 - nu^2/(n-1)]^{(n-4)/2}.$$

He suggested that any observation x_i be rejected if the associated u_i exceeded the ρ/n fractile of $f_{\mathbf{u}}(u)$, so that on average ρ observations will be discarded as "outliers", where ρ is some acceptable discard fraction. One can use this procedure both as a one-sided procedure (if we *know* that all the observations in a given direction from $\bar{x}$ are not outliers) or as a two-sided procedure, wherein upper and lower $\rho/2n$ fractiles are used.

Table 3. Percentage Points of the Skewness Statistic $\mathbf{g}_1$.

Size of sample n	Percentage points 5%	Percentage points 1%	Standard deviation	Size of sample n	Percentage points 5%	Percentage points 1%	Standard deviation	Size of sample n	Percentage points 5%	Percentage points 1%	Standard deviation
25	0.711	1.061	0.4354	200	0.280	0.403	0.1706	1000	0.127	0.180	0.0772
30	0.661	0.982	0.4052	250	0.251	0.360	0.1531	1200	0.116	0.165	0.0705
35	0.621	0.921	0.3804	300	0.230	0.329	0.1400	1400	0.107	0.152	0.0653
40	0.587	0.869	0.3596	350	0.213	0.305	0.1298	1600	0.100	0.142	0.0611
45	0.558	0.825	0.3418	400	0.200	0.285	0.1216	1800	0.095	0.134	0.0576
50	0.533	0.787	0.3264	450	0.188	0.269	0.1147	2000	0.090	0.127	0.0547
				500	0.179	0.255	0.1089				
60	0.492	0.723	0.3009	550	0.171	0.243	0.1039	2500	0.080	0.114	0.0489
70	0.459	0.673	0.2806	600	0.163	0.233	0.0995	3000	0.073	0.104	0.0447
80	0.432	0.631	0.2638	650	0.157	0.224	0.0956	3500	0.068	0.096	0.0414
90	0.409	0.596	0.2498	700	0.151	0.215	0.0922	4000	0.064	0.090	0.0387
100	0.389	0.567	0.2377	750	0.146	0.208	0.0891	4500	0.060	0.085	0.0365
				800	0.142	0.202	0.0863	5000	0.057	0.081	0.0346
125	0.350	0.508	0.2139	850	0.138	0.196	0.837				
150	0.321	0.464	0.1961	900	0.134	0.190	0.0814				
175	0.298	0.430	0.1820	950	0.130	0.185	0.0792				
200	0.280	0.403	0.1706	1000	0.127	0.180	0.0772				

SOURCE: Pearson and Hartley [1958]. Reprinted from *Biometrika Tables for Statisticians*, Vol. I, with the permission of the Biometrika Trustees.

(b) Multiple Outlier Indication

Ferguson [1961a] considers the multiple outliers problem in this form:

$$H_0: \quad \mathscr{E}\mathbf{x} = \mu,$$

$$H_1: \begin{cases} \text{for some } \mathscr{E}\mathbf{x}_i = \mu + \delta_i\sigma, & \delta_i > 0, \\ \text{for the rest } \mathscr{E}\mathbf{x}_i = \mu. \end{cases}$$

Let Δ be the n vector $\Delta = (\delta_1, \delta_2, \ldots, \delta_n)$. Then under H_0, $\Delta = 0$. Ferguson shows that the test procedure which maximizes the power in the neighborhood of $\delta = 0$ is one based on the sample skewness coefficient

$$\mathbf{g}_1 = \left[\sum_{i=i}^{n} (\mathbf{x}_i - \bar{\mathbf{x}})^3/\mathbf{s}^3 \right] \Big/ n.$$

Critical values of $\mathbf{g}_1$ are given in Table 3. The procedure is optimal for any

Table 4. Data Drawn from Various Normal Distributions.

Row	Data x_i	Studentized deviation from mean u_i
1	−4.43648	−5.89615
2	−1.41342	−2.87309
3	−0.89776	−2.35743
4	−0.77651	−2.23618
5	−0.70526	−2.16493
6	−0.67469	−2.13436
7	−0.62287	−2.08254
8	−0.24410	−1.70377
9	−0.19527	−1.65494
10	0.00963	−1.45004
11	0.02599	−1.43368
12	0.32503	−1.13464
13	0.55170	−0.90797
14	0.64365	−0.81602
15	0.85041	−0.60926
16	0.87574	−0.58393
17	0.99265	−0.46702
18	1.04666	−0.41301
19	1.20143	−0.25824
20	1.21123	−0.24844
21	1.30871	−0.15096
22	1.72238	0.26271
23	5.90736	4.44769
24	9,29919	7.83952
25	20.48644	19.02677

number of outliers up to $n/2$, but if only a single outlier is present and Δ is far from 0 then $\mathbf{u} = \max_i \mathbf{u}_i$ is more powerful (although the difference in powers in a simulation study between the two procedures was not very large).

To illustrate this procedure, suppose the data are as give in Table 4, i.e., the aforementioned data along with sixteen other data points. Then the skewness coefficient of these data is $\mathbf{g}_1 = 2.8037$. If we compare this with the critical value of $\mathbf{g}_1$ from Table 3 for a sample size 25, namely 0.711, we accept the hypothesis of the existence of outliers.

(c) Multiple Outlier Detection—Known Number

The Ferguson procedure merely indicates the presence or absence of outliers, but does not detect them. To detect the outliers a number of procedures have been adduced. First, suppose that we *know* that there are k outliers and moreover that they are *all* from a $N(\mu + \delta\sigma, \sigma^2)$ distribution, with $\delta > 0$. Then Murphy [1951] has shown that the optimal procedure (maximizing the probability of rejecting the outliers) is one based on

$$\mathbf{v} = \mathbf{u}_{(n-k+1)} + \cdots + \mathbf{u}_{(n)},$$

where $\mathbf{u}_{(1)} \leq \cdots \leq \mathbf{u}_{(n)}$. A table of fractiles of $\mathbf{v}$ is given as Table 5. (In our example, if we take $k = 4$ and assume that all outliers are from a common

Table 5. Fractiles of Distribution of Sum of k Largest Studentized Deviations from Mean.

	$k = 2$		$k = 3$		$k = 4$	
n	5%	1%	5%	1%	5%	1%
5	2.10	2.16				
6	2.41	2.50				
7	2.66	2.79	2.97	3.08		
8	2.87	3.02	3.29	3.42		
9	3.04	3.22	3.58	3.73	3.82	3.98
10	3.18	3.40	3.82	4.00	4.17	4.34
12	3.44	3.70	4.24	4.44	4.72	4.92
14	3.66	3.92	4.57	4.83	5.20	5.42
16	3.83	4.10	4.85	5.14	5.60	5.85
18	3.96	4.25	5.08	5.38	5.91	6.20
20	4.11	4.41	5.30	5.60	6.22	6.54
30	4.56	4.92	6.03	6.41	7.26	7.64
40	4.84	5.29	6.49	6.98	7.93	8.38
50	5.06	5.51	6.82	7.34	8.38	8.88
100	5.62	6.06	7.77	8.27	9.71	10.3

SOURCE: Barnett and Lewis [1978]. Reprinted from *Outliers in Statistical Data* with permission from John Wiley and Sons.

Table 6. Fracticles of $\mathbf{e}_k$: The Tietjen–Moore Statistic.

n \ k	1	1*	2	3	4	5	6	7	8	9	10
3	0.001	0.001									
4	0.025	0.025	0.001								
5	0.081	0.081	0.010								
6	0.146	0.145	0.034	0.004							
7	0.208	0.207	0.065	0.016							
8	0.265	0.262	0.099	0.034	0.010						
9	0.314	0.310	0.137	0.057	0.021						
10	0.356	0.352	0.172	0.083	0.037	0.014					
11	0.386	0.390	0.204	0.107	0.055	0.026					
12	0.424	0.423	0.234	0.133	0.073	0.039	0.018				
13	0.455	0.453	0.262	0.156	0.092	0.053	0.028				
14	0.484	0.479	0.293	0.179	0.112	0.068	0.039	0.021			
15	0.509	0.503	0.317	0.206	0.134	0.084	0.052	0.030			
16	0.526	0.525	0.340	0.227	0.153	0.102	0.067	0.041	0.024		
17	0.544	0.544	0.362	0.248	0.170	0.116	0.078	0.050	0.032		
18	0.562	0.562	0.382	0.267	0.187	0.132	0.091	0.062	0.041	0.026	
19	0.581	0.579	0.398	0.287	0.203	0.146	0.105	0.074	0.050	0.033	
20	0.597	0.594	0.416	0.302	0.221	0.163	0.119	0.085	0.059	0.041	0.028
25	0.652	0.654	0.493	0.381	0.298	0.236	0.186	0.146	0.114	0.089	0.068
30	0.698		0.549	0.443	0.364	0.298	0.246	0.203	0.166	0.137	0.112
35	0.732		0.596	0.495	0.417	0.351	0.298	0.254	0.214	0.181	0.154
40	0.758		0.629	0.534	0.458	0.395	0.343	0.297	0.259	0.223	0.195
45	0.778		0.658	0.567	0.492	0.433	0.381	0.337	0.299	0.263	0.233
50	0.797		0.684	0.599	0.529	0.468	0.417	0.373	0.334	0.299	0.258

* From Grubbs (1950, Table I).
SOURCE: Tietjen and Moore (1972). Reprinted from *Technometrics* with the permission of the ASA.

distribution,[2] then $\mathbf{v} = \mathbf{u}_4 + \mathbf{u}_{22} + \mathbf{u}_{24} + \mathbf{u}_{25} = 31.57669$, as compared with the critical value which is between 6.22 and 7.26. Thus we would reject these four observations as outliers. Clearly, our example is contrived to show that inferences based on this model, by its very restrictiveness, can lead to erroneous results when the model does not hold.)

When it is known that there are k outliers and that the ith outlier is from a $N(\mu + a_i\delta, \sigma^2)$ distribution, it has been shown (Murphy [1951]) that the

[2] In point of fact, x_{22} is from a $N(5, 1)$, x_{23} is from a $N(-5, 1)$, x_{24} is from a $N(10, 1)$, and x_{25} is from a $N(20, 1)$ distribution, while all the remaining 21 observations are random draws from a $N(0, 1)$ distribution.

optimal procedure is based on a weighted sum of the $\mathbf{u}_{(i)}$, $i = n - k + 1, \ldots, n$, where the weights are the a_i. The distribution of this weighted sum is not tabulated except for some very special cases,[3] and in any event the model, which assumes knowledge of the magnitude of the a_i's and the identity of the associated observations, is unrealistic.

The following procedure suggested by Tietjen and Moore [1972], has been effective.

1. Assume the data are ranked in ascending order of $|\mathbf{x}_i - \bar{\mathbf{x}}|$.
2. Let

$$\bar{\mathbf{x}}_k = \sum_{i=1}^{n-k} \mathbf{x}_i/(n-k),$$

$$\mathbf{s}_k^2 = \sum_{i=1}^{n-k} (\mathbf{x}_i - \bar{\mathbf{x}}_k)^2,$$

$$\mathbf{e}_k = \frac{\mathbf{s}_k^2}{(n-1)\mathbf{s}^2}.$$

3. Reject H_0 if $\mathbf{e}_k$ is smaller than the αth fractile of its sampling distribution. (A table of these fractiles for $\alpha = 0.05$ is given as Table 6.)[4]

For the example of Table 4, where the data are already ranked in order of $|\mathbf{x}_i - \bar{\mathbf{x}}|$, the following are the values of the variables of step 2 for various values of k:

k	$\bar{x}_k$	s_k^2	e_k	Critical values
10	0.688	4.975	0.0095	0.068
9	0.606	6.587	0.0125	0.089
8	0.531	8.132	0.0155	0.114
7	0.462	9.576	0.0182	0.146
6	0.397	11.030	0.0210	0.186
5	0.332	12.623	0.0241	0.236
4	0.249	15.526	0.0296	0.298
3	0.506	46.084	0.0879	0.381
2	0.292	69.454	0.1325	0.493
1	0.667	147.211	0.2808	0.654
0	1.460	524.313		

Comparing these values of e_k with the critical values of Table 6 for $n = 25$

[3] If $a_1 = 1$, $a_n = -1$, and $a_i = 0$, $i = 2, \ldots, n-1$, the Murphy statistic is the studentized range, whose distribution is given in Pearson and Stephens [1964]. The case $a_1 = a_2 = 1$, $a_i = 0$, $i = 3, \ldots, n$, is tabulated in Appendix 5 of Hawkins [1980].

[4] An analogous procedure, assuming that we know that the outliers are in one direction, ranks the data in order of the x_i's. A table of the fractiles of $\mathbf{e}_k$ for this case is given as Table 6(a).

Table 6(a). Fractiles of e_k: The Tietjen–Moore Statistic When Outliers are Known To Be in One Direction.

	k									
n	1	2	3	4	5	6	7	8	9	10
5	0.1239	0.0187								
6	0.1971	0.0561	0.0096							
7	0.2656	0.1025	0.0341							
8	0.3195	0.1446	0.0635	0.0218						
9	0.3772	0.1922	0.0985	0.0460						
10	0.4187	0.2297	0.1311	0.0706	0.0331					
15	0.5613	0.3850	0.2742	0.1964	0.1398	0.0975	0.0658			
20	0.6381	0.4812	0.3761	0.2971	0.2366	0.1889	0.1483	0.1153	0.0886	0.0666
25	0.6916	0.5492	0.4500	0.3746	0.3129	0.2627	0.2206	0.1852	0.1545	0.1270
30	0.7321	0.6013	0.5098	0.4381	0.3782	0.3265	0.2835	0.2457	0.2132	0.1848
35	0.7615	0.6428	0.5544	0.4848	0.4277	0.3790	0.3363	0.2990	0.2666	0.2362
40	0.7855	0.6721	0.5901	0.5256	0.4701	0.4224	0.3813	0.3441	0.3119	0.2821
45	0.8035	0.7000	0.6220	0.5584	0.5050	0.4594	0.4195	0.3834	0.3513	0.3223
50	0.8181	0.7209	0.6479	0.5878	0.5366	0.4911	0.4516	0.4160	0.3850	0.3565
60	0.8428	0.7553	0.6884	0.6335	0.5861	0.5448	0.5074	0.4743	0.4434	0.4154
70	0.8610	0.7833	0.7222	0.6715	0.6276	0.5892	0.5536	0.5215	0.4918	0.4642
80	0.8746	0.8021	0.7453	0.6980	0.6576	0.6209	0.5873	0.5569	0.5291	0.5033
90	0.8863	0.8195	0.7668	0.7227	0.6834	0.6486	0.6163	0.5880	0.5611	0.5365
100	0.8957	0.8333	0.7844	0.7419	0.7057	0.6729	0.6432	0.6150	0.5898	0.5656

SOURCE: Hawkins [1980]. Reprinted from *Identification of Outliers* with permission from Chapman and Hall.

indicates that not only are the true outliers, x_{25}, x_{24}, x_1, and x_{23}, detected, but also x_2 through x_7 are judged to be outliers, since e_k is smaller than the critical value for these values of k. Of course, this artifact is a manifestation of the masking effect due to the nature of the process underlying the generation of this data. Only four of the data points are outliers, but because one of them (x_{23}) comes from a distribution with $\mu = -5$ and the others come from one with a zero or positive mean, four other data points are implicated as outliers.

A way to see this procedure in action in a less artifactual manner is to delete some of the spurious observations and consider the procedure with the smaller sample size. Let $e_k^{(l)}$ denote the computation of e_k when the last l observations are deleted from the sample (so that $e_k^{(5)}$, for example, is based on the first 20 observations). Following is a table of the $e_k^{(l)}$:

k	$e_k^{(1)}$	k	$e_k^{(2)}$	k	$e_k^{(3)}$	k	$e_k^{(4)}$	k	$e_k^{(5)}$
7	0.059	6	0.124	5	0.187	4	0.556	3	0.650
6	0.069	5	0.146	4	0.220	3	0.653	2	0.763
5	0.079	4	0.168	3	0.253	2	0.750	1	0.876
4	0.090	3	0.191	2	0.288	1	0.856		
3	0.105	2	0.224	1	0.337				
2	0.313	1	0.664						
1	0.472								

We see that, since $e_k^{(5)}$ exceeds the critical values for $n = 20$, $k = 1, 2, 3$ of Table 6, we do not reject any of the first 20 observations as outliers. Similarly, $e_k^{(4)}$ exceeds the critical values for $n = 25$, $k = 1, 2, 3, 4$, and so *a fortiori* would exceed those for $n = 21$ if they were tabulated. But when we study $e_k^{(3)}$ we find that, since $e_k^{(3)}$ is smaller even than the critical values for $n = 20$, we would have to conclude that all of the five observations x_{18} through x_{22} are outliers. And the same type of conclusion is reached from a study of $e_k^{(1)}$ and $e_k^{(2)}$.

Another problem with this procedure, which, along with the defect illustrated above, motivated a suggested modification by Rosner [1975] (see also Hawkins [1978]), is the following. Let $\mathbf{x}_1, \ldots, \mathbf{x}_{n-2}$ come from a $N(0, 1)$ population, let $\mathbf{x}_{n-1} = 10$, and let $\mathbf{x}_n$ be greater than 10. The Tietjen–Moore procedure for $k = 2$ will take $\mathbf{x}_n$ and $\mathbf{x}_1$ as outliers (where $\mathbf{x}_1$ is the smallest of the $\mathbf{x}$'s) rather than $\mathbf{x}_n$ and $\mathbf{x}_{n-1}$. The modified procedure is a step-down procedure, as follows:

0. Set $l = 0$.
1. Let, as before,

$$\bar{\mathbf{x}}_l = \sum_{i=1}^{n-l} \mathbf{x}_i/(n - l),$$

$$\mathbf{s}_l^2 = \sum_{i=1}^{n-l} (\mathbf{x}_i - \bar{\mathbf{x}}_l)^2,$$

$$\mathbf{e}_l = \frac{\mathbf{s}_l^2}{(n - 1)\mathbf{s}^2}.$$

Table 7. Fractiles of the Rosner Statistic for $\alpha = 0.05$.

k	n		k	n		k	n	
2	5	0.0061	4	50	0.5235	7	40	0.2907
2	10	0.1640	4	75	0.6366	7	50	0.3668
2	15	0.3104	4	100	0.7041	7	75	0.4992
2	20	0.4136				7	100	0.5838
2	25	0.4886	5	10	0.0096			
2	30	0.5455	5	15	0.0763	8	20	0.0546
2	40	0.6262	5	20	0.1560	8	25	0.1070
2	50	0.6810	5	25	0.2283	8	30	0.1591
2	75	0.7639	5	30	0.2906	8	40	0.2517
2	100	0.8109	5	40	0.3895	8	50	0.3275
			5	50	0.4634	8	75	0.4628
3	10	0.0740	5	75	0.5854	8	100	0.5510
3	15	0.1967	5	100	0.6600	9	20	0.0365
3	20	0.2972				9	25	0.0818
3	25	0.3758	6	15	0.0451	9	30	0.1294
3	30	0.4379	6	20	0.1123	9	40	0.2178
3	40	0.5297	6	25	0.1786	9	50	0.2924
3	50	0.5942	6	30	0.2382	9	75	0.4290
3	75	0.6948	6	40	0.3964	9	100	0.5200
3	100	0.7534	6	50	0.4121			
			6	75	0.5403	10	20	0.0230
4	10	0.0299	6	100	0.6205	10	25	0.0611
4	15	0.1239				10	30	0.1043
4	20	0.2154	7	15	0.0249	10	40	0.1881
4	25	0.2923	7	20	0.0794	10	50	0.2612
4	30	0.3558	7	25	0.1389	10	75	0.3986
4	40	0.4530	7	30	0.1949	10	100	0.4920

SOURCE: Hawkins [1980]. Reprinted from *Identification of Outliers* with permission from Chapman and Hall.

2. Let k be the index corresponding to

$$\max_i |\mathbf{x}_i - \bar{\mathbf{x}}_l|.$$

3. Reject H_0 if $\mathbf{e}_l$ is smaller than the αth fractile of its sampling distribution. (A table of the sampling distribution of $\mathbf{e}_l$ for this procedure is given as Table 7.)
4. Renumber the observations so that $\mathbf{x}_k$ now becomes $\mathbf{x}_{n-l}$.
5. Set $l = l + 1$.
6. Go to step 1. (Step 1 is so constructed that none of the identified outliers are ever used again in the computation of $\bar{\mathbf{x}}_l$ and $\mathbf{s}_l^2$.)

Instead of keeping the fixed ordering based on the $|\mathbf{x}_i - \bar{\mathbf{x}}|$'s, we reorder at each successive value of l based on the $|\mathbf{x}_i - \bar{\mathbf{x}}_l|$'s (or equivalently, based on the standardized values $|\mathbf{x}_i - \bar{\mathbf{x}}_l|\sqrt{n - l - 1/\mathbf{s}_l}$). As an example, Table 8 sum-

Table 8. Successive Orderings of Observations in the Rosner Procedure.

1	1.30871	0.15096	0.25369	0.57245	0.96534	1.20241	1.19776	1.36823	1.34091	1.31035	1.28223
2	1.21123	0.24844	0.21516	0.51759	0.89138	1.09177	1.07816	1.24113	1.20822	1.17157	1.13554
3	1.20143	0.25824	0.21129	0.51208	0.88395	1.08065	1.06615	1.22836	1.19489	1.15763	1.12080
4	1.72238	0.26271	0.41720	0.80527	1.27918	1.67192	1.70528				
5	1.04666	0.41301	0.15011	0.42497	0.76653	0.90498	0.87625	1.02656	0.98422	0.93728	0.88789
6	0.99265	0.46702	0.12876	0.39457	0.72555	0.84369	0.80999	0.95615	0.91071	0.86040	0.80662
7	0.87574	0.58393	0.08255	0.32878	0.63686	0.71100	0.66657	0.80373	0.75159	0.69397	0.63070
8	0.85041	0.60926	0.07254	0.31452	0.61764	0.68225	0.63549	0.77071	0.71711	0.65791	0.59258
9	0.64365	0.81602	0.00919	0.19815	0.46077	0.44758	0.38182	0.50113	0.43568	0.36356	0.28144
10	0.55170	0.90797	0.04553	0.14640	0.39101	0.34322	0.26901	0.38125	0.31052	0.23266	0.14308
11	0.32503	1.13464	0.13513	0.01883	0.21904	0.08595	0.00909	0.08571	0.00199	0.09004	0.19802
12	0.02599	1.43368	0.25333	0.14948	0.00783	0.25346	0.37598	0.30417	0.40504	0.51576	0.64802
13	0.00963	1.45004	0.25980	0.15868	0.02024	0.27202	0.39604	0.32549	0.42730	0.53904	0.67263
14	−0.19527	1.65494	0.34078	0.27400	0.17569	0.50458	0.64742	0.59263	0.70619	0.83073	0.98096
15	−0.24410	1.70377	0.36009	0.30148	0.21274	0.56001	0.70734	0.65631	0.77266	0.90026	1.05445
16	−0.62287	2.08254	0.50980	0.51466	0.50011	0.98990	1.17204	1.15014	1.28821	1.43948	1.62442
17	−0.67469	2.13436	0.53029	0.54382	0.53942	1.04872	1.23562	1.21770	1.35874	1.51325	1.70240
18	−0.70526	2.16493	0.54237	0.56102	0.56261	1.08341	1.27312	1.25755	1.40035	1.55676	
19	−0.77651	2.23618	0.57053	0.60112	0.61667	1.16428	1.36053	1.35044	1.49733		
20	−0.89776	2.35743	0.61846	0.66937	0.70866	1.30190	1.50930	1.50854			
21	−1.41342	2.87309	0.82228	0.95958	1.09988	1.88716					
22	5.90736	4.44769	2.07140	3.16061							
23	−4.43648	5.89615	2.01721	2.66099	3.39341						
24	9.29919	7.83952	3.41209								
25	20.48644	19.02677									

marizes these orderings for $l = 1, \ldots, 8$. Note that the successive deleted observations are $x_{25}, x_{24}, x_{22}, x_{23}, x_{21}, x_4, x_{20}, x_{19}, x_{18}$ in this procedure. The values of the $\bar{x}_l$, s_l^2, and e_l for this example are:

l	$\bar{x}_l$	s_l^2	e_l
1	0.666892	147.211	0.2808
2	0.291575	69.454	0.1325
3	0.036312	36.484	0.0696
4	0.249302	15.526	0.0296
5	0.332438	12.623	0.0241
6	0.259283	10.589	0.0202
7	0.323563	9.176	0.0175
8	0.388273	7.895	0.0151
9	0.456619	6.624	0.0126

Comparing these values of e_l with the critical values of Table 7 for $n = 25$ indicates that all nine observations tagged by this procedure are outliers. Thus, although this procedure compensates for some of the masking problems of the Tietjen–Moore procedure, it is not foolproof.

(d) Multiple Outlier Detection—Unknown Number

Now suppose that the number of outliers is *unknown*. A heuristic rule due to Tietjen and Moore is to estimate k as that value for which the "gap" $\mathbf{x}_{n-k+1} - \mathbf{x}_{n-k}$ is maximum (where it is assumed that $\mathbf{x}_1 \leq \cdots \leq \mathbf{x}_n$). They show by simulation that if one selects k in this way and uses their procedures for known k with $\alpha = 0.025$, then an overall α of about 0.05 results. In our example (see Table 8, where the data are ordered by magnitude of the x's) $x_{n-k+1} - x_{n-k}$ is maximum for $k = 1$, since $x_{25} - x_{24} = 11.18725$. Thus this procedure is "masked" by the magnitude of x_{25}. If x_{25} were not in the data set, then k would equal 2, since $x_{23} - x_{22} = 4.18498$ is maximum, and so x_{23} and x_{24} would be classified as outliers. Note that in this case x_1 is not deleted, being masked by the two large values of x.

An alternative approach when the number of outliers is unknown is the sequential examination of potential outliers. Two general tacks are the "backward elimination" and the "forward selection" procedures. In forward selection one first tests for a single outlier, then, if it exists, eliminates it from the sample and once more tests for a single outlier. This process continues until one can no longer reject an observation as being an outlier. It is known (Hawkins [1980]) that this procedure is highly vulnerable to masking.

In backward elimination, which is not vulnerable to masking, one sets an upper limit k_0 on the number of outliers to be tested, deletes the most outlying k_0 observations, and sequentially includes those suspected outliers one-at-a-time, starting with the most inlying of these observations, and tests whether the observation should be included. Unfortunatey, the determination of the level of significance for such procedures is tricky.

Hawkins [1980] considers the statistics

$$\mathbf{t}_{ni} = |\mathbf{x}_i - \overline{\mathbf{x}}_{i-1}|/\mathbf{s}_{i-1},$$

where

$$\overline{\mathbf{x}}_{i-1} = \sum_{j=i}^{n} \mathbf{x}_j/(n - i + 1),$$

$$\mathbf{s}_{i-1}^2 = \sum_{j=i}^{n} (\mathbf{x}_j - \overline{\mathbf{x}}_{i-1})^2,$$

for $i = k_0, \ldots, n$, as the test statistics for outliers, wherein for each i the observations are reordered so that the most aberrant is dubbed "$\mathbf{x}_i$". In this procedure one concludes that there are k outliers by setting k equal to the largest value of i such that $\mathbf{t}_{ni}$ exceeds an appropriate critical value of $\mathbf{t}_{ni}$. Tables of the critical values of $\mathbf{t}_{ni}$ are given in Table 9 for $n = 10, 15, 20, 30, 50$, $k_0 = 2, 3, 5$, and 5. Rosner [1983] provides tables of critical values for this procedure, based instead on the statistic

$$\mathbf{t}_{ni}^* = \sqrt{n - i - 1}\,\mathbf{t}_{ni},$$

for $k_0 = 10$, $n = 25(1)50(10)100(50)500$. These are given in Table 9(a).

Table 9. Critical Values of t_{ni}, Backward Elimination Statistic.

K	i	n 10	15	20	30	50
2	1	0.783	0.695	0.632	0.547	0.450
2	2	0.789	0.698	0.633	0.546	0.450
3	1	0.804	0.711	0.643	0.554	0.454
3	2	0.806	0.714	0.640	0.552	0.453
3	3	0.796	0.705	0.632	0.546	0.450
4	1	0.825	0.727	0.655	0.561	0.458
4	2	0.832	0.734	0.653	0.559	0.458
4	3	0.820	0.723	0.642	0.554	0.454
4	4	0.811	0.711	0.634	0.546	0.450
5	1	0.844	0.745	0.668	0.568	0.461
5	2	0.854	0.752	0.666	0.567	0.461
5	3	0.837	0.744	0.653	0.559	0.458
5	4	0.839	0.729	0.643	0.552	0.454
5	5	0.841	0.717	0.635	0.546	0.450

SOURCE: Hawkins [1980]. Reprinted from *Identification of Outliers* with permission from Chapman and Hall.

Table 9(a). Critical Values of $t^*_{ni} = \sqrt{n-i-1}\,t_{ni}$, the Backward Elimination Statistic.

n	k		n	k		n	k		n	k		n	k	
25	1	2.82	33	1	2.95	41	1	3.05	49	1	3.12	200	1	3.61
	2	2.80		2	2.94		2	3.04		2	3.11		2	3.60
	3	2.78		3	2.92		3	3.03		3	3.10		3	3.60
	4	2.76		4	2.91		4	3.01		4	3.09		4	3.60
	5	2.73		5	2.89		5	3.00		5	3.09		5	3.60
	10	2.59		10	2.80		10	2.94		10	3.04		10	3.59
26	1	2.84	34	1	2.97	42	1	3.06	50	1	3.13	250	1	3.67
	2	2.82		2	2.95		2	3.05		2	3.12		5	3.67
	3	2.80		3	2.94		3	3.04		3	3.11		10	3.66
	4	2.78		4	2.92		4	3.03		4	3.10			
	5	2.76		5	2.91		5	3.01		5	3.09			
	10	2.62		10	2.82		10	2.95		10	3.05			
27	1	2.86	35	1	2.98	43	1	3.07	60	1	3.20	300	1	3.72
	2	2.84		2	2.97		2	3.06		2	3.19		5	3.72
	3	2.82		3	2.95		3	3.05		3	3.19		10	3.71
	4	2.80		4	2.94		4	3.04		4	3.18			
	5	2.78		5	2.92		5	3.03		5	3.17			
	10	2.65		10	2.84		10	2.97		10	3.14			
28	1	2.88	36	1	2.99	44	1	3.08	70	1	3.26	350	1	3.77
	2	2.86		2	2.98		2	3.07		2	3.25		5	3.76
	3	2.84		3	2.97		3	3.06		3	3.25		10	3.76
	4	2.82		4	2.95		4	3.05		4	3.24			
	5	2.80		5	2.94		5	3.04		5	3.24			
	10	2.68		10	2.86		10	2.98		10	3.21			
29	1	2.89	37	1	3.00	45	1	3.09	80	1	3.31	400	1	3.80
	2	2.88		2	2.99		2	3.08		2	3.30		5	3.80
	3	2.86		3	2.98		3	3.07		3	3.30		10	3.80
	4	2.84		4	2.97		4	3.06		4	3.29			
	5	2.82		5	2.95		5	3.05		5	3.29			
	10	2.71		10	2.88		10	2.99		10	3.26			
30	1	2.91	38	1	3.01	46	1	3.09	90	1	3.35	450	1	3.84
	2	2.89		2	3.00		2	3.09		2	3.34		5	3.83
	3	2.88		3	2.99		3	3.08		3	3.34		10	3.83
	4	2.86		4	2.98		4	3.07		4	3.34			
	5	2.84		5	2.97		5	3.06		5	3.33			
	10	2.73		10	2.89		10	3.00		10	3.31			
31	1	2.92	39	1	3.03	47	1	3.10	100	1	3.38	500	1	3.86
	2	2.91		2	3.01		2	3.09		2	3.38		5	3.86
	3	2.89		3	3.00		3	3.09		3	3.38		10	3.86
	4	2.88		4	2.99		4	3.08		4	3.37			
	5	2.86		5	2.98		5	3.07		5	3.37			
	10	2.76		10	2.91		10	3.01		10	3.35			
32	1	2.94	40	1	3.04	48	1	3.11	150	1	3.52			
	2	2.92		2	3.03		2	3.10		2	3.51			
	3	2.91		3	3.01		3	3.09		3	3.51			
	4	2.89		4	3.00		4	3.09		4	3.51			
	5	2.88		5	2.99		5	3.08		5	3.51			
	10	2.78		10	2.92		10	3.03		10	3.50			

SOURCE: Rosner [1983]. Reprinted from *Technometrics* with the permission of the ASA.

Continuing with our example, following are the values of $\mathbf{t}^*_{25,i}$, as read off from Table 8.

i	$\mathbf{t}^*_{25,i}$
1	19.02677
2	3.41209
3	3.16061
4	3.39341
5	1.88716
6	1.70528
7	1.50854
8	1.49733
9	1.55676
10	1.70240

Comparing these values with Table 9(a), we see that there are four outliers.

2. Nonparametric Procedures

Prescription

Suppose $\mathbf{x}_1 \le \cdots \le \mathbf{x}_n$, and suppose we know that, if there are outliers, there are r of them, namely, $\mathbf{x}_1, \ldots, \mathbf{x}_r$. A procedure due to Walsh [1958], applicable under very mild restrictions, for making the proper determination is the following: Let

$$k = r + [\sqrt{2n}],$$

$$b^2 = 1/\alpha,$$

$$c = [\sqrt{2n}],$$

$$a = \frac{1 + b\sqrt{(c - b^2)/(c - 1)}}{c - b^2 - 1}.$$

Then reject the smallest r observations if

$$x_r - (1 + a)x_{r+1} + ax_k < 0.$$

Similarly, if we suspect that the r largest x's are outliers, namely $x_{n+1-r}, \ldots, x_n$, then they are rejected if

$$x_{n+1-r} - (1 + a)x_{n-r} + a_{n+1-k} > 0.$$

Finally, if both these inequalities are satisfied, then both the r smallest and the r largest x's are rejected as being outliers. This test is of course applicable only if $c - b^2 - 1 > 0$, i.e.,

$$\sqrt{2n} > 1 + \frac{1}{\alpha}.$$

For $\alpha = 0.05$. this implies that $n > 220$; for $\alpha = 0.10$, n must exceed 60.

We illustrate this procedure on a small sample, but with $\alpha = 0.36$

x_1	−4.43648
x_2	−1.41342
x_2	−0.77651
x_4	−0.77651
x_5	0.55170
x_6	0.64365
x_7	1.04666
x_8	1.04666
x_9	1.72238
x_{10}	5.90736
x_{11}	9.29919
x_{12}	20.48644

If $r = 3$, $k = 3 + [\sqrt{24}] = 7$, $b^2 = 1/0.36 = 2.777$, $c = [\sqrt{24}] = 4$,

$$a = \frac{1 + 1.666\sqrt{(4 - 2.777)/(4 - 1)}}{4 - 2.777 - 1}$$

$$= 9.288.$$

We thus consider

$$x_{10} - 10.288x_9 + 9.288x_6 = -5.834,$$

which, being negative, leads us not to take x_{10}, x_{11}, and x_{12} as outliers even at level of significance $\alpha = 0.36$. If we consider the other tail,

$$x_3 - 10.288x_4 + 9.288x_7 = 8.846$$

and so again we cannot conclude that the extreme values x_1, x_2, x_3 are outliers.

Another nonparametric procedure due to Walsh [1950], which has no such restrictive sample size requirements, is applicable if one knows that the distribution from which the x's are sampled is symmetric. The test is based on a comparison of a subset of s of the r largest observations with some of the smallest observations. The procedure is as follows.

1. Select a convenient value of s. (For example, $s = 2$.)
2. For that value of s find a level of significance α from Table 10 that you wish to use in this procedure. (For example, $\alpha = 0.0469$.)
3. Calculate w, where w is the smallest integer satisfying

$$\Pr\{\mathbf{x}_w < \phi\} \leq \alpha,$$

where ϕ is the median of the distribution of $\mathbf{x}$ and α is the level of significance for the outlier procedure. For large n, w is approximately given by $w = n/2 + \sqrt{n}z_\alpha/2$, where z_α is the upper $100\alpha\%$ point of the $N(0, 1)$ distribution. (In our example, $w = n/2 + \sqrt{n} \times 1.987/2$.)

Table 10. Significance Levels for the Walsh Procedure.
Some Values of α for $s \leq 5$.

α	s	i_1	i_2	i_3	i_4	i_5	j_1	j_2	j_3	j_4	j_5
.0625	1	4					1				
.0312	1	5					1				
.0156	1	6					1				
.0078	1	7					1				
.0039	1	8					1				
.0352	1	7					2				
.0195	1	8					2				
.0107	1	9					2				
.0469	2	4	5				1	2			
.0234	2	5	6				1	2			
.0117	2	6	7				1	2			
.0059	2	7	8				1	2			
.0391	3	4	5	6			1	2	3		
.0195	3	5	6	7			1	2	3		
.0098	3	6	7	8			1	2	3		
.0459	4	4	5	6	7		1	2	3	4	
.0229	4	5	6	7	8		1	2	3	4	
.0115	4	6	7	8	9		1	2	3	4	
.0308	5	4	5	6	7	8	1	2	3	4	5
.0154	5	5	6	7	8	9	1	2	3	4	5
.0077	5	6	7	8	9	10	1	2	3	4	5

SOURCE: Walsh [1950]. Reprinted from *Annals of the Institute of Statistical Mathematics*, Vol. 10, with the permission of the IMS.

4. Calculate the set of statistics $\mathbf{u}_k = \mathbf{x}_{n+1-i_k} + \mathbf{x}_{j_k}$ for a prespecified set of pairs of integers (i_k, j_k), $k = 1, \ldots, s$; where the value of s and the prespecified set of integers are given in Table 10. (For example, $\mathbf{u}_1 = \mathbf{x}_{n-3} + \mathbf{x}_1$, $\mathbf{u}_2 = \mathbf{x}_{n-4} + \mathbf{x}_2$, corresponding to the values (i_k, j_k) in Table 10.)
5. Compare $\min_k u_k$ with $2x_w$. If $\min_k \mathbf{u}_k > 2\mathbf{x}_w$, reject the r largest observations as outliers.

Let us see how this test performs on our data of Table 4. Here $w = 17.4675$, and so $x_w = x_{17} = 0.99265$, $u_1 = x_{22} + x_1 = -2.71410$, $u_2 = x_{21} + x_2 = -0.10471$, and, since $2x_w = 1.9953$, we would not reject the five largest observations. If we had eliminated x_1 from our data, so that $n = 24$, then w would equal 16.867, $x_w = x_{16} = 0.87574$, $u_1 = x_{22} + x_2 = 0.30896$, $u_2 = x_{21} + x_3 = 0.41095$ and, since $2x_w = 1.74148$, we would still not reject the five largest observations. Thus this nonparametric procedure is seen to be conservative.

Theoretical Background

The significance level α is a function of s, the i_k, and the j_k, but *not* a function of either r (except that r must be at least 4) or n. It is given by the formidable formula

$$\alpha = 2^{-w}\left\{1 + m_1 + \sum_{h_1=1}^{m_2} (m_1 - h_1) + \sum_{h_2=1}^{m_3} \sum_{h_1=1}^{m_2-h_2} (m_2 - h_1 - h_2) + \cdots + \sum_{h_{u-1}}^{m_u} \sum_{h_{u-2}=1}^{m_{u-1}-h_{u-1}} \cdots \sum_{h_1=1}^{m_2-h_2-\cdots-h_{u-1}} (m_1 - h_1 - \cdots - h_{u-1}),\right.$$

where

$$w = i_s + j_s - 1, \qquad u = j_s - 1,$$

$$m_{j_t+v_t-1} = i_s + j_s - i_t - j_t - v_t + 1, \qquad t = 0, 1, \ldots, s-1,$$

$$v_t = 1, \ldots, j_{t+1} - j_t, \qquad i_0 = 0, \quad j_0 = 1.$$

For example, if $s = 2$, then, from Table 10, if we take $i_1 = 4$, $i_2 = 5$, $j_1 = 1$, $j_2 = 2$ we see that v_0 is not defined, and that for $t = 1$ $v_1 = 1$, $m_1 = 2$, and $\alpha = 2^{-6}[1 + 2] = 3/64 = 0.046875$, which checks with the tabulated value of α in Table 10.

3. Outliers in Regression

To check whether $(\mathbf{y}_i, X_i')$ is an outlying point in a multiple regression, the standard procedure is to begin by fitting the regression with this point deleted from the data set used in estimating the regression coefficients. Let $\hat{\mathbf{B}}_{(i)}$ denote the estimator of the vector of regression coefficients when the ith data point is deleted, and let $\hat{\boldsymbol{\sigma}}^2_{(i)}$ denote the associated estimator of the common variance of the true residuals. Recall that the hat matrix P is $X(X'X)^{-1}X'$. Let $X_{(i)}$ be the $n - 1 \times p + 1$ submatrix of X with its ith row deleted. Since

$$(X_{(i)}'X_{(i)})^{-1} = (X'X)^{-1} + \frac{(X'X)^{-1}X_iX_i'(X'X)^{-1}}{1 - p_{ii}},$$

with a bit of algebra one can see that

$$X_i'(X_{(i)}'X_{(i)})^{-1}X_i = p_{ii}/(1 - p_{ii}).$$

Also

$$\hat{\mathbf{B}}_{(i)} = \hat{\mathbf{B}} - \frac{\hat{\mathbf{u}}_i(X'X)X_i}{1 - p_{ii}},$$

and

$$\hat{\boldsymbol{\sigma}}^2_{(i)} = \frac{\hat{\boldsymbol{\sigma}}^2(n - p - 1 - \mathbf{w}_i^2)}{n - p - 2},$$

where $\mathbf{w}_i$ is the ith studentized residual. Thus one need not actually fit a new regression to obtain $\hat{\mathbf{B}}_{(i)}$, $\boldsymbol{\sigma}^2_{(i)}$, and $X_i'(X_{(i)}'X_{(i)})^{-1}X_i$; one merely fits the original regression and calculates as by-products the associated hat matrix and studentized residuals.

A comparison of $\hat{\mathbf{B}}_{(i)}$ and $\hat{\mathbf{B}}$ reveals the influence of a single data point on the results of the regression. Cook [1977, 1979] defined

$$\mathbf{d}_i = \frac{(\mathbf{B}_{(i)} - \hat{\mathbf{B}})'(X'X)(\hat{\mathbf{B}}_{(i)} - \hat{\mathbf{B}})}{(p+1)\sigma^2}$$

as the distance between $\hat{\mathbf{B}}_{(i)}$ and $\hat{\mathbf{B}}$. The $\mathbf{d}_i$ can be computed directly from the regression, as

$$\mathbf{d}_i = \frac{\mathbf{w}_i^2}{p+1}\left(\frac{p_{ii}}{1-p_{ii}}\right).$$

As a rule of thumb, observations for which $\mathbf{d}_i$ is greater than 1 may be judged as influential observations. MINITAB calculates the d_i in its regression output using the subcommand COOKD. It is recommended that the value of d_i be compared with the 50% point of an $F(p+1, n-p-1)$ distribution. For large n, this value is approximately 1.

In Chapter 5, Section 1, Table 1, we will consider a data set in which a regression is performed and the residuals will be found to be heteroscedastic. The reader is referred forward to Chapter 5 for details about this data set. Here we examine the data to see if there are any significant outliers. For those data the matrix $X'X$ is given by

$$\begin{bmatrix} 100 & 14774 & 48.2233 & 112.8 \\ & 32691150 & 3148.6747 & 15664.9 \\ & & 74.319 & 84.4485 \\ & & & 269.5400 \end{bmatrix},$$

and the resulting computations of the studentized residuals, the p_{ii}, and the d_i are given in Table 11. Note that a number of the d_i are larger than 1, especially those for observations 77, 95, and 96.

A more formal procedure is based on the following arguments. Let $\tilde{\mathbf{y}}_i = \hat{\mathbf{B}}_{(i)}'X_i$ be the estimator of $\mathbf{y}_i$ based on the remaining $n-1$ data points. Then $\mathbf{y}_i$ and $\tilde{\mathbf{y}}_i$ are independent and, under the null hypothesis that $\mathbf{y}_i$ is not an outlier, $\mathscr{E}\mathbf{y}_i = \mathscr{E}\tilde{\mathbf{y}}_i = B'X_i$. Finally, $\mathbf{y}_i - \tilde{\mathbf{y}}_i$ is normally distributed with

$$\mathscr{V}(\mathbf{y}_i - \tilde{\mathbf{y}}_i) = \sigma^2[1 + X_i'(X_{(i)}'X_{(i)})^{-1}X_i].$$

Thus

$$\mathbf{z}_i = \frac{\mathbf{y}_i - \tilde{\mathbf{y}}_i}{\hat{\boldsymbol{\sigma}}_{(i)}\sqrt{1 + X_i'(X_{(i)}'X_{(i)})^{-1}X_i}}$$

Table 11. Residuals, Studentized Residuals, and Influence Statistics for Grinding Steel Data.

	Residual	Studentized residual	Influence statistic d_i	z_i
1	0.01491	-1.09377	-0.22541	-0.22431
2	0.01216	-1.36781	-0.28150	-0.28016
3	0.01320	-1.23030	-0.25333	-0.25210
4	0.14113	-5.44620	-1.20204	-1.20484
5	0.01587	-1.42198	-0.29320	-0.29181
6	0.13868	-0.55864	-0.12312	-0.12250
7	0.01687	2.94058	0.60662	0.60463
8	0.01247	-1.92856	-0.39696	-0.39523
9	0.01573	1.51240	0.31182	0.31036
10	0.01260	-0.65318	-0.13445	-0.13377
11	0.01140	3.24258	0.66707	0.66515
12	0.01195	-2.11671	-0.43557	-0.43375
13	0.01682	-0.10994	-0.02268	-0.02256
14	0.01489	-1.23786	-0.25511	-0.25387
15	0.01136	-2.22552	-0.45783	-0.45595
16	0.01374	-1.77986	-0.36659	-0.36495
17	0.01136	-1.73683	-0.35730	-0.35568
18	0.01178	-2.19115	-0.45085	-0.44900
19	0.01562	-0.53113	-0.10950	-0.10894
20	0.01260	-1.80730	-0.37203	-0.37037
21	0.01321	-2.00972	-0.41382	-0.41205
22	0.01362	-1.66692	-0.34331	-0.34174
23	0.07277	-2.72895	-0.57969	-0.57769
24	0.01155	-1.22533	-0.25210	-0.25088
25	0.02881	4.03239	0.83695	0.83565
26	0.01309	2.79876	0.57626	0.57426
27	0.01560	1.00022	0.20621	0.20519
28	0.01132	-2.45385	-0.50479	-0.50284
29	0.01481	-0.33756	-0.06956	-0.06921
30	0.01554	-1.49220	-0.30762	-0.30618
31	0.01355	-1.85064	-0.38113	-0.37945
32	0.01380	-1.31424	-0.27070	-0.26940
33	0.03761	-0.84628	-0.17645	-0.17557
34	0.03676	-4.23074	-0.88174	-0.88072
35	0.01726	-2.04717	-0.42240	-0.42061
36	0.01497	5.94960	1.22617	1.22940
37	0.23507	-3.09654	-0.72419	-0.72241
38	0.01505	-2.04527	-0.42154	-0.41974
39	0.01440	-1.31612	-0.27117	-0.26987
40	0.02008	-1.68930	-0.34906	-0.34748
41	0.05522	7.87952	1.65815	1.67347
42	0.01199	-1.90953	-0.39295	-0.39123
43	0.01216	-1.86781	-0.38440	-0.38270
44	0.01458	0.93784	0.19324	0.19228
45	0.01412	2.13166	0.43913	0.43730
46	0.01492	-0.33163	-0.06835	-0.06799
47	0.01295	-0.85425	-0.17587	-0.17499
48	0.01685	0.17007	0.03508	0.03490
49	0.01145	-1.67182	-0.34394	-0.34237
50	0.01574	9.22517	1.90199	1.92846

(*continued*)

Table 11 (*continued*)

	Residual	Studentized residual	Influence statistic d_i	z_i
51	0.02301	-1.83259	-0.37924	-0.37756
52	0.01747	-0.09596	-0.01980	-0.01970
53	0.01189	-1.53324	-0.31550	-0.31403
54	0.01376	-0.58437	-0.12036	-0.11975
55	0.01328	-0.91487	-0.18839	-0.18745
56	0.01715	0.51718	0.10671	0.10616
57	0.01449	-0.18410	-0.03793	-0.03774
58	0.01498	0.78992	0.16280	0.16198
59	0.01699	10.61778	2.19050	2.23517
60	0.01500	-1.82826	-0.37680	-0.37512
61	0.42412	2.23947	0.60363	0.60164
62	0.01455	-0.23850	-0.04914	-0.04889
63	0.01356	-0.34374	-0.07079	-0.07043
64	0.01050	1.00108	0.20585	0.20483
65	0.01752	-1.31561	-0.27149	-0.27019
66	0.01598	-1.73993	-0.35877	-0.35716
67	0.01573	-0.73760	-0.15207	-0.15131
68	0.01559	-0.54880	-0.11314	-0.11256
69	0.01500	5.57172	1.14832	1.15023
70	0.01592	-0.23994	-0.04947	-0.04922
71	0.01833	-0.70272	-0.14507	-0.14434
72	0.02438	-2.61934	-0.54243	-0.54045
73	0.01397	-1.93669	-0.39894	-0.39720
74	0.01308	-1.61261	-0.33203	-0.33050
75	0.15508	-5.07724	-1.12982	-1.13146
76	0.02288	-2.76585	-0.57233	-0.57033
77	0.87115	-9.37256	-5.34086	-6.32383
78	0.01033	1.04916	0.21572	0.21466
79	0.01135	-2.23159	-0.45907	-0.45720
80	0.03843	-2.18126	-0.45500	-0.45313
81	0.13723	-5.08913	-1.12069	-1.12219
82	0.02129	-2.42238	-0.50085	-0.49890
83	0.01754	-0.31554	-0.06512	-0.06478
84	0.01606	-3.70359	-0.76371	-0.76206
85	0.02155	-2.19012	-0.45288	-0.45102
86	0.01317	-0.70125	-0.14439	-0.14366
87	0.01622	-1.47850	-0.30491	-0.30347
88	0.01547	-0.37825	-0.07798	-0.07758
89	0.01493	-0.48675	-0.10031	-0.09980
90	0.01568	-0.59554	-0.12278	-0.12216
91	0.02144	-2.83740	-0.58670	-0.58471
92	0.01237	-0.87068	-0.17920	-0.17831
93	0.01674	12.87097	2.65502	2.74284
94	0.07358	-0.78199	-0.16618	-0.16535
95	0.06518	27.24009	5.76280	7.06957
96	0.05459	22.24057	4.67870	5.28948
97	0.19123	-6.33858	-1.44168	-1.44985
98	0.01501	-0.80118	-0.16512	-0.16429
99	0.01518	-1.46582	-0.30213	-0.30071
100	0.01608	9.55797	1.97095	2.00125

has a t distribution with $n - p - 2$ degrees of freedom, and is the statistic used to check on whether the ith observation is an outlier. Algebraic manipulation of the formula for $\mathbf{z}_i$ reveals that

$$\mathbf{z}_i = \mathbf{w}_i \left(\frac{n - p - 2}{n - p - 1 - \mathbf{w}_i^2} \right)^{1/2},$$

so that a separate regression deleting the ith observation need not be computed.

If one suspects in advance that the ith data point is an outlier, the aforementioned procedure is an appropriate test. If, however, one looks *ex post* at the set of $\hat{\mathbf{z}}_i$, selects the largest in absolute value, and asks whether the associated data point is an outlier, the t test based on that residual is inappropriate. This is because

$$\Pr\{\max |\hat{\mathbf{z}}_i| > t_\alpha\} \gg \alpha,$$

where t_α is the two-tailed $100\alpha\%$ point of the appropriate t distribution. However, if the $|\hat{\mathbf{z}}_i|$ were independent

$$\begin{aligned} \Pr\{|\hat{\mathbf{z}}_i| > t_\alpha \quad \text{for some} \quad i\} &= \bigcup_{i=1}^{n} \Pr\{|\hat{\mathbf{z}}_i| > t_\alpha\} \\ &\leq \sum_{i=1}^{n} \Pr\{|\hat{\mathbf{z}}_i| > t_\alpha\} = n\alpha. \end{aligned}$$

They are not independent, as the above inequality is not strictly correct. However, if one wants a rough guide as to whether an outlier exists, one might use the above inequality as if it were correct and construct a significance test using as a critical value $t_{\alpha'}$, where $\alpha' = \alpha/n$. Specially prepared t tables for $\alpha = 0.05$ and 0.01 and various values of p and n are given in Weisberg [1980], one of which is reproduced as Table 12.

Looking at the z_i in Table 11, we see that observations 77, 95, and 96 had the largest values of this outlier statistic. Using Table 12 for $n = 100$ and $p = 3$ indicates that 3.60 is the critical value for z_i. Thus these three observations would be judged to be outliers in this data set.

Another measure, due to Belsley, Kuh, and Welsch [1980], is called DFITS (or sometimes DFFITS) and is implemented in MINITAB using that subcommand. The quantity

$$\mathbf{e}_i = \frac{\mathbf{u}_i \sqrt{p_{ii}}}{\hat{\boldsymbol{\sigma}}(1 - p_{ii})}$$

is to be compared with $2\sqrt{(p + 1)/n}$. Observations for which $\mathbf{e}_i$ exceeds this quantity are suspect. Cook [1977] and Weisberg [1980] suggests comparing $\mathbf{e}_i^2/(p + 1)$ to the 50% point of an $F(p + 1, n - p - 1)$ distribution.

Table 12. Critical Values for the Outlier Test, $\alpha = 0.05$.

$n/$ p'	1	2	3	4	5	6	7	8	9	10	11	12	13	14	15	20	25	30
6	4.85	6.23	10.89	76.39														
7	4.38	5.07	6.58	11.77	89.12													
8	4.12	4.53	5.26	6.90	12.59	101.9												
9	3.95	4.22	4.66	5.44	7.18	13.36	114.6											
10	3.83	4.03	4.32	4.77	5.60	7.45	14.09	127.3										
11	3.75	3.90	4.10	4.40	4.88	5.75	7.70	14.78	140.1									
12	3.69	3.81	3.96	4.17	4.49	4.98	5.89	7.94	15.44	152.8								
13	3.65	3.74	3.86	4.02	4.24	4.56	5.08	6.02	8.16	16.08	165.5							
14	3.61	3.69	3.79	3.91	4.07	4.30	4.63	5.16	6.14	8.37	16.69	178.2						
15	3.58	3.65	3.73	3.83	3.95	4.12	4.36	4.70	5.25	6.25	8.58	17.28	191.0					
16	3.56	3.62	3.68	3.77	3.87	4.00	4.17	4.41	4.76	5.33	6.36	8.77	17.85	203.7				
17	3.54	3.59	3.65	3.72	3.80	3.90	4.04	4.21	4.46	4.82	5.40	6.47	8.95	18.40	216.4			
18	3.53	3.57	3.62	3.68	3.75	3.83	3.94	4.08	4.26	4.51	4.88	5.47	6.57	9.13	18.93			
19	3.52	3.56	3.60	3.65	3.71	3.78	3.86	3.97	4.11	4.30	4.55	4.93	5.54	6.67	9.30			
20	3.51	3.54	3.58	3.62	3.67	3.73	3.81	3.89	4.00	4.15	4.33	4.59	4.98	5.60	6.76			
21	3.50	3.53	3.57	3.60	3.65	3.70	3.76	3.83	3.92	4.03	4.18	4.37	4.64	5.03	5.67			
22	3.50	3.52	3.55	3.59	3.63	3.67	3.72	3.78	3.86	3.95	4.06	4.21	4.40	4.68	5.08	280.1		
23	3.49	3.52	3.54	3.57	3.61	3.65	3.69	3.75	3.81	3.88	3.98	4.09	4.24	4.44	4.71	21.41		
24	3.49	3.51	3.53	3.56	3.59	3.63	3.67	3.71	3.77	3.83	3.91	4.00	4.12	4.27	4.47	10.07		
25	3.48	3.50	3.53	3.55	3.58	3.61	3.65	3.69	3.73	3.79	3.85	3.93	4.02	4.14	4.30	7.17		
26	3.48	3.50	3.52	3.54	3.57	3.60	3.63	3.66	3.70	3.75	3.81	3.87	3.95	4.05	4.17	5.95		
27	3.48	3.50	3.52	3.54	3.56	3.58	3.61	3.65	3.68	3.72	3.77	3.83	3.89	3.97	4.07	5.29	343.8	
28	3.48	3.50	3.51	3.53	3.55	3.58	3.60	3.63	3.66	3.70	3.74	3.79	3.84	3.91	3.99	4.88	23.63	

(continued)

Table 12 (*continued*)

n/ p'1		2	3	4	5	6	7	8	9	10	11	12	13	14	15	20	25	30
29	3.48	3.49	3.51	3.53	3.55	3.57	3.59	3.62	3.64	3.68	3.71	3.76	3.81	3.86	3.93	4.61	10.74	
30	3.48	3.49	3.51	3.52	3.54	3.56	3.58	3.60	3.63	3.66	3.69	3.73	3.77	3.82	3.88	4.42	7.53	
31	3.48	3.49	3.50	3.52	3.54	3.55	3.57	3.59	3.62	3.64	3.67	3.71	3.74	3.79	3.84	4.28	6.18	
32	3.48	3.49	3.50	3.52	3.53	3.55	3.57	3.59	3.61	3.63	3.66	3.69	3.72	3.76	3.80	4.17	5.47	407.4
33	3.48	3.49	3.50	3.52	3.53	3.54	3.56	3.58	3.60	3.62	3.64	3.67	3.70	3.74	3.77	4.08	5.03	25.66
34	3.48	3.49	3.50	3.51	3.53	3.54	3.56	3.57	3.59	3.61	3.63	3.66	3.68	3.71	3.75	4.01	4.74	11.34
35	3.48	3.49	3.50	3.51	3.52	3.54	3.55	3.57	3.58	3.60	3.62	3.64	3.67	3.70	3.73	3.96	4.53	7.84
36	3.48	3.49	3.50	3.51	3.52	3.54	3.55	3.56	3.58	3.60	3.61	3.63	3.66	3.68	3.71	3.91	4.37	6.39
37	3.48	3.49	3.50	3.51	3.52	3.53	3.55	3.56	3.57	3.59	3.61	3.62	3.65	3.67	3.69	3.87	4.26	5.62
38	3.48	3.49	3.50	3.51	3.52	3.53	3.54	3.56	3.57	3.58	3.60	3.62	3.64	3.66	3.68	3.84	4.16	5.16
39	3.49	3.49	3.50	3.51	3.52	3.53	3.54	3.55	3.57	3.58	3.59	3.61	3.63	3.65	3.67	3.81	4.09	4.84
40	3.49	3.49	3.50	3.51	3.52	3.53	3.54	3.55	3.56	3.58	3.59	3.60	3.62	3.64	3.66	3.79	4.03	4.62
50	3.51	3.51	3.51	3.52	3.53	3.53	3.54	3.54	3.55	3.56	3.57	3.57	3.58	3.59	3.60	3.66	3.75	3.88
60	3.53	3.53	3.53	3.54	3.54	3.54	3.55	3.55	3.56	3.56	3.57	3.57	3.58	3.58	3.59	3.62	3.67	3.73
70	3.55	3.55	3.55	3.55	3.56	3.56	3.56	3.56	3.57	3.57	3.57	3.58	3.58	3.59	3.59	3.61	3.64	3.67
80	3.57	3.57	3.57	3.57	3.57	3.58	3.58	3.58	3.58	3.58	3.59	3.59	3.59	3.60	3.60	3.61	3.63	3.66
90	3.58	3.59	3.59	3.59	3.59	3.59	3.59	3.60	3.60	3.60	3.60	3.60	3.60	3.61	3.61	3.62	3.63	3.65
100	3.60	3.60	3.60	3.60	3.61	3.61	3.61	3.61	3.61	3.61	3.61	3.62	3.62	3.62	3.62	3.63	3.64	3.65
200	3.73	3.73	3.73	3.73	3.73	3.73	3.73	3.73	3.73	3.73	3.73	3.73	3.73	3.73	3.74	3.74	3.74	3.74
300	3.81	3.81	3.81	3.81	3.81	3.81	3.81	3.81	3.81	3.81	3.82	3.82	3.82	3.82	3.82	3.82	3.82	3.82
400	3.87	3.87	3.87	3.87	3.87	3.87	3.87	3.88	3.88	3.88	3.88	3.88	3.88	3.88	3.88	3.88	3.88	3.88
500	3.92	3.92	3.92	3.92	3.92	3.92	3.92	3.92	3.92	3.92	3.92	3.92	3.92	3.92	3.92	3.92	3.92	3.92

SOURCE: Weisberg [1980]. Reprinted from *Applied Linear Regression* with permission from John Wiley and Sons.

References

Barnett, V. and Lewis, T., 1978. *Outliers in Statistical Data.* New York: Wiley.

Beckman, R. J. and Cook, R. D. 1983. Outlier..........s. *Technometrics* **25** (May): 119–49.

Belsley, D. A., Kuh, E., and Welsch, R. E. 1980. *Regression Diagnostics.* New York: Wiley.

Cook, R. D. 1977. Detection of influential observations in linear regression. *Technometrics* **19** (February): 15–18.

Cook, R. D. 1979. Influential observations in linear regression. *Journal of the American Statistical Association* **74** (March): 169–74.

Cook, R. D. and Weisberg, S. 1982. *Residuals and Influence in Regression.* New York: Chapman and Hall.

Ferguson, T. S. 1961a. On the rejection of outliers. In *Proceedings of the Fourth Berkeley Symposium on Mathematical Statistics and Probability*, Vol. 1, ed. J. Neyman. Berkeley: University of California Press, pp. 253–87.

Ferguson, T. S. 1961b. Rules for rejection of outliers. *Revue de l'Institut International de Statistique* **29** (No. 3): 29–43.

Grubbs, F. E. 1950. Sample criteria for testing outlying observations. *Annals of Mathematical Statistics* **21** (March): 27–58.

Hawkins, D. M. 1978. Fractiles of an extended multiple outlier test. *Journal of Statistical Computation and Simulation* **8**: 227–36.

Hawkins, D. M. 1980. *Identification of Outliers.* New York: Chapman and Hall.

Huber, P. 1981. *Robust Statistics.* New York: Wiley.

Karlin, S. and Truax, D. R. 1960. Slippage problems. *Annals of Mathematical Statistics* **31** (June): 296–324.

Kudo, A. 1956. On the testing of outlying observations. *Sankhya* **17** (June): 67–76.

Murphy, R. B. 1951. *On Tests for Outlying Observations.* Ph. D. thesis, Princeton University. Ann Arbor: University Microfilms.

Pearson, E. S. and Chandra Sekar, C. 1936. The efficiency of statistical tools and a criterion for the rejection of outlying observations. *Biometrika* **28**: 308–20.

Pearson, E. S. and Stephens, M. A. 1964. The ratio of range to standard deviation in the same normal sample. *Biometrika* **51**: 484–87.

Rosner, B. 1975. On the detection of many outliers. *Technometrics* **17** (May): 221–27.

Rosner, B. 1983. Percentage points for a generalized ESD many-outlier procedure. *Technometrics* **25** (May): 165–72.

Thompson, W. R. 1935. On a criterion for the rejection of observations and the distributions of the ratio of the deviation to the sample standard deviation. *Annals of Mathematical Statistics* **6**: 214–19.

Tietjen, G. L. and Moore, R. M. 1972. Some Grubbs-type statistics for the detection of several outliers. *Technometrics* **14** (August): 583–97.

Walsh, J. E. 1950. Some nonparametric tests of whether the largest observations of a set are too large or too small. *Annals of Mathematical Statistics* **21** (December): 583–92. (see also correction 1953 *Annals of Mathematical Statistics* **24** (March): 134–35).

Walsh, J. E. 1958. Large sample nonparametric rejection of outlying observations. *Annals of the Institute of Statistical Mathematics* **10**: 223–32.

Weisberg, S. 1980. *Applied Linear Regression.* New York: Wiley.

CHAPTER 5

Transformations

"The simple family often does quite well in transforming to normality!"

Tukey [1957]

0. Introduction

The transformation of data takes on great importance in statistical analysis, espccially as one wishes to apply inferential procedures which are highly sensitive to deviations from their underlying assumptions to the data. In this chapter we discuss a number of such transformations, each focusing on a different ill. In Section 1 we describe the Glejser suggestions for searching for a deflator to apply to a regression equation, i.e., to each of the variables in the regression (including the constant), which will rid the regression residuals of their heteroscedasticity. In Section 2 we discuss the variance stabilizing transformation, a transformation procedure which produces a random variable whose variance is functionally independent of its expected value. Section 3 describes the Box–Cox transformations, power transformations which best transform data to normality. Section 4 is an exposition of Tukey's "letter values" and "box plots", which provide a quick graphic way of finding the appropriate power transformation. In Section 5, we describe the Box–Tidwell procedures for systematically determining the appropriate power transformation of an independent variable in a regression.

1. Deflating Heteroscedastic Regressions

If indeed the heteroscedasticity is a function of a single independent variable, x_l, say, then to rid ourselves of this problem we must also know the functional form that the heteroscedasticity takes in order to estimate the regression equation properly. One approach to estimation of the regression is *deflation.*

That is, if $\mathscr{V}\mathbf{u}_i = \phi(x_{li})\sigma^2$, then the functional form of ϕ must be known so that we can deflate the dependent variable and all the independent variables by dividing each of them (including the unit constant if there is an intercept in the model) by $\sqrt{\phi(x_{li})}$. For if $\mathbf{y}_i = \alpha + \beta_1 x_{1i} + \cdots + \beta_k x_{ki} + \mathbf{u}_i$ and $\mathscr{V}\mathbf{u}_i = \sigma^2\phi(x_{li})$, then

$$\frac{\mathbf{y}_i}{\sqrt{\phi(x_{li})}} = \frac{\alpha}{\sqrt{\phi(x_{li})}} + \beta_1 \frac{x_{1i}}{\sqrt{\phi(x_{li})}} + \cdots + \beta_k \frac{x_{ki}}{\sqrt{\phi(x_{li})}} + \frac{\mathbf{u}_i}{\sqrt{\phi(x_{li})}}.$$

This can be rewritten as

$$\mathbf{y}_i^* = \alpha x_{0i}^* + \beta_1 x_{1i}^* + \cdots + \beta_{ki}^* x_{ki}^* + \mathbf{u}_i^*,$$

where $x_{0i}^* = 1/\sqrt{\phi(x_{li})}$ and the other variables are the original variables deflated by $\sqrt{\phi(x_{li})}$. Then $\mathscr{V}\mathbf{y}_i^* = \mathscr{V}\mathbf{u}_i^* = \sigma^2$. Thus the variance of all the resulting true residuals will be the hoped for σ^2.

Another approach to estimation of the regression equation is the use of *weighted least squares*. Briefly, if the covariance matrix of $\mathbf{U}$ is known up to a scale factor σ^2 to be $\sigma^2\Omega$, then the weighted least squares estimator of B is

$$\hat{\mathbf{B}} = (X\Omega^{-1}X^1)^{-1}X\Omega^{-1}\mathbf{y}.$$

In particular, if the $\mathbf{u}_i$ are independent then Ω is a diagonal matrix, and this estimator can be seen to be equivalent to that of deflating $\mathbf{y}_i$ and each of the x_{ki}'s by the square root of the reciprocal of the variance of $\mathbf{u}_i$. Glejser [1969] suggests that the search for the functional form of ϕ be carried out by regressing the $|\hat{\mathbf{u}}_i|$ onto some candidate functions ϕ and testing the coefficients of these resulting regressions for significance.

Let us consider his proposal in some detail. Suppose $\phi(x_l)$ is of the form

$$\phi(x_l) = \alpha_0 + \alpha_1\psi(x_l),$$

where $\psi(x_l)$ is a prespecified mathematical function of x_l involving no additional parameters (e.g., $\psi(x_l) = \sqrt{x_l}$, or $\psi(x_l) = \sin^{-1} x_l$). Then Glejser suggests regressing the $|\hat{u}_i|$ on the $\psi(x_{li})$ and testing the hypothesis that $\alpha_0 = 0$ and $\alpha_1 = 0$. There are four possibilities:

(1) $\alpha_0 = 0, \alpha_1 = 0$ } both imply that $\psi(x_l)$ is not an
(2) $\alpha_0 \neq 0, \alpha_1 = 0$ } appropriate form on which to base a deflator.
(3) $\alpha_0 = 0, \alpha_1 \neq 0$, this implies that $\sqrt{\psi(x_l)}$ should be used as a deflator.
(4) $\alpha_0 \neq 0, \alpha_1 \neq 0$, this implies that $\sqrt{\hat{\alpha}_0 + \alpha_1\hat{\psi}(x_l)}$ should be used as a deflator.

To verify that such deflation has indeed eliminated the heteroscedasticity, one can use any of the test procedures of Chapter 2, Section 4.

The approach taken by Glejser is to model the variance as a function of one of the independent variables and determine the validity of the model based on a significance test (in his example, the Goldfeld–Quandt test, probably because it was the only published procedure at the time of publication of

Table 1. Grinding of Stainless Steel Data.

Row	HOURS	NUMBER	DIAMET	TOLERA	Residual	Row	HOURS	NUMBER	DIAMET	TOLERA	Residual
1	1.75	126	.25500	.3	-1.09378	51	1.75	74	.09800	2.0	-1.83259
2	1.25	9	.22300	1.0	-1.36781	52	1.75	6	.12900	.2	-0.09596
3	1.50	44	.40625	.5	-1.23030	53	1.25	36	.22400	1.0	-1.53324
4	0.25	6	.25500	5.0	-5.44620	54	3.25	280	.19200	.5	-0.58437
5	0.50	1	.09900	.4	-1.42198	55	2.50	202	.07813	.8	-0.91485
6	6.75	265	.28125	5.0	-0.55864	56	2.50	24	.14700	.2	0.51718
7	5.00	25	.19300	.2	2.94057	57	2.25	62	.19150	.4	-0.18410
8	0.50	4	.21875	.8	-1.92856	58	3.00	34	.15625	.4	0.78991
9	5.75	33	.68750	2.0	1.51240	59	15.00	101	1.00950	1.0	1.61778
10	2.75	4	.75500	1.0	-.65318	60	0.25	3	.19300	.4	-1.82826
11	7.00	198	.21800	1.0	3.24258	61	11.75	1	4.7500	1.0	2.23946
12	0.50	2	.25000	1.0	-2.11671	62	2.00	15	.25000	.4	-0.23850
13	2.00	2	.31700	.2	-0.10994	63	2.00	4	.31250	.5	-0.34375
14	1.75	151	.25000	.3	-1.23787	64	4.25	43	.50000	1.0	1.00108
15	0.50	3	.31700	1.0	-2.22552	65	0.50	1	.12900	.2	-1.31561
16	0.50	8	.25500	.5	-1.77986	66	0.25	1	.19300	.3	-1.73993
17	1.00	6	.31250	1.0	-1.73683	67	3.50	33	.68750	2.0	-0.73760
18	0.50	13	.25500	1.0	-2.19115	68	3.75	380	.09900	.5	-0.54880
19	1.75	3	.37500	.3	-0.53113	69	8.25	100	.25000	.3	5.57172
20	0.75	8	.18750	1.0	-1.80730	70	4.50	436	.16500	.5	-0.23994
21	0.50	12	.14063	1.0	-2.00971	71	1.00	4	.04350	.2	-0.70272
22	1.50	140	.06250	1.0	-1.66692	72	0.50	4	.07300	2.0	-2.61934
23	3.75	167	.62500	4.0	-2.72895	73	0.50	12	.09300	1.0	-1.93669

24	1.50	12	.28125	1.0	-1.22533	74	1.25	104	.25500	.5	-1.61261
25	9.25	5	1.43750	2.0	4.03239	75	1.75	1	3.00000	1.0	-5.07724
26	5.50	55	.34375	.5	2.79876	76	0.50	15	.12500	2.0	-2.76585
27	4.75	15	.93750	1.0	1.00022	77	24.50	5261	.06300	.5	-9.37256
28	0.75	4	.62500	1.0	-2.45385	78	4.50	75	.50500	1.0	1.04916
29	2.00	55	.15625	.4	-0.33756	79	0.50	4	.31700	1.0	-2.23159
30	0.50	6	.12500	.4	-1.49220	80	6.50	1050	.25500	.6	-2.18126
31	0.50	4	.31700	.5	-1.85064	81	2.50	1	3.00000	2.0	-5.08913
32	1.00	30	.14063	.6	-1.31422	82	1.00	25	.18750	2.0	-2.42238
33	7.75	1036	.25500	.6	-0.84628	83	1.50	2	.12500	.2	-0.31554
34	0.50	3	.63100	3.0	-4.23074	84	0.25	2	.62500	2.0	-3.70359
35	1.75	5	1.00800	1.0	-2.04717	85	1.50	77	.15625	2.0	-2.19012
36	9.25	111	.56250	.4	5.94960	86	2.25	95	.10000	1.0	-0.70125
37	7.75	4	3.12500	6.0	-3.09654	87	1.25	1	.62500	.4	-1.47851
38	0.50	1	.50550	.4	-2.04527	88	2.00	19	.37500	.3	-0.37825
39	1.00	12	.31250	.4	-1.31612	89	2.75	192	.25000	.3	-0.48675
40	3.25	267	.21875	2.0	-1.68930	90	1.50	4	.25000	.3	-0.59554
41	13.00	4	1.87500	1.0	7.87952	91	0.50	11	.18750	2.0	-2.83740
42	0.75	13	.23438	1.0	-1.90953	92	2.00	90	.26600	.6	-0.87068
43	0.75	9	.22300	1.0	-1.86781	93	18.00	243	.43750	2.0	12.87097
44	3.50	92	.15625	.4	0.93784	94	4.75	11	.62500	4.0	-0.78199
45	4.75	46	.37500	.4	2.13166	95	38.00	1053	.40625	3.0	27.24009
46	2.50	124	.25500	.3	-0.33163	96	31.75	847	.40625	3.0	22.24057
47	2.25	155	.16100	.6	-0.85425	97	3.25	2	2.31250	6.0	-6.33858
48	3.25	13	.75700	.5	0.17007	98	1.50	59	.06700	.5	-0.80118
49	1.25	51	.25500	1.0	-1.67182	99	1.00	65	.25000	.3	-1.46582
50	12.50	91	.62500	.4	9.22517	100	12.50	156	.25000	.2	9.55797

Glejser's paper). We are of course muddying the inferential procedures to be used with the final regression model by using the same data to model the heteroscedasticity, test the heteroscedasticity model, and create deflators prior to fitting the final regression model. One should view this approach then as a more formalized bit of data exploration than as a rigorous procedure for finding the source of heteroscedasticity of residuals.

To illustrate this method of hunting down and eliminating heteroscedasticity of residuals, we will study the data set of 100 observations on time required for machining of grinding orders for stainless steel. The variables are $\mathbf{y}$ = HOURS, x_1 = NUMBER of bars in an order, x_2 = DIAMET, diameter in inches of bar prior to grinding, and x_3 = TOLERA, maximum allowable spread in 1000ths of an inch. The data and the regression residuals are given in Table 1. First, we can learn a bit by plotting the residuals of the regression

```
  HORZ OBS. IN ROW  77 AN OUTLIER; PLOTTED AS  3.4
  VERT OBS. IN ROW  95 AN OUTLIER; PLOTTED AS  3.4
  VERT OBS. IN ROW  96 AN OUTLIER; PLOTTED AS  3.4

  -                                              1   1
  -
 3
  -
  -                                    1
  -
  -                                  1
 2                                 1 1
  -
  -                                1
  -
  -                                  2
 1
  -                                1
  -                                2 1
  -                                3
  -                                6
 0                                 9 1     1
  -                                F 6 2 1           1
  -                                * 2 1             1
  -                                6 1
  -                                2
-1                                 2
  -                                1
  -                                1
  -
  -
-2                                                                    1
  -
  -
  -
  -
-3
  -
  -
    - - : - - - - - : - - - - - : - - - - - : - - - - - : - - - - - : - - - - - : - -
       -3          -2          -1           0           1           2           3

                    MEAN        STD.DEV.
 VERT.  VAR.    -1.4901E-08     4.8142E+00
 HORIZ. VAR.     1.4774E+02     5.5513E+02
 SAMPLE SIZE = 100
```

Figure 1. Scatter Plot of 100 Standardized Values of RESID versus NUMBER.

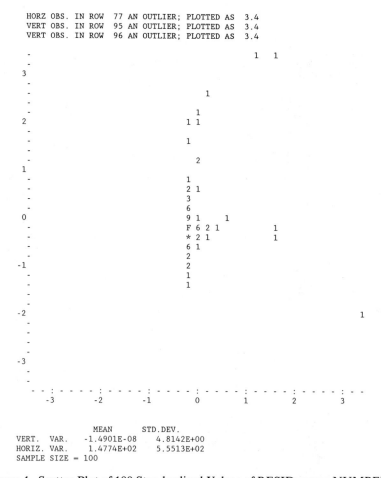

```
HORZ OBS. IN ROW  37 AN OUTLIER; PLOTTED AS  3.4
HORZ OBS. IN ROW  61 AN OUTLIER; PLOTTED AS  3.4
HORZ OBS. IN ROW  75 AN OUTLIER; PLOTTED AS  3.4
HORZ OBS. IN ROW  81 AN OUTLIER; PLOTTED AS  3.4
VERT OBS. IN ROW  95 AN OUTLIER; PLOTTED AS  3.4
VERT OBS. IN ROW  96 AN OUTLIER; PLOTTED AS  3.4

  -                                2
  -
 3
  -
  -                                  1
  -
  -                                          1
 2                               1     1
  -
  -                                                     1
  -
  -                              1     1
 1
  -                                                1
  -                              2 1
  -                                1   1                               1
  -                              3   2     1
 0                               7 3     1
  -                            6 B 5   2 1
  -                            3 * 4 1 1     1
  -                            1 3     2                               1
  -                                    2
-1                                                                     2
  -                              1
  -                                                            1
  -
  -
-2                             1
  -
  -
  -
  -
-3
  -
  -
    - - : - - - - - : - - - - - : - - - - - : - - - - - : - - - - - : - - - - - : - -
       -3        -2        -1         0         1         2         3

                     MEAN        STD.DEV.
VERT.  VAR.    -1.4901E-08     4.8142E+00
HORIZ. VAR.     4.8223E-01     7.1819E-01
SAMPLE SIZE =  100
```

Figure 2. Scatter Plot of 100 Standardized Values of RESID versus DIAMET.

of HOURS on NUMBER, DIAMET, and TOLERA against each of the these independent variables. These plots are given as Figures 1–3. Notice that in the plot of residuals versus NUMBER there is a great deal of scatter at low values of NUMBER but less scatter at the high values of NUMBER. But this may be due to the paucity of data for high values of NUMBER. The same phenomenon is true for DIAMET, but here there seems to be more data at higher levels of DIAMET than at higher levels of NUMBER. Looking at the TOLERA plot we see roughly equal scatter of the residuals at various levels of TOLERA. Thus our prime candidates for the independent variable potentially causing heterosce-dasticity are DIAMET and NUMBER.

Next we apply the Goldfeld–Quandt procedure to see if our visual per-

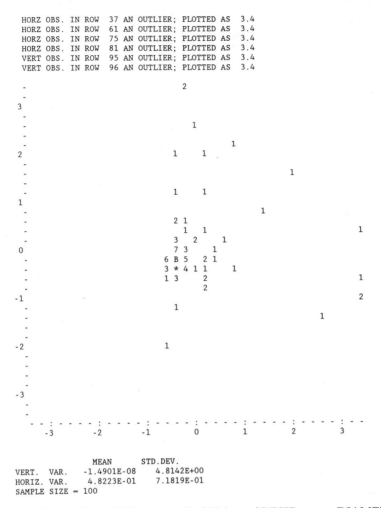

```
HORZ OBS. IN ROW  37 AN OUTLIER; PLOTTED AS  3.4
VERT OBS. IN ROW  95 AN OUTLIER; PLOTTED AS  3.4
VERT OBS. IN ROW  96 AN OUTLIER; PLOTTED AS  3.4
HORZ OBS. IN ROW  97 AN OUTLIER; PLOTTED AS  3.4
```

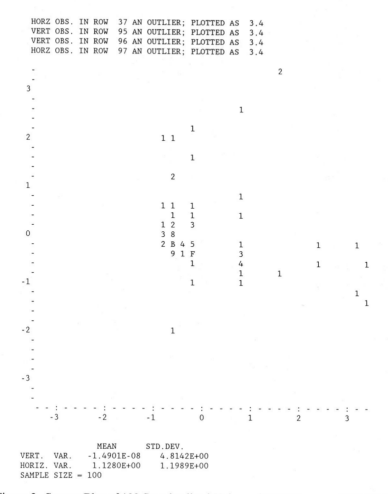

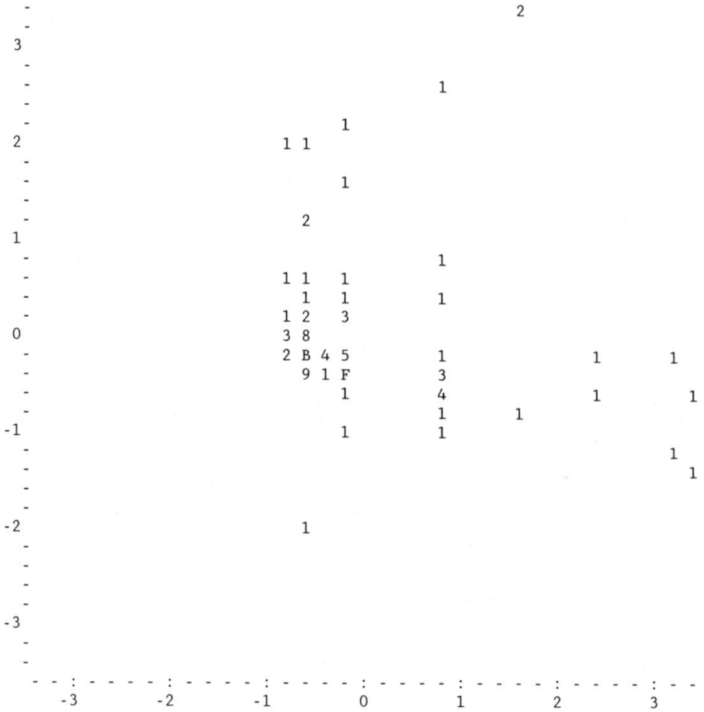

```
                MEAN        STD.DEV.
VERT.  VAR.   -1.4901E-08    4.8142E+00
HORIZ. VAR.    1.1280E+00    1.1989E+00
SAMPLE SIZE = 100
```

Figure 3. Scatter Plot of 100 Standardized Values of RESID versus TOLERA.

ception is very good. If we sort on NUMBER and use the first and last 40 observations for the test, we find that $\mathbf{q} = 4.58984/35.7366 = 0.1284$, as compared with 0.574, the 5% point of the $F(36, 36)$ distribution. So, as suspected, NUMBER is a cause of heteroscedasticity. Sorting on DIAMET and once again using the first and last 40 observations, we obtain $\mathbf{q} = 1.42792/10.6331 = 0.1343$, also significant. Finally, applying the same procedure using TOLERA we obtain $\mathbf{q} = 7.2676/11.0748 = 0.6562$, so that as we surmised TOLERA is not the root of the heteroscedasticity of residuals.

Now let us find a function of one of the independent variables that can be used to deflate the dependent variable and eliminate the heterosce-dasticity. Since the Goldfeld–Quandt procedure based on NUMBER produced a more significant departure from homoscedasticity than did that based on DIAMET, we use it as the potential deflator, and consider five functions:

(1) $|\hat{u}_i| = \alpha_1 + \beta_1 x_{1i}$.
(2) $|\hat{u}_i| = \alpha_2 + \beta_2 \sqrt{x_{1i}}$.
(3) $|\hat{u}_i| = \alpha_3 + \beta_3 / x_{1i}$.
(4) $|\hat{u}_i| = \alpha_4 + \beta_4 / \sqrt{x_{1i}}$.
(5) $|\hat{u}_i| = \alpha_5 + \beta_5 \log x_{1i}$.

The correlations of the independent variables with these $|\hat{\mathbf{u}}_i|$ will indicate which of these five functions will yield the best deflating function. The significance of these five functions is determined by the t test on the slope and intercept of the regressions. Following are the results of the regressions, with an asterisk indicating that the associated coefficient is significant at the 5% level.

	Correlation	Intercept	Slope
(1)	0.3354	2.356*	0.0023983*
(2)	0.4162	1.3871*	0.17363*
(3)	−0.0567	2.855*	−0.81384
(4)	−0.1258	3.2893*	−1.8095
(5)	0.2853	0.90161	0.58505*

The ideal deflation situation is one wherein the intercept is not significantly different from zero whereas the slope is. In that case deflation by $\sqrt{\phi(x_1)}$ will do the trick. Such is the case in this example if we take $\phi(x_1) = \log x_1$, but is not the case if we were to take $\phi(x_1) = \sqrt{x_1}$ or $\phi(x_1) = x_1$.. For this example I would deflate using $\sqrt{\log x_1}$, since that is the easiest procedure. Had I not considered $\log x_1$ as a potential deflator, or had it not turned out to behave so ideally, I would be forced to deflate using the best deflator,

$$\phi(x_1) = 1.3871 + 0.17363\sqrt{x_1},$$

instead of merely a function of x_1 above. Such a procedure introduces some "noise" in the subsequent analysis, since 1.3871 and 0.173636 are sample statistics.

Upon dividing HOURS by $\sqrt{\log \text{NUMBER}}$, we obtain a new dependent variable which we can regress on the three deflated independent variables. The next step in the process is to see if that deflator produces residuals with no heteroscedasticity. The problem with the Glejser approach is that, though it points to potential deflators, these deflators are based only on one independent variable, not on a subset of them. If the heteroscedasticity is due to an interaction between independent variables, the Glejser procedure will not detect such an interaction directly.

Such unfortunately is the case with this data set. Upon deflating using $\sqrt{\log \text{NUMBER}}$ (and even upon deflating using

$\sqrt{1.3871 + 0.17363\sqrt{\text{NUMBER}}}$), we applied the Goldfeld–Quandt procedure to the data and still rejected the hypothesis of homoscedasticity of residuals both based on sorting on NUMBER and sorting on DIAMET. That the regression based on the dependent variable deflated by a function of NUMBER failed the Goldfeld–Quandt test based on sorting on NUMBER is a strong indication that the heteroscedasticity is due to an interaction of NUMBER and DIAMET.

2. Variance Stabilizing Transformations

As has been seen earlier in the regression context, sometimes the heteroscedasticity is characterized by the variance of the variable under consideration being a function of the expected value of that variable. For example, the Glejser procedure seeks to determine the nature of that function by postulating various functional forms involving one (or more) of the independent variables. There are still other situations in which the variance of the random variable $\mathbf{x}$ is naturally a function of its mean, the simplest example being when $\mathbf{x}$ is the average of n independent Bernoulli trials each governed by the same parameter θ. Then $\mathscr{E}x = \theta$ and $\mathscr{V}\mathbf{x} = \theta(1 - \theta)/n = \mathscr{E}\mathbf{x}(1 - \mathscr{E}\mathbf{x})/n$. In order to use statistical procedures which require homoscedasticity for their validity on such data, we must first transform $\mathbf{x}$ into a random variable $\mathbf{y} = h(\mathbf{x})$ such that $\mathscr{V}\mathbf{y}$ is a constant, independent of $\mathscr{E}\mathbf{y}$.

Prescription

In general, if one can characterize the functional relationship between $\mathscr{E}\mathbf{x}$ and $\mathscr{V}\mathbf{x}$, say $\mathscr{V}\mathbf{x} = g(\mathscr{E}\mathbf{x})$, then the appropriate transformation to eliminate the heteroscedasticity is given by any linear function of

$$\int \frac{1}{\sqrt{g(\mathscr{E}\mathbf{x})}} d(\mathscr{E}\mathbf{x}).$$

The task for the data analyst, then, is to determine the relationship g between variance and mean, after which the determination of the appropriate transformation to homoscedasticity is given by the above formula.

This formidable looking expression is really quite an easy-to-manipulate and handy gadget, depending only on the nature of the function g. To see this, let us look at a simple example, where $\mathscr{E}\mathbf{x} = \theta$ and $\mathscr{V}\mathbf{x} = \theta^2/n$. Then

$$\begin{aligned} h(\theta) &= k \int \frac{1}{\sqrt{\theta^2/n}} d\theta \\ &= k\sqrt{n} \log \theta + C, \end{aligned}$$

so that, using the convention that $C = 0$ and $k = 1/\sqrt{n}$, we find that $h(\mathbf{x}) = \log \mathbf{x}$.

One situation in which this transformation is useful is that where one knows that the variance of $\mathbf{x}$ is the square of the mean, either on theoretical grounds (e.g., $\mathbf{x}$ comes from a gamma distribution in which $\alpha = 1$) or based on empirical evidence. Such empirical evidence can be built up by creating random subsamples of the data, and plotting the sample mean and variance for each of the subsamples with the mean as the abscissa and the variance as the ordinate. If the resulting plot appears to be an increasing function, increasing quadratically, then we would be concerned about the functional dependence of the variance upon the mean and seek out such a variance stabilizing transformation.

Another situation in which this transformation is useful is when $\mathbf{x}$ is the sample variance calculated from a sample of n independent observations drawn from a $N(\mu, \theta)$ distribution. Then $\mathscr{E}\mathbf{x} = \theta$ and $\mathscr{V}\mathbf{x} = 2\theta^2/n$, so that in this case $h(\mathbf{x}) = \log \mathbf{x}/\sqrt{2}$ produces a new random variable whose variance is independent of its expected value.

Theoretical Background

We require a transformation of $\mathbf{x}$ into a new random variable $\mathbf{y}$, say $\mathbf{y} = h(\mathbf{x})$, such that, even though $\mathscr{E}\mathbf{y} = h(\mathscr{E}\mathbf{x})$, $\mathscr{V}\mathbf{y}$ is a constant independent of $\mathscr{E}\mathbf{y}$. To do this, at least approximately, we rely on Taylor's theorem, which says that for any function $h(x)$ possessing first derivatives one can approximate $h(x)$ by selecting a point x_0 and expressing $h(x)$ as

$$h(x) \simeq h(x_0) + h'(x_0)(x - x_0),$$

where $h'(x_0)$ is the derivative of h with respect to x, evaluated at $x = x_0$. If $\mathbf{x}$ is a random variable and $\mathbf{y} = h(\mathbf{x})$, we can take $x_0 = \mathscr{E}\mathbf{x}$ and express $\mathbf{y}$ as

$$\mathbf{y} \simeq h(\mathscr{E}\mathbf{x}) + h'(\mathscr{E}\mathbf{x})(\mathbf{x} - \mathscr{E}\mathbf{x}),$$

so that $\mathscr{E}\mathbf{y} \simeq h(\mathscr{E}\mathbf{x})$ and

$$\mathscr{V}\mathbf{y} \simeq [h'(\mathscr{E}\mathbf{x})]^2 \mathscr{V}\mathbf{x}.$$

Now if $\mathscr{V}\mathbf{y}$ is to approximately equal k^2, an arbitrary constant independent of $\mathscr{E}\mathbf{y}$, the function h must satisfy

$$h'(\mathscr{E}\mathbf{x}) = k/\sqrt{\mathscr{E}\mathbf{x}},$$

or

$$h(\mathscr{E}\mathbf{x}) = k \int \frac{1}{\sqrt{\mathscr{E}\mathbf{x}}} d(\mathscr{E}\mathbf{x}).$$

To make this development of the nature of the appropriate transformation concrete, consider the example where $\mathbf{x}$ is the mean of n independent Bernoulli trials each governed by the same parameter θ. Then $\mathscr{E}\mathbf{x} = \theta$ and $\mathscr{V}\mathbf{x} = \theta(1 - \theta)/n$, and so

$$\begin{aligned} h(\mathscr{E}\mathbf{x}) = h(\theta) &= k \int \frac{1}{\sqrt{\theta(1 - \theta)/n}} d\theta \\ &= k\sqrt{n}\, 2 \sin^{-1} \sqrt{\theta} + C, \end{aligned}$$

where C is an arbitrary constant. We can conventionally take $C = 0$ and set $k = 1/\sqrt{n}$, in which case $h(\theta) = 2\sin^{-1}\sqrt{\theta}$ and hence $\mathbf{y} = h(\mathbf{x}) = 2\sin^{-1}\sqrt{\mathbf{x}}$. From the above, we see that $\mathscr{E}\mathbf{y} \simeq 2\sin^{-1}\sqrt{\theta}$ and $\mathscr{V}\mathbf{y} \simeq 1/n$.

As a second example, suppose $\mathscr{E}\mathbf{x} = \theta$ and $\mathscr{V}\mathbf{x} = \theta/n$ (e.g., if $\mathbf{x}$ is the mean of n independent Poisson variables each with mean θ). Then

$$
\begin{aligned}
h(\theta) &= k\int \frac{1}{\sqrt{\theta/n}}\,d\theta \\
&= k\sqrt{n}2\sqrt{\theta} + C,.
\end{aligned}
$$

so that, if $C = 0$ and $k = 1/\sqrt{n}$, $h(\mathbf{x}) = 2\sqrt{\mathbf{x}}$.

A final example, to show the wide range of applicability of this method, is one where $\mathbf{x}$ is the sample correlation coefficient. Then $\mathscr{E}\mathbf{x} = \rho$, the population correlation coefficient, and $\mathscr{V}\mathbf{x} = (1 - \rho^2)^2/n$, approximately. Then using our usual convention

$$
\begin{aligned}
h(\rho) &= \int \frac{1}{1-\rho^2}\,d\rho \\
&= \frac{1}{2}\log\frac{1+\rho}{1-\rho} \\
&= \tanh^{-1}\rho.
\end{aligned}
$$

The variance stabilizing transformation sometimes has the added property that it also transforms the distribution to one which is more nearly normal. For an in-depth study of conditions under which this occurs, see Efron [1982].

3. Power Transformations (Box–Cox)

We have seen earlier one rationale for selection of a transformation of our data, namely to create a new random variable whose variance is functionally independent of its expected value. Another, perhaps more prevalent, reason for seeking an appropriate transformation is to produce random variables that are more nearly normally distributed. One class of transformations which have received extensive study are the power transformations, usually called Box–Cox transformations because of their intensive study of a variant of the power transformation (see Box and Cox [1964] and Figure 4). Box and Cox studied transformations of $\mathbf{x}$ of the form

$$
\mathbf{y}_\lambda = \mathbf{y}_\lambda(\mathbf{x}) = \begin{cases} (\mathbf{x}^\lambda - 1)/\lambda, & \lambda \neq 0, \\ \log \mathbf{x}, & \lambda = 0, \end{cases}
$$

where $\mathbf{y}_0(\mathbf{x}) = \log \mathbf{x}$ is so defined since $\lim_{\lambda\to 0}(x^\lambda - 1)/\lambda = \log x$.

Note that for this transformation to make sense unequivocally $\mathbf{x}$ must be a positive random variable. And even if $\mathbf{x}$ is positive $\mathbf{y}_\lambda$ will have a normal

BOX & COX

BOX AND COX To play box and cox is to alternate privileges—as if, say, one were to entertain two lovers in turn, neither knowing of the other's existence. The usage arises from the one-act farce *Box and Cox* (1847), by John Maddison Morton (1811–1891). Mrs. Bouncer, a lodging-house keeper, lets her room to Box, a journeyman printer out all night, and Cox, a journeyman hatter out all day. The arrangement is upset when one arrives home out of turn.

Source: Espy [1978]. Reprinted from *O Thou Improper, Thou Uncommon Noun*, illustration © 1978 by Paul Degan, used by permission of Clarkson N. Potter.

Figure 4. Box and Cox.

distribution *exactly* only if $\lambda = 0$ or $1/\lambda$ is an even integer. But $\mathbf{y}_\lambda$ can be *approximately* normally distributed even for other values of λ, and it is the aim of the Box–Cox transformation to find the λ that does this best.

In what follows we will assume either that $\mathbf{x} > 0$ or that there is a known lower bound $-\xi$ on $\mathbf{x}$, and so we will assume that our data have been translated to $\mathbf{x} + \xi$ before determining the appropriate Box–Cox transformation.

(a) Maximum Likelihood

Prescription

Assuming that $\mathbf{y}_\lambda$ is distributed as $N(\mu, \sigma^2)$, we can estimate the best value of the power parameter λ using the maximum likelihood method on the joint density of a sample of n independent $\mathbf{x}_i$'s. For given λ the maximum likelihood estimator of μ is

$$\overline{\mathbf{y}}_\lambda = \sum_{i=1}^{n} \mathbf{y}_{\lambda i}/n,$$

and the maximum likelihood estimator of σ^2 is

$$\mathbf{s}_\lambda^2 = \sum_{i=1}^{n} (\mathbf{y}_{\lambda i} - \overline{\mathbf{y}}_\lambda)^2/n.$$

It can be shown that the log likelihood function, merely as a function of λ, is

$$l(\lambda) = -\frac{n}{2}\log(2\pi) - \frac{n}{2} - \frac{n}{2}\log s_\lambda^2 + (\lambda - 1)\sum_{i=1}^{n} \log x_i.$$

Let $\dot{x} = (\prod_{i=1}^{n} x_i)^{1/n}$, the geometric mean of the x_i's. Then $\sum_{i=1}^{n} \log x_i = n \log \dot{x}$, so that

$$l(\lambda) = -\frac{n}{2}\log(2\pi) - \frac{n}{2} - \frac{n}{2}\log s_\lambda^2 + n(\lambda - 1)\log \dot{x}.$$

Letting $\mathbf{z}_{\lambda i} = \mathbf{y}_{\lambda i}/\dot{\mathbf{x}}^{\lambda - 1}$, we can rewrite $\mathbf{s}_\lambda^2$ as

$$\mathbf{s}_\lambda^2 = \dot{\mathbf{x}}^{2(\lambda-1)} \sum_{i=1}^{n} (\mathbf{z}_{\lambda i} - \overline{\mathbf{z}}_\lambda)^2/n,$$

and thus

$$l(\lambda) = -\frac{n}{2}\log(2\pi) - \frac{n}{2} - \frac{n}{2}\log\left[\sum_{i=1}^{n} (z_{\lambda i} - \overline{z}_\lambda)^2/n\right]$$
$$- n(\lambda - 1)\log \dot{x} + n(\lambda - 1)\log \dot{x},$$

so that the maximum likelihood estimation problem reduces to the minimization with respect to λ of

$$t_\lambda^2 = \sum_{i=1}^{n} (z_{\lambda i} - \overline{z}_\lambda)^2/n.$$

Table 2. One Hundred Independent Gamma (5, 5) Observations.

1	6.48875	51	24.25881
2	7.36181	52	24.73173
3	7.70804	53	24.96917
4	8.11865	54	25.24047
5	8.46150	55	25.24985
6	12.75674	56	25.37080
7	13.08055	57	26.35537
8	13.43281	58	26.47865
9	13.93881	59	26.48096
10	14.22067	60	26.99213
11	14.25665	61	27.52403
12	14.69054	62	28.05616
13	15.02005	63	28.09613
14	15.03517	64	28.29106
15	15.14141	65	28.46719
16	15.31778	66	28.71782
17	15.44101	67	29.20037
18	15.52946	68	29.56230
19	15.75355	69	29.73896
20	15.76179	70	30.30640
21	15.78501	71	30.32239
22	16.24040	72	30.57425
23	16.62110	73	31.49794
24	16.65918	74	31.84531
25	16.71427	75	31.85266
26	17.42898	76	32.70566
27	18.76179	77	32.93997
28	18.79385	78	33.31204
29	19.07174	79	33.35007
30	19.32876	80	33.43215
31	19.62610	81	33.71671
32	19.78951	82	34.05206
33	19.96001	83	34.25993
34	20.04533	84	34.44229
35	20.14921	85	35.43620
36	20.62760	86	35.90700
37	20.99996	87	36.77312
38	21.00931	88	36.86588
39	21.24029	89	37.67851
40	21.47477	90	37.84356
41	21.81918	91	38.00723
42	22.05309	92	40.46576
43	22.33860	93	40.72330
44	22.87242	94	43.38677
45	22.94075	95	45.75185
46	23.23921	96	45.79899
47	23.45028	97	46.53636
48	23.74068	98	47.08834
49	23.81423	99	52.63478
50	24.15124	100	52.84011

Table 3. Trial and Error Iterations to Find λ.

0.4	0.48875	0.48896
0.95609336E + 02	0.95418356E + 02	0.95418353E + 02
0.5	0.4889	0.48897
0.95421189E + 02	0.95418354E + 02	0.95418353E + 02
0.6	0.48895	0.48898
0.95707911E + 02	0.95418354E + 02	0.95418353E + 02
0.45	0.48885	0.48899
0.95454870E + 02	0.95418354E + 02	0.95418353E + 02
0.475	0.4888	0.489
0.95423078E + 02	0.95418355E + 02	0.95418353E + 02
0.4875	0.48891	0.4891
0.95418412E + 02	0.95418354E + 02	0.95418353E + 02
0.4825	0.48892	0.4892
0.95419384E + 02	0.95418354E + 02	0.95418353E + 02
0.4925	0.48893	0.4893
0.95418632E + 02	0.95418354E + 02	0.95418354E + 02
0.49	0.48894	0.48925
0.95418373E + 02	0.95418354E + 02	0.95418354E + 02
0.4885	0.48895	0.48921
0.95418361E + 02	0.95418354E + 02	0.95418354E + 02

In summary, one can, at least by trial-and-error, calculate $\mathbf{t}_\lambda^2$ for a variety of values of λ and find the value of λ which minimizes $\mathbf{t}_\lambda^2$. We illustrate this with an example, a data set consisting of 100 observations from the gamma distribution

$$f_{\mathbf{x}}(x) = \frac{x^{\alpha-1}}{\Gamma(\alpha)\beta^\alpha} e^{-x/\beta}$$

with $\alpha = \beta = 5$. The ordered data are given in Table 2. We note that $\dot{x} = 23.35731$. The trial-and-error steps to estimate λ are presented in Table 3, from which we conclude that $\lambda = 0.489$ minimizes $\mathbf{t}_\lambda^2$. One can see the effect of the transformation by comparing Figures 5 and 6, the original, and the Box–Cox transformed data plotted on normal probability paper. The original data had a skewness of 0.5043, and the transformation produced a more symmetric distribution, with skewness of -0.018335. (The kurtosis increased, but from -0.11665 to -0.3134, still insignificant.)

As an alternative to this trial-and-error procedure, one can attempt to solve for λ the defining equation for the maximum likelihood estimate:

$$0 = \frac{\partial l(\lambda)}{\partial \lambda} = \sum_{i=1}^{n} \log x_i - \frac{\sum_{i=i}^{n} (y_{\lambda i} - \bar{y}_\lambda)}{\lambda^2 s_\lambda^2} [(1 + \lambda y_{\lambda i}) \log(1 + \lambda y_{\lambda i}) - \lambda y_{\lambda i}].$$

A computer program that uses Newton's method to solve this equation

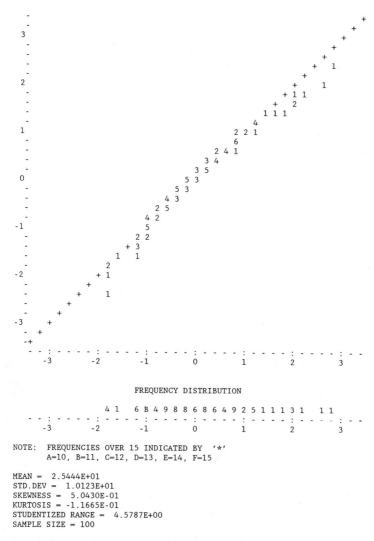

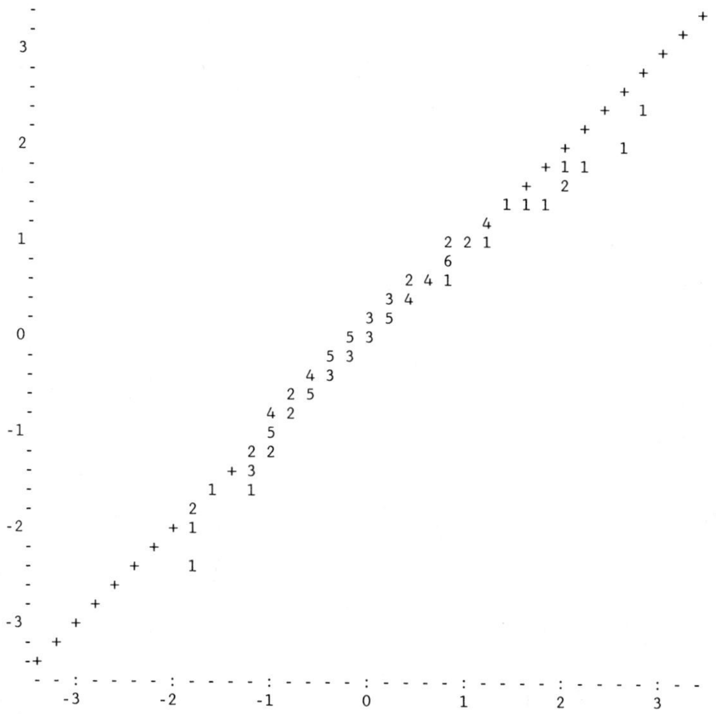


```
                          FREQUENCY DISTRIBUTION

               4 1   6 B 4 9 8 8 6 8 6 4 9 2 5 1 1 1 3 1   1 1
 - - : - - - - - : - - - - : - - - - - : - - - - - : - - - - - : - - - - - : - -
    -3          -2         -1          0           1           2           3

NOTE:   FREQUENCIES OVER 15 INDICATED BY  '*'
        A=10, B=11, C=12, D=13, E=14, F=15

MEAN =   2.5444E+01
STD.DEV =   1.0123E+01
SKEWNESS =   5.0430E-01
KURTOSIS = -1.1665E-01
STUDENTIZED RANGE =   4.5787E+00
SAMPLE SIZE = 100
```

Figure 5. Normal Probability Plot of Gamma (5, 5) Data.

iteratively is given in Appendix I. The iterative steps taken by this program to estimate λ for the gamma distributed data are given in Table 4. Note that the solution is $\lambda = 0.489$, as above.

Suppose that instead we are in a regression context and we wish to find the appropriate power transformation of the dependent variable to make the residuals normal. Similar reasoning to the above dictates that the role of ns_λ^2 is to be taken by the sum of squares of the residuals of the regression of the transformed dependent variable onto the independent variables. (As has been noted by Schlesselman [1971], the Box–Cox transformation does not produce

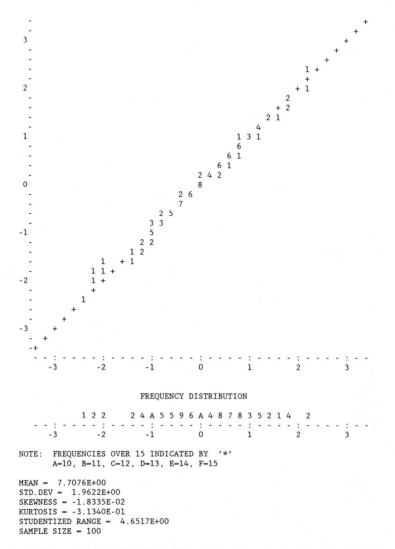

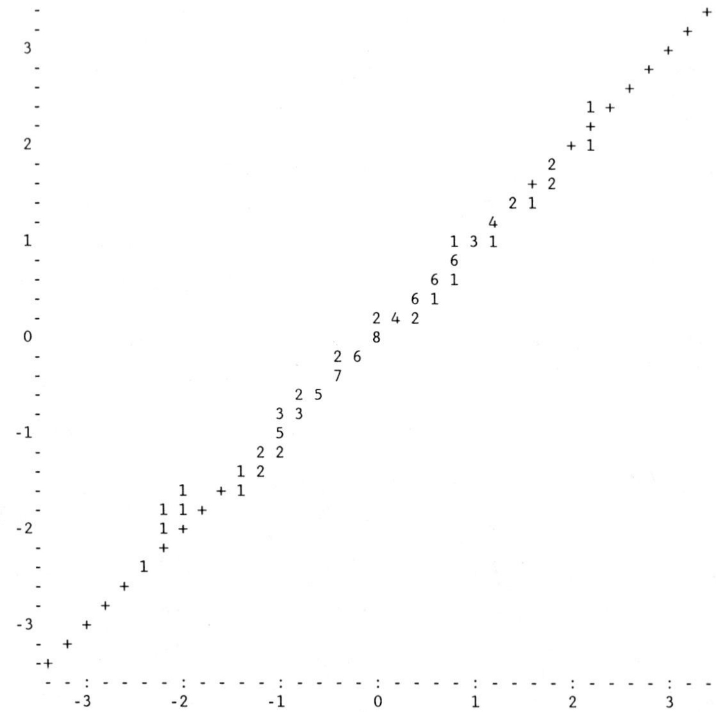

```
                              FREQUENCY DISTRIBUTION

              1 2 2    2 4 A 5 5 9 6 A 4 8 7 8 3 5 2 1 4   2
 - - : - - - - - : - - - - - : - - - - - : - - - - - : - - - - - : - - - - - : - - -
    -3          -2          -1           0           1           2           3

NOTE:  FREQUENCIES OVER 15 INDICATED BY  '*'
       A=10, B=11, C=12, D=13, E=14, F=15

MEAN =  7.7076E+00
STD.DEV =  1.9622E+00
SKEWNESS = -1.8335E-02
KURTOSIS = -3.1340E-01
STUDENTIZED RANGE =  4.6517E+00
SAMPLE SIZE = 100
```

Figure 6. Normal Probability Plot of Box–Cox Transformed Gamma (5, 5) data.

Table 4. Iterations of Newton's Method to Find λ.

Trial λ	Derivative of log likelihood	Second derivative
$0.250000E+00$	$0.620837E+01$	$-0.269809E+02$
$0.480103E+00$	$0.224628E+00$	$-0.250214E+02$
$0.489080E+00$	$0.335693E-03$	$-0.249447E+02$
$0.489093E+00$	$-0.762939E-05$	$-0.249440E+02$

Table 5. Cycles to Failure of Worsted Yarn: 3^3 Factorial Experiment.

Factor levels			Cycles to failure, y
x_1	x_2	x_3	
−1	−1	−1	674
−1	−1	0	370
−1	−1	+1	292
−1	0	−1	338
−1	0	0	266
−1	0	+1	210
−1	+1	−1	170
−1	+1	0	118
−1	+1	+1	90
0	−1	−1	1,414
0	−1	0	1,198
0	−1	+1	634
0	0	−1	1,022
0	0	0	620
0	0	+1	438
0	+1	−1	442
0	+1	0	332
0	+1	+1	220
+1	−1	−1	3,636
+1	−1	0	3,184
+1	−1	+1	2,000
+1	0	−1	1,568
+1	0	0	1,070
+1	0	+1	566
+1	+1	−1	1,140
+1	+1	0	884
+1	+1	+1	360

SOURCE: Box and Cox [1964]. Reprinted from the *Journal of the Royal Statistical Society*, Vol. 26, with permission from the Royal Statistical Society.

an $l(x)$ which is invariant to changes of scale of x nor whose maximization is equivalent to minimization of the residual sum of squares unless the regression has a nonzero intercept term.) This is illustrated by a consideration of the data of Table 4 of Box and Cox [1964], reproduced as Table 5 herein. (Box and Cox analyzed the data after dividing y by 1,000, but failed to mention that in their paper.) Consider, for example, the trial value $\lambda = 0.2$. The following annotated IDA session illustrates the procedure for determining ns_λ^2.

```
COMMAND> **** LOGE ****                      These two instructions are used to
COLUMN 5 = "LOGE" OF COLUMN 1                determine ẋ, the geometric mean
LOGY                       Y

COMMAND> **** MEAN ****

VARIABLE        MEAN        STD. DEV.
Y              0.861333     0.882075
X1             0.000000     0.832050
X2             0.000000     0.832050
X3             0.000000     0.832050
LOGY          -0.573091     0.944696     ẋ = exp(-0.573091) = 0.5637801

BASED ON     27 ACTIVE ROWS:
FIRST, LAST ACTIVE ROWS ARE 1,     27

COMMAND> **** TGEN ****

YLAM = (Y**.2-1)/.2
NEW DATA IN COL 6                            here we transform x using λ = 0.2

COMMAND> **** MULC ****

CONSTANT IS 0.63225                          that is, 1/ẋ ** (0.2 - 1)
COLUMN 7 = "MULC" OF COLUMN 6
ZLAM                       YLAM

COMMAND> **** REGR ****

  DEPENDENT VARIABLE     :  ZLAM
INDEPENDENT VARIABLE   1:  X1
INDEPENDENT VARIABLE   2:  X2
INDEPENDENT VARIABLE   3:  X3

COMMAND> **** ANOV ****

  SOURCE           SS         DF       MS          F
REGRESSION   7.40652E + 00    3   2.36884E + 00  137.98
RESIDUALS    4.11520E - 01   23   1.78922E - 02          <----and here is s²_λ
TOTAL        7.81804E + 00   26   3.00694E - 01
```

Theoretical Background

We already saw that $\mathbf{y} = \log \mathbf{x}$ is the variance stabilizing transformation if $\mathscr{V}\mathbf{x}$ is proportional to $(\mathscr{E}\mathbf{x})^2$, and $\mathbf{y} = \sqrt{\mathbf{x}}$ is the variance stabilizing transformation if $\mathscr{V}\mathbf{x}$ is proportional to $\mathscr{E}\mathbf{x}$. What relationship does $\mathscr{V}\mathbf{x}$ have to have to $\mathscr{E}\mathbf{x}$ for the power transformation $\mathbf{y} = \mathbf{x}^\lambda$ to be a variance stabilizing transformation? Clearly, the relationship must be

$$\mathscr{V}\mathbf{x} \propto \left[\left. \frac{\partial y}{\partial x} \right|_{x=\mathscr{E}\mathbf{x}} \right]^{-2},$$

or, in our case

$$\mathscr{V}\mathbf{x} \propto [\lambda(\mathscr{E}\mathbf{x})^{\lambda-1}]^{-2}$$
$$\propto (\mathscr{E}\mathbf{x})^{2(1-\lambda)}.$$

These Box–Cox transformations have the property that, for all λ,

$$\frac{\partial y_\lambda}{\partial x} = x^{\lambda-1}.$$

This property is quite a mathematical convenience, because for $\mathbf{y}_\lambda$ to be normally distributed $\mathbf{x}$ has to have probability density function

$$\begin{aligned} f_{\mathbf{x}}(x) &= \frac{1}{\sqrt{2\pi}\sigma} e^{-(y_\lambda-\mu)^2/2\sigma^2} \frac{\partial y_\lambda}{\partial x} \\ &= \begin{cases} \dfrac{x^{\lambda-1}}{\sqrt{2\pi}\sigma} e^{-([x^\lambda-1)/\lambda]-\mu)^2/2\sigma^2}, & \lambda \neq 0, \\ \dfrac{1}{\sqrt{2\pi}\sigma x} e^{-(\log x-\mu)^2/2\sigma^2}, & \lambda = 0. \end{cases} \end{aligned}$$

From this one can derive both the reduced log likelihood function $l(\lambda)$ and the defining equation for the maximum likelihood estimate $\partial l(\lambda)/\partial\lambda = 0$ given above.

Bickel and Doksum [1981] investigated the effect on estimates of μ and σ^2 of transforming the $\mathbf{x}_i$ using an estimate of λ based on the self-same $\mathbf{x}_i$'s. They note first of all that for small σ the model underlying the Box–Cox transformation is nearly unidentifiable. For moderate or large values of σ, especially as μ/σ is small, λ is relatively accurately estimable. However, when the $\mathbf{x}$'s are modeled as dependent variables in a regression or by an analysis of variance model, the effect on the other parameter estimates of first estimating λ can be large regardless of the value of σ. Even in a simple example where λ is estimated but is truly equal to zero, the variance of the estimate of μ based on the estimated λ is asymptotically (see Hinkley [1975])

$$\frac{\sigma^2}{n}\left[1 + \frac{1}{6}\left(1 + \left(\frac{\mu}{\sigma}\right)^2\right)^2\right],$$

whereas the estimate of μ using the true value of λ has variance σ^2/n. Hinkley and Runger [1984] argue, though, that inference about μ should be made with the scale determined by the estimate of λ regarded as fixed. The discussion which follows this paper, in particular that of Rubin, gives reasons for rejecting this point of view (see also Box and Cox [1982]).

The Box–Cox approach (a) disregards the form of the distribution of $\mathbf{x}$, and (b) requires that the transformed distribution *is* normal. Another view of these power transformations is taken in Hernandez and Johnson [1980]. They begin with the distribution of $\mathbf{x}$ and ask for the power transformation of $\mathbf{x}$ whose distribution has minimum Kullback–Liebler distance from the normal distribution. Let $f_{\mathbf{y}_\lambda}(y)$ be the density function of $\mathbf{y}_\lambda$, and let $\phi_{\mu,\sigma^2}(y)$ denote the density function of a normally distributed random variable with mean μ and variance σ^2. The Kullback–Liebler distance between $f_{\mathbf{y}_\lambda}$ and ϕ_{μ,σ^2} is given by

$$I[f_{\mathbf{y}_\lambda}, \phi_{\mu,\sigma^2}] = \int f_{\mathbf{y}_\lambda}(y) \log\left\{\frac{f_{\mathbf{y}_\lambda}(y)}{\phi_{\mu,\sigma^2}(y)}\right\} dy,$$

and one might seek to find values of μ, σ^2, and λ minimizing $I[f_{\mathbf{y}_\lambda}, \phi_{\mu,\sigma^2}]$. Hernandez and Johnson show that the Box–Cox estimates of μ, σ^2, and λ converge to the parameters minimizing the Kullback–Liebler distance between $f_{\mathbf{y}_\lambda}$ and ϕ_{μ,σ^2}.

In particular, suppose $\mathbf{x}$ has density $f_{\mathbf{x}}(x)$. Then the minimizing μ and σ^2 are given by

$$\mu_\lambda^* = \int y_\lambda(x) f_{\mathbf{x}}(x)\, dx,$$

$$(\sigma_\lambda^2)^* = \int y_\lambda^2(x) f_{\mathbf{x}}(x)\, dx - (\mu_\lambda^*)^2,$$

and the minimizing λ is found by minimizing

$$G(x) = \frac{1}{2}[\log(2\pi) + 1)] + \int_{-\infty}^{\infty} \log[y_\lambda(x)] f_{\mathbf{x}}(x)\, dx + (1 - \lambda) \int_{-\infty}^{\infty} \log x f_{\mathbf{x}}(x)\, dx + \frac{1}{2}\log(\sigma_\lambda^2)^*.$$

For example, if $f_{\mathbf{x}}$ is the $\Gamma(\alpha, \beta)$ density, then

$$\begin{aligned} G(\lambda) &= \tfrac{1}{2}[\log(2\pi) + 1] - 2\log[\Gamma(\alpha)] + \alpha[\psi(\alpha) - 1] - \lambda\psi(\alpha) \\ &\quad + \tfrac{1}{2}\log\left\{\frac{\Gamma(\alpha)\Gamma(2\lambda + \alpha) - [\Gamma(\lambda + \alpha)]^2}{\lambda^2}\right\} \\ &\simeq G(\tfrac{1}{3}) + \frac{3}{4\alpha}(\lambda - \tfrac{1}{3})^2 + O(\alpha^{-2}), \end{aligned}$$

where ψ is the digamma function $\psi(x) = d\log(\Gamma(x))/dx$, so that $\lambda = \frac{1}{3}$ is the best power transformation for this distribution. Note that this differs from the $\hat{\lambda} = 0.489$ found based on 100 observations from this distribution.

(b) Hinkley Estimation Procedure

Hinkley [1975] noted that a paramount property of the normal distribution, which the power transformation wishes to achieve, is the property of symmetry. Suppose the $\mathbf{x}$'s are ordered so that $\mathbf{x}_1 \le \mathbf{x}_2 \le \cdots \le \mathbf{x}_n$. Then, since the power transformation is a monotonic transformation, the $\mathbf{y}_\lambda$'s also satisfy $\mathbf{y}_{\lambda 1} \le \mathbf{y}_{\lambda 2} \le \cdots \le \mathbf{y}_{\lambda n}$. By symmetry,

$$\tilde{\mathbf{y}}_\lambda - \mathbf{y}_{\lambda r} = \mathbf{y}_{\lambda, n-r+1} - \tilde{\mathbf{y}}_\lambda,$$

i.e., the rth and $(n - r + 1)$st ordered transformed data points should be equidistant from the sample median, $\tilde{\mathbf{y}}_\lambda$. This equation, rearranged, is (when

Table 6. Hinkley Estimates of λ.

Row	$x_r/\tilde{x}$	$x_{n-r+1}/\tilde{x}$	Lambda
1	0.26807	2.18302	0.50803
2	0.30414	2.17454	0.44286
3	0.31845	1.94540	0.61192
4	0.33541	1.92259	0.59904
5	0.34958	1.89213	0.60180
6	0.52703	1.89018	0.05415
7	0.54041	1.79247	0.11296
8	0.55496	1.68243	0.23723
9	0.57586	1.67179	0.15618
10	0.58751	1.57022	0.34714
11	0.58900	1.56346	0.35929
12	0.60692	1.55664	0.27315
13	0.62053	1.52307	0.29745
14	0.62116	1.51923	0.30690
15	0.62555	1.48345	0.41584
16	0.63283	1.46400	0.44934
17	0.63793	1.42294	0.59712
18	0.64158	1.41541	0.61154
19	0.65084	1.40682	0.58718
20	0.65118	1.39296	0.67180
21	0.65214	1.38121	0.74262
22	0.67095	1.37782	0.59883
23	0.68668	1.37624	0.48678
24	0.68825	1.36087	0.55129
25	0.69053	1.35119	0.60483
26	0.72006	1.31595	0.57664
27	0.77512	1.31565	0.04389
28	0.77644	1.30130	0.06712
29	0.78792	1.26314	0.18312
30	0.79854	1.25273	0.13673
31	0.81083	1.25207	0.05603
32	0.81758	1.22863	0.11008
33	0.82462	1.22133	0.09319
34	0.82815	1.20638	0.18168
35	0.83244	1.18644	0.46501
36	0.85220	1.17609	0.15723
37	0.86759	1.16881	0.06481
38	0.86797	1.16076	0.10323
39	0.87752	1.15910	0.05511
40	0.88720	1.13712	0.09676
41	0.90143	1.11515	0.14542
42	0.91110	1.09403	0.49993
43	0.92289	1.09393	0.09737
44	0.94495	1.08884	0.03462
45	0.94777	1.04816	2.45260
46	0.96010	1.04317	0.43607
47	0.96882	1.04278	0.09637
48	0.98082	1.03157	0.08480
49	0.98385	1.02176	0.18773
50	0.99778	1.00222	0.50000

$\lambda \neq 0$) of the form

$$\left[\frac{\mathbf{x}_r}{\tilde{\mathbf{x}}}\right]^\lambda + \left[\frac{\mathbf{x}_{n-r+1}}{\tilde{\mathbf{x}}}\right]^\lambda = 2$$

for all r, where $\tilde{\mathbf{x}}$ is the median of the $\mathbf{x}$'s.

One can use this relation as the basis for a quick-and-dirty procedure to determine λ. First, $\lambda = 0$ always satisfies this, so we exclude $\lambda = 0$ as a possibility. When $\lambda = 0$, the relation

$$\frac{\tilde{\mathbf{x}}}{\mathbf{x}_r} = \frac{\mathbf{x}_{n-r+1}}{\tilde{\mathbf{x}}}$$

should hold for all r. This, then, is a way of checking the appropriateness of the logarithmic (i.e., $\lambda = 0$) transformation.

Second, if this relation holds for each r, it holds for all r, and so one might want to solve this equation for λ for a number of different values of r and examine the various estimates of λ, possibly taking their average. Finally, one might consider a weighted combination of m such estimating equations, e.g., an equation of the form

$$\sum_{i=i}^{m} w_i \left[\left(\frac{\mathbf{x}_i}{\tilde{\mathbf{x}}}\right)^\lambda + \left(\frac{\mathbf{x}_{n-i+1}}{\tilde{\mathbf{x}}}\right)^\lambda\right] = 2 \sum_{i=1}^{m} w_i = 2,$$

where $\sum_{i=1}^{m} w_i = 1$. In an example, Hinkley has shown that if $i/n > 0.05$ for all values of i (i.e., $n > 20$) then little improvement, measured by the variance of the estimate of λ, can be obtained by taking $m > 1$. He does not study properties of the average estimator of λ.

Table 6 presents the $\mathbf{x}_r/\tilde{\mathbf{x}}$ and $\mathbf{x}_{n-r+1}/\tilde{\mathbf{x}}$ based on the 100 gamma distributed observations, along with the Hinkley estimate of λ associated with each $r = 1, \ldots, 50$. One notes how variable the estimates are, with the estimates closest to the maximum likelihood estimates $\hat{\lambda} = 0.489$ associated with lower values of r, i.e., ones wherein $\mathbf{x}_r/\tilde{\mathbf{x}}$ is quite distinct from $\mathbf{x}_{n-r+1}/\tilde{\mathbf{x}}$ and wherein both differ from 1.

(c) Graphic Procedure

Prescription

The graphic procedure of Emerson and Stoto is motivated by the following reasoning. If one plots $(\mathbf{x}_r + \mathbf{x}_{n-r+1})/2 - \tilde{\mathbf{x}}$ against $\{[\mathbf{x}_{n-r+1} - \tilde{\mathbf{x}}]^2 + [\mathbf{x}_r - \tilde{\mathbf{x}}]^2\}/4\tilde{\mathbf{x}}$ and the plot is linear, its slope will be $1 - \lambda$. This suggestion by Emerson and Stoto [1982] gives a graphic procedure for estimating λ. Cameron [1984] (see also Emerson and Stoto [1984]) points out that this series expansion is not very accurate for highly skewed distributions, especially for quantiles far from the median.

We illustrate this procedure on our gamma distributed data. Table 7

Table 7. Computations of the Emerson–Stoto Variables.

Row	$x_r - \tilde{x}$	$x_{n-r+1} - \tilde{x}$	LIN	QUAD
1	-17.71627	28.63508	5.45940	11.71073
2	-16.84321	28.42975	5.79327	11.27808
3	-16.49698	22.88331	3.19316	8.21933
4	-16.08637	22.33133	3.12248	7.82338
5	-15.74353	21.59396	2.92522	7.37613
6	-11.44828	21.54682	5.04927	6.14882
7	-11.12448	19.18174	4.02863	5.07842
8	-10.77222	16.51828	2.87303	4.01667
9	-10.26621	16.26073	2.99726	3.81952
10	-9.98435	13.80220	1.90892	2.99719
11	-9.94837	13.63853	1.84508	2.94339
12	-9.51449	13.47348	1.97950	2.80996
13	-9.18498	12.66086	1.73794	2.52697
14	-9.16985	12.56809	1.69912	2.49993
15	-9.06361	11.70198	1.31918	2.26281
16	-8.88724	11.23118	1.17197	2.11859
17	-8.76401	10.23726	0.73662	1.87574
18	-8.67557	10.05491	0.68967	1.82159
19	-8.45148	9.84704	0.69778	1.73922
20	-8.44324	9.51169	0.53423	1.67073
21	-8.42001	9.22712	0.40355	1.61161
22	-7.96462	9.14505	0.59021	1.51897
23	-7.58393	9.10701	0.76154	1.45067
24	-7.54585	8.73495	0.59455	1.37615
25	-7.49076	8.50063	0.50494	1.32588
26	-6.77605	7.64763	0.43579	1.07830
27	-5.44323	7.64029	1.09853	0.90893
28	-5.41118	7.29291	0.94087	0.85176
29	-5.13329	6.36923	0.61797	0.69115
30	-4.87627	6.11736	0.62055	0.63210
31	-4.57892	6.10137	0.76122	0.60105
32	-4.41551	5.53394	0.55921	0.51767
33	-4.24501	5.35728	0.55613	0.48255
34	-4.15969	4.99534	0.41782	0.43644
35	-4.05581	4.51279	0.22849	0.38024
36	-3.57743	4.26217	0.34237	0.31981
37	-3.20507	4.08604	0.44049	0.27854
38	-3.19571	3.89110	0.34769	0.26186
39	-2.96474	3.85114	0.44320	0.24397
40	-2.73026	3.31901	0.29437	0.19077
41	-2.38584	2.78711	0.20063	0.13902
42	-2.15194	2.27593	0.06200	0.10133
43	-1.86643	2.27362	0.20360	0.08937
44	-1.33261	2.15035	0.40887	0.06610
45	-1.26428	1.16578	-0.04925	0.03055
46	-0.96582	1.04483	0.03951	0.02091
47	-0.75475	1.03544	0.14035	0.01696
48	-0.46434	0.76414	0.14990	0.00826
49	-0.39079	0.52671	0.06796	0.00444
50	-0.05379	0.05379	0.00000	0.00006

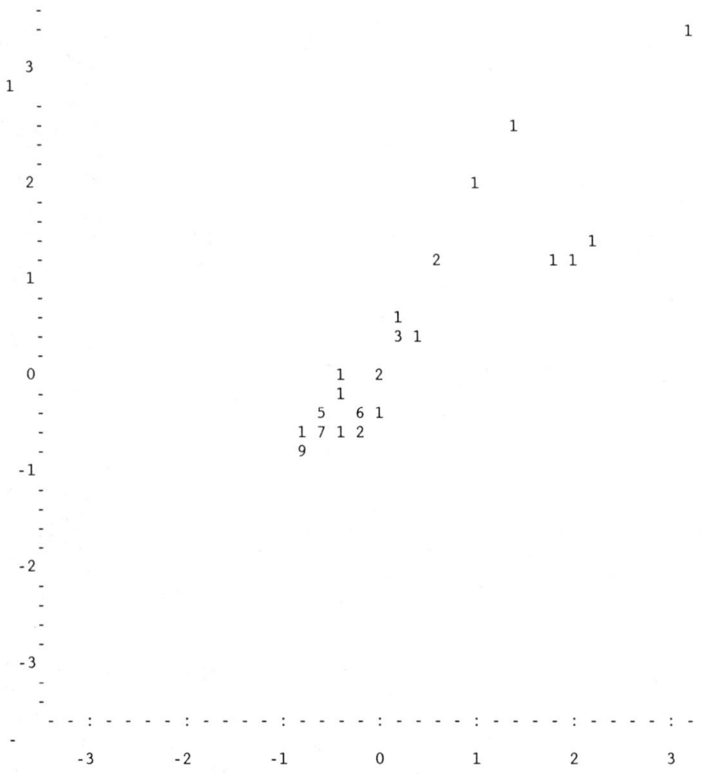

	MEAN	STD.DEV.
VERT. VAR.	1.2389E+00	1.4503E+00
HORIZ. VAR.	2.1275E+00	2.8277E+00
SAMPLE SIZE =	50	

Figure 7. Scatter Plot of 50 Standardized Values of LIN versus QUAD.

provides the data, with FIRST denoting $\mathbf{x}_r - \tilde{\mathbf{x}}$, LAST denoting $\mathbf{x}_{n-r+1} - \tilde{\mathbf{x}}$, LIN denoting $(\mathbf{x}_r + \mathbf{x}_{n-r+1})/2 - \tilde{\mathbf{x}}$, and QUAD denoting

$$\{[\mathbf{x}_{n-r+1} - \tilde{\mathbf{x}}]^2 + [\mathbf{x}_r - \tilde{\mathbf{x}}]^2\}/4\tilde{\mathbf{x}}.$$

Figure 7 depicts the scatter plot of LIN versus QUAD. The slope of the regression of LIN on QUAD is 0.482, yielding an estimate of λ of $1 - 0.482 = 0.518$. Thus we have the pair of variables to be plotted expressed as being in the ratio $1 - \lambda$.

Theoretical Background

Expanding $\mathbf{x}_r^\lambda$ and $\mathbf{x}_{n-r+1}^\lambda$ around $\tilde{\mathbf{x}}$ in second-order Taylor series yields the equations

$$\mathbf{x}_r^\lambda \approx \tilde{\mathbf{x}}^\lambda + \lambda\tilde{\mathbf{x}}^{\lambda-1}(\mathbf{x}_r - \tilde{\mathbf{x}}) + \frac{\lambda(\lambda-1)}{2}\tilde{\mathbf{x}}^{\lambda-2}(\mathbf{x}_r - \tilde{\mathbf{x}})^2,$$

$$\mathbf{x}_{n-r+1}^{\lambda} \approx \tilde{\mathbf{x}}^{\lambda} + \lambda\tilde{\mathbf{x}}^{\lambda-1}(\mathbf{x}_{n-r+1} - \tilde{\mathbf{x}}) + \frac{\lambda(\lambda-1)}{2}\tilde{\mathbf{x}}^{\lambda-2}(\mathbf{x}_{n-r+1} - \tilde{\mathbf{x}})^2,$$

or

$$\frac{\mathbf{x}_r + \mathbf{x}_{n-r+1}}{2} - \tilde{\mathbf{x}} \approx (1-\lambda)\frac{[\mathbf{x}_{n-r+1} - \tilde{\mathbf{x}}]^2 + [\mathbf{x}_r - \tilde{\mathbf{x}}]^2}{4\tilde{\mathbf{x}}}.$$

4. Letter-Values and Boxplots

One of the aims of exploratory data analysis is to develop data description techniques for ascertaining normality of the distribution underlying a data set and, if not normal, determining the appropriate power transformation to transform the data to normality. To this end, Tukey [1977] invented a set of descriptive statistics called "letter-values" and a pictorial representation of the distribution of the data, simpler to look at and to construct than the sample histogram, called the "boxplot".

First we introduce the concept of "depth" of a data value. Suppose our n observations are ordered from lowest to the highest value. The *depth* of the ith datum is the smaller of i and $n + 1 - i$. We define the M-value or *median* of our data as the $(n + 1)/2$th observation when n is odd and as the average of the $n/2$ and $(n + 2)/2$th observation when n is even. The M-value splits our data set into two halves, of size $(n - 1)/2$ if n is odd and size $n/2$ if n is even. Thus the depth of the median is $(n + 1)/2$, where $(n + 1)/2$ is to be interpreted as halfway between the two data values whose depths are $(n + 1)/2 - \frac{1}{2}$ when n is even.

Now consider each of these halves of the data set separately. We can find the "middle" of each of these sets over again using the above definition. These values are called H-values or *hinges*, and are similar to quartiles in that they (along with the M-value) split the data set into four roughly equally sized data sets. More precisely, the depth of each of the hinges is given by $([(n + 1)/2] + 1)/2$, where $[x]$ is the greatest integer in x.

Similarly, we can take each of the outer quarters of the data set, i.e., the subset from the smallest (largest) observation to the nearest hinge, and find the median of each of these subsets. These values are called E-values or *eighths*. Again, more precisely, the depth of the eighths is given by ([depth of the hinges] + 1)/2.

This process can be continued many more times, each time producing a median of the associated data subset. These medians have not been dubbed with names; instead they are called D-values, C-values, B-values, A-values, Z-values, Y-values, X-values, and so on. In general, the depth of a letter-value is given by ([depth of the predecessor letter value] + 1)/2.

Aside from the median, each of the letter-values is associated with two values. We can define a *letter-spread* as the absolute difference between the two letter-values. Now we can compare these letter-spreads with their theoretical values if the data were from an infinite standard normally distributed population, namely:

Letter	Theoretical letter-spread
H	1.349
E	2.301
D	3.068
C	3.726
B	4.308
A	4.836
Z	5.320

If the ratio of the letter-spreads to the theoretical letter-spread are all approximately equal, then we can say that the data come from a normal distribution. If not, then, since the power transformation is a monotonic transformation, each letter-value of a power-transformed data set will equal the power-transform of the letter-values of the original data set. Thus one can use these letter-values and their associated letter-spreads to ascertain the appropriate power transformation to make the data set more normally distributed.

To illustrate these concepts, consider our example data of Table 2. Following are the depths, letter-values, letter-spreads, and ratio of letter-spread to theoretical letter-spread for this data set:

	Depth	Letter-values		Letter-spread	Ratio
M	101/2 = 50.5	24.205025		—	—
H	50/2 = 25.5	17.071625	32.27916	15.207535	11.273
E	26/2 = 13	15.02005	36.86588	19.84583	8.625
D	14/2 = 7	13.08055	43.38677	30.30622	9.878
C	8/2 = 4	8.11865	46.53636	38.41771	10.311
B	5/2 = 2.5	7.534925	49.86156	42.326635	9.825

Each of these ratios is, in the case of normally distributed data, an estimate of σ. (Recall that $\sigma = \sqrt{125} = 11.18$ in this example.) Since these ratios are somewhat variable, one might want to find some power transformation to reduce the variability of these ratios. For example, the Box–Cox power 0.489 produces the following:

		Letter-values		Letter-spread	Ratio
	H	4.00481	5.46843	1.46362	1.085
	E	3.76177	5.83552	2.07375	0.901
Power = 0.489	*D*	3.51585	6.31928	2.80343	0.914
	C	2.78444	6.53959	3.75515	1.008
	B	2.68467	6.76406	4.07939	0.947

Clearly the ratio is less variable.

The boxplot is a pictorial way of exhibiting the median, hinges, and extreme observations. First, a "box" or rectangle is created with sides at the hinges. A vertical line is drawn through the rectangle at the median. A line perpendicular to the hinge line, called a "whisker", is drawn from the hinge line to the largest (smallest) data point within 1.5 times the H-spread. Finally, "notches" are placed symmetrically around the median at values given by median $\pm$ 1.58 $\times$ H-spread/$\sqrt{n}$. The notches are used in informal comparisons of boxplots. Two boxplots are said to be significantly different if the intervals between the notches of the two plots do not overlap.

In our example, the whiskers end at the largest data points within upper hinge + 1.5 $\times$ H-spread (32.27916 + 1.5 $\times$ 15.207535 = 55.09046) and lower hinge − 1.5 $\times$ H-spread (17.071625 − 1.5 $\times$ 15.207535 = −5.73968), i.e., to the endpoints of the data. The notches are at 24.205025 $\pm$ 1.58 $\times$ 15.207535/$\sqrt{100}$ = (21.80223, 26.60782). The boxplot then looks like

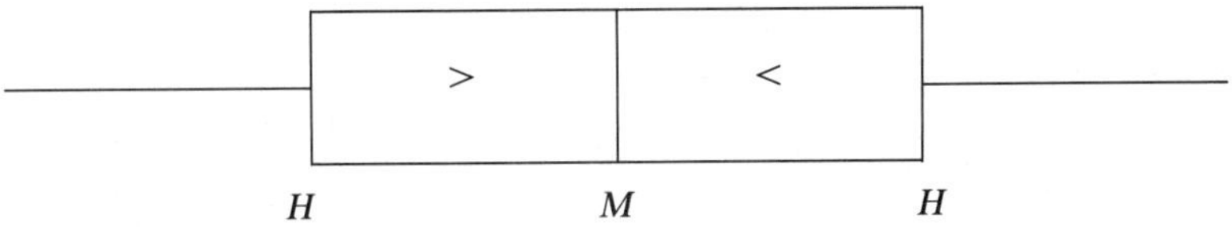

5. Power Transformations of Regression Independent Variables

Prescription

Suppose that our model is

$$y = \alpha + \beta x^{\gamma},$$

but we do not know the appropriate value of γ. Based on a suggestion of Box and Tidwell [1962], following is a procedure for determining γ.

Step 0. Select an initial value γ_0 of γ.

Step 1. Regress y on x^{γ_0} to obtain an estimated slope $\hat{\beta}_0$.

Step 2. Regress y on the two independent variables x^{γ_0} and $x^{\gamma_0} \log x$ to obtain $\hat{\delta}_0$ as coefficient of $x^{\gamma_0} \log x$.

Step 3. Calculate $\gamma_1 = \gamma_0 + \hat{\delta}_0/\hat{\beta}_0$.

Step 4. If $\hat{\delta}_0/\hat{\beta}_0$ is small, terminate the procedure. Otherwise, go back to Step 0, replacing γ_0 with γ_1 in the procedure, and continue the procedure.

For example, suppose our data are as follows:

x	y
1	11.000
2	11.414
3	11.732
4	12.000
5	12.236

Let $\gamma_0 = 1$. A regression of y on x^{γ_0} produces $\hat{\beta}_0 = 0.30580$; a regression of y on x^{γ_0} and $x^{\gamma_0} \log x$ produces $\hat{\delta}_0 = -0.15914$. Thus $\gamma_1 = 1 - 0.15914/0.30580 = 0.52041$. Now a regression of y on x^{γ_1} produces $\hat{\beta}_1 = 0.94300$; a regression of y on $x^{\gamma_1} \log x$ produces $\hat{\delta}_1 = -0.018897$. Thus $\gamma_2 = 0.52041 - 0.018897/0.94300 = 0.50037$. If we regress y on x^{γ_2} we obtain $\hat{\beta}_2 = 0.099894$ and an $R^2 = 1$, so we can no longer proceed.

It is transparent from the example that the true relation is $y = 10 + \sqrt{x}$, i.e., the true value of γ is 0.5. We see how this procedure has in short order converged to a reasonable approximation to the true vaue. This procedure is a slight variant of that given in Box and Tidwell [1962], as is this example (except that there are some slight computational errors in their presentation of this example). It can be used more generally when there are many independent variables, each appearing in the regresssion in some power transformation form. The procedure described is modified merely by regressing y first onto the p variables $x_i^{\gamma_i}$ and then onto the $2p$ variables $(x_i^{\gamma_{ti}}, x_i^{\gamma_{ti}} \log x_i)$ at iteration t. The coefficients are used for each independent variable separately in the same manner as above.

Theoretical Background

Once again we use Taylor's theorem to expand the relation $y = \alpha + \beta x^{\gamma}$ to first order around the value $\gamma = \gamma_0$. Then we can approximate this model by

$$\begin{aligned} y &\approx \alpha + \beta[x^{\gamma_0} + x^{\gamma_0} \log x(\gamma - \gamma_0)] \\ &= \alpha + \beta x^{\gamma_0} + \beta(\gamma - \gamma_0)x^{\gamma_0} \log x. \end{aligned}$$

On the one hand, we can pretend that $\gamma = \gamma_0$, and so we can regress y upon x^{γ_0} and estimate β by the slope of this regression, say $\hat{\beta}_0$. On the other hand, we can regress y on x^{γ_0} and $x^{\gamma_0} \log x$ and estimate the parameter $\delta = \beta(\gamma - \gamma_0)$ by the coefficient of $x^{\gamma_0} \log x$, say $\hat{\delta}_0$. We can now approximate γ by

$$\gamma_1 = \gamma_0 + \frac{\hat{\delta}_0}{\hat{\beta}_0}.$$

Now we can repeat this procedure with the relation

$$y \approx \alpha + \beta x^{\gamma_1} + \beta(\gamma - \gamma_1)x^{\gamma_1} \log x$$

to produce

$$\gamma_2 = \gamma_1 + \frac{\hat{\delta}_1}{\hat{\beta}_1} = \gamma_0 + \frac{\hat{\delta}_0}{\hat{\beta}_0} + \frac{\hat{\delta}_1}{\hat{\beta}_1},$$

etc., until the procedure converges.

Let us consider once again the power relation $y = \alpha + \beta x^{\gamma}$. Now $y' = dy/dx = \beta\gamma x^{\gamma-1}$, and $y'' = d^2y/dx^2 = \beta\gamma(\gamma - 1)x^{\gamma-2}$, so that $y'/y'' = x/(\gamma - 1)$, and

$$\frac{d}{dx}\left(\frac{y'}{y''}\right) = \frac{1}{\gamma - 1}.$$

Thus to estimate γ one need only estimate the quantity $d(y'/y'')/dx$. Assuming, as we do, that there is no error in the relation, one can estimate the functions $y'$and y'' using methods of finite differences (see Milne-Thompson [1951]), and hence develop an estimate of the function $w = y'/y''$ by taking the ratio of these functions, after which one can once again apply finite differences methods to w to estimate dw/dx. Hopefully, this resulting function is a constant, which will be equal to $(\gamma - 1)^{-1}$.

Dolby [1963] used this argument to develop this approach as a procedure for estimating the appropriate power transformation of the independent variable. Recognizing that one merely needs four points to accomplish this finite differences program, Dolby provided a rule for selection of the "best" four points. He suggests using the smallest and largest x value and the two interior x values that most nearly subdivide the range of x's into three equal intervals. He notes that there is little gain from estimating γ from each of the $\binom{n}{4}$ selections of four x's and averaging these estimates. Moreover, using more than four x's leads to a great deal of extra computational effort.

The general form of the finite differences procedure is the following. Let the selected pairs of points be denoted by (x_1, y_1), (x_2, y_2), (x_3, y_3), (x_4, y_4). Using standard notation from the calculus of finite differences, let

$$[i, j] = \frac{y_i - y_j}{x_i - x_j},$$

and

$$[i, j, k] = \frac{[i, j] - [j, k]}{x_i - x_k}.$$

Then

$$\frac{1}{\gamma - 1} = \frac{1}{(x_4 + x_3 - x_2 - x_1)}\left(\frac{[4, 3] + [3, 2]}{[4, 3, 2]} - \frac{[3, 2] + [2, 1]}{[3, 2, 1]}\right).$$

If the x's are equally spaced, $(\gamma - 1)^{-1}$ reduces to

$$\frac{1}{\gamma - 1} = \frac{1}{4}\left(\frac{y_4 - y_2}{y_4 - 2y_3 + y_2} - \frac{y_3 - y_1}{y_3 - 2y_2 + y_1}\right).$$

Applying this formula to our example, if we used the four points associated with $x = 1, 2, 3, 4$, we would find that

$$\frac{1}{\gamma - 1} = \frac{1}{4}\left(\frac{12 - 11.414}{12 - 23.464 + 11.414} - \frac{11.732 - 11}{11.732 - 22.828 + 11}\right) = -1.02375,$$

or $\gamma = 0.023$. If instead we used $x = 1, 2, 3, 5$, we obtain the following table:

x	y	$[\cdot, \cdot]$	$[\cdot, \cdot, \cdot]$
1	11		
		0.414	
2	11.414		−0.048
		0.318	
3	11.732		−0.022
		0.252	
5	12.236		

so that

$$\frac{1}{\gamma - 1} = \frac{1}{5+3-2-1}\left(\frac{0.252+0.318}{-0.022} - \frac{0.314+0.414}{-0.048}\right) = -2.13,$$

or $\gamma = 0.531$. Thus this "quick-and-dirty" procedure has produced rather rapidly a "ball-park" estimate of the true $\gamma(0.5)$.

Appendix I

```
      dimension x(100),yl(100),dy(100)
      open(unit=21,file='gamma.dat')
      read(21)n,k
      yx=0.
      do 21 i=1,n
      read(21)x(i)
21    yx=yx+alog(x(i))
      xl=.25
30    continue
10    continue
      do 1 i=1,n
1     yl(i)=(x(i)**xl-1.)/xl
      ym=0.
      ys=0.
      do 2 i=1,n
      ym=ym+yl(i)/n
2     ys=ys+yl(i)*yl(i)
      ys=(ys-n*ym*ym)/n
      z=0.
      xlx=0.
      yxlx=0.
      do 3 i=1,n
      xlx=xlx+alog(x(i))*x(i)**xl
      yxlx=yxlx+yl(i)*alog(x(i))*x(i)**xl
3     z=z+(yl(i)-ym)*((1.+xl*yl(i))*alog(1.+xl*yl(i))
     +           -xl*yl(i))
      z=yx-z/(xl*xl*ys)
      do 4 i=1,n
4     dy(i)=(-yl(i)+x(i)**xl*alog(x(i)))/xl
      dm=0.
      do 5 i=1,n
```

```
5      dm=dm+dy(i)/n
       dsig=-2.*ys/xl+2.*(yxlx-xlx*ym)/(xl*n)
       g=-(z-yx)*(2.*xl*ys+xl*xl*dsig)/(xl*xl*ys)
       s2=0.
       s1=0.
       do 6 i=1,n
       wy=1.+xl*yl(i)
       s1=s1+(yl(i)-ym)*alog(wy)*(yl(i)+xl*dy(i))
       s2=s2+(dy(i)-dm)*(wy*alog(wy)-xl*yl(i))
6      continue
       g=g-(s1+s2)/(xl*xl*ys)
       write(5,500)xl,z,g
500    format(1h 3(x,e12.6))
       xl=xl-z/g
       wz=abs(z)
       if(wz.gt.10.**-5) go to 30
       end
```

References

Bickel, P. J. and Doksum, K. A. 1981. An analysis of transformations revisited. *Journal of the American Statistical Association* **76** (June): 296–311.

Box, G. E. P. and Cox, D. R. 1964. An analysis of transformations. *Journal of the Royal Statistical Society, Series B* **26**: 211–52.

Box, G. E. P. and Cox, D. R. 1982. An analysis of transformations, revisited, rebutted. *Journal of the American Statistical Association* **77** (March): 209–10.

Box, G. E. P. and Tidwell, P. W. 1962. Transformation of the independent variables. *Technometrics* **4** (November): 531–50.

Cameron, M. A. 1984. Choosing a symmetrizing power transformation. *Journal of the American Statistical Association* **79** (March): 107–8.

Dolby, J. L. 1963. A quick method for choosing a transformation. *Technometrics* **5** (August): 317–25.

Efron, B. 1982. Transformation theory: How normal is a family of distributions? *Annals of Statistics* **10** (June): 323–39.

Emerson, J. D. and Stoto, M. A. 1982. Exploratory methods for choosing power transformations. *Journal of the American Statistical Association* **77** (March): 103–8.

Emerson, J. D. and Stoto, M. A. 1984. Rejoinder. *Journal of the American Statistical Association* **79** (March): 108–9.

Espy, W. R. 1978. *O Thou Improper, Thou Uncommon Noun.* New York: Clarkson N. Potter.

Glejser, H. 1969. A new test for heteroskedasticity. *Journal of the American Statistical Association* **64** (March): 316–23.

Hernandez, F. and Johnson, R. A. 1980. The large-sample behavior of transformations to normality. *Journal of the American Statistical Association* **75** (December): 855–61.

Hinkley, D. V. 1975. On power transformations to symmetry. *Biometrika* **62** (April): 101–11.

Hinkley, D. V. and Runger, G. 1984. The analysis of transformed data. *Journal of the American Statistical Association* **79** (June): 302–28.

Milne-Thompson, L. M. 1951. *The Calculus of Finite Differences.* New York: Macmillan.

Schlesselman, J. 1971. Power families: a note on the Box and Cox transformation. *Journal of the Royal Statistical Society, Series B* **33**: 307–11.
Tukey, J. W. 1957. On the comparative anatomy of transformations. *Annals of Mathematical Statistics* **28** (September): 602–32.
Tukey, J. W. 1977. *Exploratory Data Analysis.* Reading, MA: Addison-Wesley.

CHAPTER 6

Independent Variable Selection in Multiple Regression

"Stepwise regression can lead to confusing results ..."
Daniel and Wood ([1980], p. 85)

0. Introduction

The data analyst confronted with a large number of variables from which he must select a parsimonious subset, as independent variables in a multiple regression, is faced with a number of technical issues. What criterion should he use to judge the adequacy of his selection? What procedure should he use to select the subset of independent variables? How should he check for/guard against/correct for possible multicollinearity in his chosen set of independent variables?

These questions will be discussed in turn in this chapter. We consider the regression model

$$\mathbf{y}_i = \beta_0 + X_i^{*\prime}B^* + \mathbf{u}_i, \qquad i = 1, \ldots, n,$$

where the $\mathbf{u}_i$ are independent $N(0, \sigma^2)$ random variables and X_i^* is a p vector of (nonrandom) independent variables. Section 1 describes the various criteria adduced for judging adequacy of the set of independent variables, including the classical multiple correlation coefficient, the Mallows C_p criterion, and the Akaike criterion. Section 2 contains a description of both stepwise procedures and subset procedures for selection of independent variables. Finally, Section 3 describes the multicollinearity problem and methods for circumventing it, including ridge regression.

1. Criteria for Goodness of Regression Model

Without the introduction of any independent variables, our model is of the form $\mathbf{y}_i = \beta_0 + \mathbf{u}_i$, our best estimate of β_0 is $\bar{\mathbf{y}} = \sum_{i=1}^n \mathbf{y}_i/n$, so that our estimate of each $\mathbf{y}_i$ is also $\bar{\mathbf{y}}$, and $\sum_{i=1}^n (\mathbf{y}_i - \bar{\mathbf{y}})^2$ is a measure of how far off the "estimates"

$\hat{\mathbf{y}}_i = \bar{\mathbf{y}}$ of the actual $\mathbf{y}_i$'s are from the observed values. Generalizing this, let $\hat{\boldsymbol{\beta}}_0$ and $\hat{\mathbf{B}}^*$ be least squares estimates of β_0 and B^*, so that our best estimate of $\mathbf{y}_i$ is

$$\hat{\mathbf{y}}_i = \hat{\boldsymbol{\beta}}_0 + X_i^{*\prime}\hat{\mathbf{B}}^*.$$

Then $\sum_{i=1}^{n} (\mathbf{y}_i - \hat{\mathbf{y}}_i)^2$, sometimes called the "sum of squares of residuals", is a measure of how far off the estimates $\hat{\mathbf{y}}_i$ are from the observed values. It is intuitively clear, based on the principle of "the more you know, the better", that

$$\sum_{i=1}^{n} (\mathbf{y}_i - \hat{\mathbf{y}}_i)^2 \le \sum_{i=1}^{n} (\mathbf{y}_i - \bar{\mathbf{y}})^2.$$

(Remember that the least squares criterion finds $\hat{\beta}_0$ and $\hat{B}^*$ which minimizes $\sum_{i=1}^{n} (y_i - \beta_0 - X_i^{*\prime}B^*)^2$, so that using least squares estimates of β_0 and B^* to determine $\hat{\mathbf{y}}_i$ will produce the minimum value of $\sum_{i=1}^{n} (\mathbf{y}_i - \hat{\mathbf{y}}_i)^2$. Remember also that one possible value of B^* is $B^* = 0$, at which point the best estimate of β_0 is $\bar{y}$. These observations are the basis for the above inequality). Moreover, augmenting the regression model with more independent variables will produce a still smaller value of the sum of squares of residuals.

A classic index of goodness of a regression model is the squared multiple correlation coefficient $\mathbf{r}^2$ (usually capitalized), given by

$$\mathbf{r}^2 = 1 - \frac{\sum_{i=1}^{n} (\mathbf{y}_i - \hat{\mathbf{y}}_i)^2}{\sum_{i=1}^{n} (\mathbf{y}_i - \bar{\mathbf{y}})^2}.$$

Note that when $\hat{\mathbf{y}}_i = \mathbf{y}_i$ for all i then $\mathbf{r}^2 = 1$, and when $\hat{\mathbf{y}}_i = \bar{\mathbf{y}}$ for all i, i.e., the independent variables do not contribute to the prediction of $\mathbf{y}_i$, then $\mathbf{r}^2 = 0$. Clearly, as one augments the model with more independent variables one increases the value of $\mathbf{r}^2$.

One would like a criterion which balances the goodness of predictability of the y_i via the regression against the number of independent variables used in the regression, docking the criterion for each additional independent variable it uses. A number of criteria have been adduced. In this section we will motivate and describe seven such criteria, $\bar{\mathbf{r}}^2$ (the adjusted multiple correlation coefficient), Γ_p, C_p, J_p, S_p, the Akaike criteria, and PRESS. We depart from our format of preceding the theoretical background discussion by a prescription section, since the theoretical background discussion in this instance provides the basic motivation behind the various criteria. Instead, the criteria will be summarized and exemplified at the end of this section.

One such proposed criterion is the adjusted squared multiple correlation coefficient $\bar{\mathbf{r}}^2$, given by

$$\bar{\mathbf{r}}^2 = 1 - \frac{\sum_{i=1}^{n} (\mathbf{y}_i - \hat{\mathbf{y}}_i)^2/(n - p - 1)}{\sum_{i=1}^{n} (\mathbf{y}_i - \bar{\mathbf{y}})^2/(n - 1)}.$$

The quantity $\sum_{i=1}^{n}(\mathbf{y}_i - \bar{\mathbf{y}})^2/(n-1)$ is the maximum likelihood estimate of σ^2 without recourse to use of the independent variables, and the quantity $\sum_{i=1}^{n}(\mathbf{y}_i - \bar{\mathbf{y}})^2/(n-p-1)$ is the maximum likelihood estimate of σ^2 based on the regression. Thus $1 - \bar{\mathbf{r}}^2$ is the ratio of two separate (though not statistically independent) estimates of σ^2, one based on the regression and one not so based. (Theil [1961] provides a less *ad hoc* argument for the use of $\bar{\mathbf{r}}^2$, namely the suggestion that, in comparing two regression models, one should select the model with the smaller associated maximum likelihood estimate of σ^2.)

It can be shown that $\mathbf{r}^2$ has a beta distribution with parameters $((n-p+1)/2, (p-1)/2)$. Since $1 - \bar{\mathbf{r}}^2 = [(n-1)/(n-p-1)](1-\mathbf{r}^2)$, the distribution of $\bar{\mathbf{r}}^2$ is directly ascertainable from that of $\mathbf{r}^2$. Alternatively, since

$$\sum_{i=1}^{n}(y_i - \bar{y})^2 = \sum_{i=1}^{n}(y_i - \hat{y}_i)^2 + \sum_{i=1}^{n}(\hat{y}_i - \bar{y})^2,$$

a monotonic function of $\mathbf{r}^2$, namely

$$\mathbf{v} = \frac{\mathbf{r}^2/p}{(1-\mathbf{r}^2)/(n-p-1)} = \frac{\sum_{i=1}^{n}(\hat{\mathbf{y}}_i - \bar{\mathbf{y}})^2/p}{\sum_{i=1}^{n}(\mathbf{y}_i - \hat{\mathbf{y}})^2/(n-p-1)}$$

has an $F(p, n-p-1)$ distribution. Typically, this calculation is presented the form of an analysis of variance table, namely

Source of variation	Sum of squares	df
Regression	$\sum_{i=1}^{n}(\hat{\mathbf{y}}_i - \bar{\mathbf{y}})^2$	p
Residual	$\sum_{i=1}^{n}(\mathbf{y}_i - \hat{\mathbf{y}}_i)^2$	$n-p-1$
Total	$\sum_{i=1}^{n}(\mathbf{y}_i - \bar{\mathbf{y}})^2$	$n-1$

Then $\mathbf{v}$ is the F test statistic for testing for the existence of a significant regression equation (apart from β_0), that is, the ratio of the regression mean square to the residual mean square.

An alternative approach toward development of a criterion of goodness of a regression is the following. Suppose we only used the first p_1 independent variables, i.e., express X_i^* as $X_i^{*\prime} = (X_{1i}^{*\prime}, X_{2i}^{*\prime})$ and B^* as $B^{*\prime} = (B_1^{*\prime} B_2^{*\prime})$, where B_1^* and X_{1i}^* are p_1 vectors, and estimate β_0 and B_1^* using only the X_{1i}^*'s. Assume for the moment that the model is

$$\mathbf{y}_i = \beta_0 + X_{1i}^{*\prime} B_1^* + \mathbf{u}_i,$$

where, as before, $\mathbf{u}_i$ is distributed as $N(0, \sigma^2)$. That is, we assume that the error distribution for this model is identical with that of the more extensive regression model involving the entire X_i^* vector.

Let $X'_{1i} = (1, X^{*\prime}_{1i})$, $B'_1 = (\beta_0, B^{*\prime}_1)$, $\mathbf{Y}' = [\mathbf{y}_1, \ldots, \mathbf{y}_n]$, and

$$X_1 = \begin{bmatrix} X'_{11} \\ \vdots \\ X'_{1n} \end{bmatrix}.$$

Then $\hat{\mathbf{B}}_1 = (X'_1X_1)^{-1}X'_1\mathbf{Y}$ is the least squares estimator of B_1.

Now suppose one observes another vector X, say X_P, and wishes to predict the associated value of $\mathbf{y}$. One can either use the full vector, obtaining the prediction $\hat{\mathbf{y}}_P = X'_P\hat{\mathbf{B}} = X'_P(X'X)^{-1}X'\mathbf{Y}$ or the $p_1 + 1$ subvector X_{1P} obtaining the prediction $\hat{\mathbf{y}}_{1P} = X'_{1P}\hat{\mathbf{B}}_1 = X'_{1P}(X'_1X_1)^{-1}X_1\mathbf{Y}$. The mean square error of the prediction if one uses the subvector is

$$\begin{aligned}\mathscr{E}(\hat{\mathbf{y}}_{1P} - \mathbf{y})^2 &= \mathscr{E}(\hat{\mathbf{y}}_{1P} - \mathscr{E}\hat{\mathbf{y}}_{1P} + \mathscr{E}\hat{\mathbf{y}}_{1P} - \mathbf{y})^2 \\ &= \mathscr{V}\hat{\mathbf{y}}_{1P} + \mathscr{E}(\mathbf{y} - \mathscr{E}\mathbf{y} + \mathscr{E}\mathbf{y} - \mathscr{E}\hat{y}_{1P})^2 \\ &= \mathscr{V}\hat{\mathbf{y}}_{1P} + \mathscr{V}\mathbf{y} + (\mathscr{E}\hat{\mathbf{y}}_{1P} - \mathscr{E}\mathbf{y})^2 \\ &= \sigma^2X'_{1P}(X'_1X_1)^{-1}X_{1P} + \sigma^2 + (X'_{1P}(X'_1X_1)^{-1}X'_1XB - X'_PB)^2.\end{aligned}$$

This expression is useful as a criterion for selection of the best subset of independent variables for predicting y from a given vector X_p. But in itself it does not provide a general criterion for assessing goodness of a regression model. However, if we evaluate $\mathscr{E}(\hat{\mathbf{y}}_P - \hat{\mathbf{y}})^2$ based on X_P equal to each column of the data matrix X, average these values, and normalize by dividing this average by σ^2, we obtain an overall measure of goodness of prediction by any subset of independent variables. This measure, dubbed Γ_{p_1} in Gorman and Toman [1966], is given by

$$\begin{aligned}\Gamma_{p_1} &= \sum_{i=1}^n (1 + X'_{1i}(X'_1X_1)^{-1}X_{1i}) + \frac{1}{\sigma^2}\sum_{i=1}^n (\mathscr{E}\hat{\mathbf{y}}_{1i} - X'_iB)^2 \\ &= n + \sum_{i=1}^n X'_{1i}(X'_1X_1)^{-1}X_{1i} + \frac{1}{\sigma^2}\sum_{i=1}^n (\mathscr{E}\hat{\mathbf{y}}_{1i} - X'_iB)^2 \\ &= n + \operatorname{tr}\sum_{i=1}^n X'_{1i}(X'_1X_1)^{-1}X_{1i} + \frac{1}{\sigma^2}\sum_{i=1}^n (\mathscr{E}\hat{\mathbf{y}}_{1i} - X'_iB)^2 \\ &= n + \operatorname{tr}(X'_1X_1)^{-1}\sum_{i=1}^n X_{1i}X'_{1i} + \frac{1}{\sigma^2}\sum_{i=1}^n (\mathscr{E}\hat{\mathbf{y}}_{1i} - X'_iB)^2 \\ &= (n + p_1 + 1) + \frac{1}{\sigma^2}[\mathscr{E}\hat{\mathbf{Y}}_1 - XB]'[\mathscr{E}\hat{\mathbf{Y}}_1 - XB],\end{aligned}$$

where $\hat{\mathbf{Y}}'_1 = [\hat{\mathbf{y}}_{1i}, \ldots, \hat{\mathbf{y}}_{1n}]$ and $\hat{\mathbf{y}}_{1i}$ is the predicted value of $\mathbf{y}_i$ based on X_{1i}.

Since σ^2 is unknown, it must be replaced by an estimate of it if the criterion is to be usable, and a natural estimate of σ^2 is the residual mean square based on the full regression, i.e.,

$$\hat{\boldsymbol{\sigma}}^2 = \sum_{i=1}^n \frac{(\mathbf{y}_i - \hat{\boldsymbol{\beta}}_0 - \hat{\boldsymbol{\beta}}_1x_{1i} - \cdots - \hat{\boldsymbol{\beta}}_px_{pi})^2}{n - p - 1}.$$

Also, the expected value of the sum of squares due to regression using only X_1 is given by

$$[\mathscr{E}\hat{\mathbf{Y}}_1 - XB]'[\mathscr{E}\hat{\mathbf{Y}}_1 - XB] + (n - p_1 - 1)\sigma^2,$$

so that $[\mathscr{E}\hat{\mathbf{Y}}_1 - XB]'[\mathscr{E}\hat{\mathbf{Y}}_1 - XB]$ can be estimated by

$$\sum_{i=1}^{n} (\mathbf{y}_i - \hat{\mathbf{y}}_{1i})^2 - (n - p_1 - 1)\hat{\boldsymbol{\sigma}}^2.$$

Thus Γ_p is estimated by

$$\sum_{i=1}^{n} \frac{(\mathbf{y}_i - \hat{\mathbf{y}}_{1i})^2}{\hat{\boldsymbol{\sigma}}^2} - (n - p_1 - 1) + (n + p_1 + 1) = \sum_{i=1}^{n} \frac{(\mathbf{y}_i - \hat{\mathbf{y}}_{1i})^2}{\hat{\boldsymbol{\sigma}}^2} + 2(p_1 + 1).$$

This is the C_{p_1} statistic suggested by Mallows [1967].

A related criterion, which neglects the bias term in $\mathscr{E}(\hat{\mathbf{y}}_{1p} - \mathbf{y})^2$, is that based on $\mathscr{E}(\hat{\mathbf{y}}_{1p} - \mathbf{y})^2$. Since

$$\frac{1}{n}\sum_{i=1}^{n} \mathscr{E}(\hat{\mathbf{y}}_{1i} - \mathbf{y}_i) = \frac{n + p_1 + 1}{n}\sigma^2,$$

where only X_1 is used, one can estimate this quantity by using as an estimate of σ^2 the residual mean square

$$\hat{\boldsymbol{\sigma}}^2_{p_1} = \frac{\sum_{i=1}^{n} (\mathbf{y}_i - \hat{\boldsymbol{\beta}}_0 - \hat{\boldsymbol{\beta}}_1 x_{1i} - \cdots - \boldsymbol{\beta}_{p_1} x_{p_1 i})^2}{n - p_1 - 1}.$$

Then, neglecting the divisor n, Hocking [1972] suggests the criterion

$$\mathbf{J}_{p_1} = (n + p_1 + 1)\hat{\boldsymbol{\sigma}}^2_{p_1}.$$

(In most literature the residual sum of squares is referred to by the symbol RSS_p, where p is used to denote what we have been calling p_1, the number of independent variables, excluding the constant 1, used in the regression. Thus the literature refers to Γ_p, C_p, and J_p, rather than the Γ_{p_1}, C_{p_1}, and J_{p_1} of this section.)

These criteria use the values of the independent variables as the basis for prediction of the associated values of the $\mathbf{y}_i$. If instead we view X as the value of a random vector drawn from a multivariate normal distribution, we can instead consider as a criterion the expected value (taken with respect to the distribution of $\mathbf{X}$) of $\mathscr{E}(\hat{\mathbf{y}}_{1p} - \mathbf{y})^2$. This is given by

$$\varepsilon_{p_1} = \frac{\sigma^2_{p_1}}{n}\left(1 + n + \frac{(p_1 + 1)(n + 1)}{n - p_1 - 3}\right),$$

where $\sigma^2_{p_1} = \mathscr{V}(\hat{\mathbf{y}}|\mathbf{X}_{p_1} = X_{p_1})$, i.e., the variance of the prediction based on the observed values of the subset of p_1 independent variables. This criterion can be estimated by estimating $\sigma^2_{p_1}$ as above, and simplified by omitting factors only involving n, to produce the criterion

$$S_{p_1} = \frac{\text{RSS}_{p_1}}{(n - p_1)(n - p_1 - 2)}.$$

This criterion is due to Hocking [1976], who motivates it via a different argument. The aforegiven motivation is based on Thompson [1978a, b], who argues that this criterion is the "most suitable" of those presented so far, in the case when X is the value of a multivariate normal random vector.

Akaike [1972] suggests that one compare models by calculating the Kullback–Leibler information number for discrimination between the maximized likelihood functions of the two models (see Kullback and Leibler [1951]). In the case of comparing two regression models, one based on X and one based only on X_1, this reduces to the criterion

$$\mathbf{a} = \log\{(\mathbf{Y} - X_1\hat{\mathbf{B}}_1)'(\mathbf{Y} - X_1\hat{\mathbf{B}}_1)/n\} + 2(p_1 + 1)/n$$

(see Amemiya [1976] for a derivation of this result. Amemiya also shows that if σ^2 were assumed known, the Akaike criterion would instead be

$$\mathbf{a} = (\mathbf{Y} - X_1\hat{\mathbf{B}}_1)'(\mathbf{Y} - X_1\hat{\mathbf{B}}_1)/n + 2(p_1 + 1)\sigma^2/n,$$

which is quite similar to Γ_p.)

In all these criteria we consider the prediction of $\mathbf{y}_i$ using a regression based on all the values of X, including X_i. Allen [1971, 1974] suggests that we simulate "prediction" more closely by predicting $\mathbf{y}_i$ based on a regression using all the values of X except for X_i. Thus, let $\hat{\mathbf{B}}_{(i)}$ denote the regression coefficients based on all data except for ith observation, and let $\hat{\mathbf{y}}_{(i)} = X_i'\hat{\mathbf{B}}_{(i)}$. Allen's criterion, called PRESS, is given by $\sum_{i=1}^{n} (\mathbf{y}_i - \hat{\mathbf{y}}_{(i)})^2$. This quantity is related to RSS, the residual sum of squares (which forms the cornerstone for the other criteria described above), by the equation

$$\text{PRESS} = \sum_{i=1}^{n} (y_i - \hat{y}_{1i})^2/(1 - q_i)^2,$$

where

$$q_i = X_{1p}'(X_1'X_1)^{-1}X_{1p}.$$

That is, PRESS is a weighted sum of squares of residuals. (The residuals $\mathbf{y}_i - \hat{\mathbf{y}}_{(i)}$ are produced by MINITAB using the subcommand TRESIDS.)

Summary of Criteria

$$\bar{\mathbf{r}}^2 = 1 - \frac{\sum_{i=1}^{n} (\mathbf{y}_i - \hat{\mathbf{y}}_i)^2/(n - p - 1)}{\sum_{i=1}^{n} (\mathbf{y}_i - \bar{\mathbf{y}})^2/(n - 1)},$$

$$\Gamma_p = (n + p + 1) + \frac{1}{\sigma^2}[\mathscr{E}\hat{\mathbf{Y}} - BX]'[\mathscr{E}\hat{\mathbf{Y}} - XB],$$

$$\mathbf{C}_p = \frac{\sum_{i=1}^{n} (\mathbf{y}_i - \hat{\mathbf{y}}_{1i})^2}{\hat{\boldsymbol{\sigma}}^2} + 2(p + 1),$$

$$\mathbf{J}_p = (n + p + 1)\hat{\boldsymbol{\sigma}}_p^2,$$

$$\mathbf{S}_p = \frac{(n - p - 1)\hat{\boldsymbol{\sigma}}_p^2}{(n - p)(n - p - 2)}.$$

$$\text{Akaike } \mathbf{a} = \log\{(\mathbf{Y} - X_1\hat{B}_1)(\mathbf{Y} - X_1\hat{B}_1)'/n\} + 2(p_1 + 1)/n$$

$$\text{PRESS} = \sum_{i=1}^{n} (\mathbf{y}_i - \mathbf{y}_{1i})^2/(1 - q_i)^2,$$

where

$$\boldsymbol{\sigma}^2 = \sum_{i=1}^{n} \frac{(\mathbf{y}_i - \hat{\boldsymbol{\beta}}_0 - \hat{\boldsymbol{\beta}}_1 x_{1i} - \cdots - \hat{\boldsymbol{\beta}}_p x_{pi})^2}{n - p - 1},$$

$$\hat{\boldsymbol{\sigma}}_{p_1}^2 = \sum_{i=1}^{n} \frac{(\mathbf{y}_i - \hat{\boldsymbol{\beta}}_0 - \hat{\boldsymbol{\beta}}_1 x_{1i} - \cdots - \hat{\boldsymbol{\beta}}_{p_1} x_{p_1 i})^2}{n - p_1 - 1},$$

$$q_i = X'_{1p}(X'_1X_1)^{-1}X_{1p}.$$

Table 1. Artificial Multiple Regression Data.

y	x_1	x_2	x_3
13.69	4	8	26
18.22	4	12	17
17.68	6	10	22
22.43	7	13	11
26.70	8	16	8
14.04	8	4	21
23.48	10	10	14
19.62	10	7	15
29.46	12	16	9
17.08	12	4	19
28.61	13	13	11
22.08	13	7	17
25.45	14	10	15
33.44	15	15	5
24.81	16	8	10

SOURCE: Abt [1967]. Reprinted from *Metrika* with the permission of Springer-Verlag.

Let us illustrate the use of all these criteria with a data set due to Abt [1967] given in Table 1. The following table summarizes the criteria calculations:

Independent variables	p_1	RSS_{p_1}	r^2	$\bar{r}^2$	C_{p_1}	J_{p_1}	S_{p_1}	$\log \text{RSS}_{p_1}$	a
x_1	1	260.076	0.4427	0.3988	459.56	22.67	1.55	5.56097	3.76
x_2	1	205.904	0.5588	0.5248	361.55	17.95	1.23	5.32741	3.52
x_3	1	74.4054	0.8406	0.8283	123.62	6.49	0.44	4.30953	2.51
x_1, x_2	2	7.07505	0.9848	0.9823	3.80	0.71	0.049	1.95657	0.286
x_1, x_3	2	61.7816	0.8676	0.8455	102.78	6.18	0.43	4.12458	2.45
x_2, x_3	2	64.8416	0.8610	0.8379	108.32	6.48	0.45	4.17195	2.50
x_1, x_2, x_3	3	6.07958	0.9870	0.9834	4.00	0.70	0.051	1.80494	0.267
none	0	466.6536	—	—	—	—	2.39	6.14559	4.21

where $\hat{\sigma}^2 = 6.07958/11 = 0.552689$.

We see from this that the Akaike criterion and the J_p, r^2, and $\bar{r}^2$ criteria would select (by a slight margin) the regression with all three independent variables, whereas the C_p and S_p criterion would select (by a slight margin) the regression with only x_1 and x_2 as independent variables.

To illustrate the use of the PRESS criterion in our example, we record the PRESS calculations herewith. One should note that each value of PRESS is based on the computation of n separate regressions with one observation deleted, followed by the computation of appropriate $\hat{y}_{(i)}$ from each regression.

Table 2 arrays the 15 sets of regression coefficients for each of the seven regressions based on deletion of a single observation from the data set. Table 3 arrays the associated $\hat{y}_{(i)}$ for each of these seven regressions. The resulting values of PRESS are:

Independent variable	PRESS
x_1	336.147
x_2	260.087
x_3	98.3407
x_1, x_2	10.9709
x_1, x_3	96.875
x_2, x_3	90.7032
x_1, x_2, x_3	10.4601

We can see from this that once again all these independent variables are selected, but by a slight margin.

Table 2. Coefficients in Each Possible Regression Based on Deletion of a Single Observation from Data of Table 1.

	Deleted observation							
	1	2	3	4	5	6	7	8
x_1	0.89334	1.07110	0.98200	1.05000	1.06900	0.92639	0.99801	0.99536
1	13.63500	11.43100	12.55300	11.57900	11.15300	13.52500	12.25700	12.55900
x_2	1.02690	1.15390	1.09320	1.14240	1.17070	1.03680	1.09890	1.10900
1	12.44300	11.13300	11.62700	11.03100	10.69300	12.01900	11.15400	11.08900
x_3	−0.96961	−0.90096	−0.94867	−0.94121	−0.94318	−0.87051	−0.91162	−0.91034
1	36.51500	35.81900	36.21000	36.50500	36.43200	35.42700	35.79300	35.98500
x_1	0.96314	0.97129	0.96651	0.97875	0.97936	0.98352	0.97940	0.97880
x_2	1.07640	1.08360	1.08110	1.08120	1.07920	1.09610	1.08280	1.09410
1	1.74870	1.57410	1.67200	1.50540	1.51520	1.27160	1.38570	1.31750
x_1	0.34503	0.27516	0.32975	0.21554	0.28175	0.30954	0.30410	0.30237
x_3	−0.85913	−0.80283	−0.83002	−0.84986	−0.81247	−0.74994	−0.79444	−0.79401
1	31.33600	31.48700	31.09400	32.90900	31.55400	30.53000	30.98500	31.21500
x_2	0.26368	0.39021	0.25665	0.33580	0.37696	0.23087	0.29914	0.24267
x_3	−0.81294	−0.70857	−0.81006	−0.78394	−0.77987	−0.77192	−0.76859	−0.79506
1	31.61300	29.09900	31.62900	30.79900	30.27000	31.56900	30.63300	31.76600
x_1	0.88572	0.86986	0.88907	0.86076	0.87492	0.89353	0.90063	0.90331
x_2	0.96876	0.96917	0.97833	0.95631	0.96969	0.98389	0.98830	0.99917
x_3	−0.11038	−0.11294	−0.10042	−0.12854	−0.11717	−0.10229	−0.09203	−0.08815
1	5.21130	5.43600	4.94920	5.89510	5.43530	4.84560	4.50370	4.36370

(continued)

Table 2 (*continued*)

	Deleted observation							
	9	10	11	12	13	14	15	NONE
x_1	0.94676	1.06810	0.94628	1.04710	1.01580	0.82181	1.12570	0.99721
1	12.48400	12.15500	12.61800	12.08300	12.22600	13.62600	11.34900	12.34800
x_2	1.07560	1.15200	1.05320	1.15010	1.10090	0.94430	1.15100	1.09770
1	11.42600	10.57600	11.48100	11.48500	10.99400	12.36000	10.36400	11.25600
x_3	−0.88672	−0.89765	−0.88808	−0.92169	−0.91476	−0.85174	−0.93346	−0.91226
1	35.31600	35.72400	35.27100	35.84400	35.63300	34.74800	36.28600	35.83200
x_1	0.99067	0.98484	0.97334	0.97377	0.99081	0.94519	1.01690	0.97847
x_2	1.11960	1.06210	1.07660	1.08650	1.08060	1.04990	1.06700	1.08140
1	1.08970	1.68430	1.58280	1.48110	1.43450	2.06900	1.35450	1.50700
x_1	0.30755	0.38185	0.27433	0.28191	0.22833	0.28159	0.42294	0.30247
x_3	−0.76706	−0.73686	−0.78597	−0.80753	−0.82622	−0.75243	−0.78996	−0.79589
1	30.43700	29.59700	31.02300	31.38900	32.08600	30.46700	30.02500	31.06100
x_2	0.26256	0.28614	0.26928	0.34773	0.29754	0.29131	0.26242	0.29621
x_3	−0.77267	−0.77322	−0.76196	−0.75959	−0.77273	−0.71419	−0.79522	−0.77085
1	31.02800	30.89200	30.69500	29.86700	30.51400	29.76400	31.49500	30.73700
x_1	0.91984	0.89004	0.86788	0.87040	0.90081	0.86272	0.90239	0.88622
x_2	1.03140	0.94581	0.95096	0.96915	0.98179	0.95058	0.91297	0.97078
x_3	−0.08084	−0.11157	−0.11967	−0.11744	−0.09682	−0.09860	−0.14635	−0.10789
1	3.87960	5.47170	5.67290	5.43350	4.74850	5.36960	6.25780	5.15270

Table 3. Fitted Dependent Variable in Each Possible Regression for Data of Table 1.

	$\hat{y}$ based on independent variables						
y	x_1	x_2	x_3	x_1, x_2	x_1, x_3	x_2, x_3	x_1, x_2, x_3
13.69	17.20836	20.65820	11.30514	14.21246	10.37874	12.58600	13.63438
18.22	15.71540	24.97980	20.50268	18.46246	18.93953	21.73583	18.62550
17.68	18.44500	22.55900	15.33926	18.28206	14.81206	16.37418	17.85768
22.43	18.92900	25.88220	26.15169	22.41225	25.06932	26.54106	22.93851
26.70	19.70500	29.42420	28.88656	26.61728	27.30824	30.06240	27.01234
14.04	20.93612	16.16620	17.14629	13.52416	17.25758	16.28216	13.78131
23.48	22.23710	22.14300	23.03032	22.00770	22.90384	22.86414	22.10455
19.62	22.51260	18.85200	22.32990	18.76420	22.32855	21.53879	19.06878
29.46	23.84512	28.63560	27.33552	30.89134	27.22406	28.27493	30.69248
17.08	24.97220	15.18400	18.66865	17.75078	20.17886	17.34538	17.81559
28.61	24.91964	25.17260	25.50212	28.23202	25.94362	25.81408	28.00145
22.08	25.69530	18.53570	20.17527	21.74561	21.32582	19.38808	21.53627
25.45	26.44720	22.00300	21.91160	26.11184	22.88932	21.89845	25.72546
33.44	25.95315	26.52450	30.48930	31.99535	30.92870	30.56270	32.07612
24.81	29.36020	19.57200	26.95140	26.16090	28.89244	25.64216	26.53630
13.69	16.33684	20.03760	12.11324	14.07208	11.57774	13.06458	13.65868
18.22	16.33684	24.42840	20.32358	18.39768	18.74075	21.18707	18.51281
17.68	18.33126	22.23300	15.76228	18.19182	15.36624	16.74040	17.80424
22.43	19.32847	25.52610	25.79714	22.41449	24.42350	26.10838	22.78959
26.70	20.32568	28.81920	28.53392	26.63716	27.11364	29.30956	26.91182
14.04	20.32568	15.64680	16.67454	13.66036	16.76707	15.73399	13.85989
23.48	22.32010	22.23300	23.06036	22.10570	22.94324	22.90720	22.21224
19.62	22.32010	18.93990	22.14810	18.86150	22.14735	21.24772	19.19201
29.46	24.31452	28.81920	27.62166	30.55014	27.52763	28.53871	30.34881
17.08	24.31452	15.64680	18.49906	17.57424	19.56873	17.27569	17.62055
28.61	25.31173	25.52610	25.79714	28.28531	26.23832	26.10838	28.10691
22.08	25.31173	18.93990	20.32358	21.79691	21.46298	19.70602	21.63489
25.45	26.30894	22.23300	22.14810	26.01958	23.35723	22.13635	25.64923
33.44	27.30615	27.72150	31.27070	32.40505	31.61860	31.32590	32.46825
24.81	28.30336	20.03760	26.70940	25.81372	27.94162	25.39818	26.01956

2. Stepwise Procedures

Suppose one has computed a multiple regression involving p independent variables and then decides to augment the set of independent variables with one more variable or to eliminate one of the p variables from consideration. The basic numerical procedure for determining the new set of regression coefficients from the previous set, without recourse to a complete recalculation of the solution of the new so-called "normal equations" defining the least squares estimator of B, has been known for a long time. (An early reference to this procedure is Section 29.1 of the fifth edition of Fisher's *Statistical Methods for Research Workers* (Fisher [1934].) The development of a computer program for the implementation of this procedure was a natural consequence of the advent of electronic computers as an aid in statistical calculations. The earliest reference to such a program is Efroymson [1962].

At its first step the forward stepwise procedure selects that independent variable whose squared correlation with y is largest for inclusion in the regression. After that it selects at each step that independent variable which, when used in conjunction with *all* previously included independent variables, produces the largest value of the squared multiple correlation coefficient with the dependent variable. (As a variant one can select as the next variable for inclusion the one that maximizes C_p, or some other criterion for goodness of regression.)

(a) Order of Selection of Variables

One of the fascinating by-products of the forward stepwise procedure is the order in which potential independent variables become included in the independent variable set. Let me hasten to point out to the user of this procedure that this order does not necessarily coincide with the order of "importance" of the independent variables. One should recognize that at each step one merely seeks the additional independent variable that adds most to r^2 when it is combined with the previously selected variables. This in no way guarantees a global optimum.

Let the correlation matrix of the variables be given by

$$R = \left|\begin{matrix} 1 & \cos\theta_1 & \cos\theta_2 & \cos\theta_3 \\ & 1 & \cos\theta_{12} & \cos\theta_{13} \\ & & & \cos\theta_{23} \\ & & & 1 \end{matrix}\right| \begin{matrix} y \\ x_1 \\ x_2 \\ x_3 \end{matrix},$$

where the variables defining the rows of R are given on the right-hand side of the matrix. The squared multiple correlation coefficient of y with x_i and x_j can be expressed as

$$r_{ij}^2 = \frac{\cos^2\theta_i + \cos^2\theta_j - 2\cos\theta_i\cos\theta_j\cos\theta_{ij}}{1 - \cos^2\theta_{ij}}.$$

For x_1 to be selected first in the stepwise procedure, we require that $|\cos\theta_1|$ be greater than $|\cos\theta_j|$ for $j = 2, 3$. At step 2, for x_2 to be selected we require merely that $r_{12}^2 > r_{13}^2$. But is it possible that r_{23}^2 is even greater than r_{12}^2? And if so, how large can the discrepancy between the squared multiple correlation coefficient based on the best pair of independent variables and one selected by the stepwise procedure be?

First,

$$\cos \lim_{\theta_{ij}\to 1} r_{ij}^2 = \cos_i\theta\cos_j\theta,$$

so that if x_i and x_j are highly collinear and at least one of them is highly uncorrelated with y, then r_{ij}^2 will be quite close to 0. Second, $r_{ij}^2 = 1$ if and only if

$$\cos\theta_{ij} = \cos\theta_i\cos\theta_j \pm \sqrt{(1 - \cos^2\theta_i)(1 - \cos^2\theta_j)}.$$

So if we let x_1 and x_2 be highly collinear (e.g., $\cos\theta_{12} = 0.999$) and let x_2 be highly uncorrelated with y (e.g., $\cos\theta_2 = 0.0001$), then r_{12}^2 will be close to 0. Meanwhile, if we let $\cos\theta_{23}$ satisfy

$$\cos\theta_{23} = \cos\theta_2 \cos\theta_3 \pm \sqrt{(1-\cos^2\theta_2)(1-\cos^2\theta_3)},$$

we can have $r_{23}^2 = 1$. In this case, since $\cos\theta_2$ is small, $\cos\theta_{23}$ will be small, i.e., x_2 and x_3 must be highly uncorrelated. Thus, the covariance matrix

$$\begin{bmatrix} 1 & 0.011 & 0.01 & 0.0001 \\ & 1 & 0.91 & 0.008 \\ & & 1 & 0.99995099 \\ & & & 1 \end{bmatrix}$$

will produce $r_{12}^2 = 0.000121001$, $r_{13}^2 = 0.000121000$, and $r_{23}^2 = 0.999901$. The discrepancy between the value of r^2 for the best pair of independent variables and that selected by stepwise regression is almost 1! Fortunately, such situations are theoretical constructs and not cases which ordinarily occur in real problems. But still the phenomenon should be checked for by the user of stepwise regression.

(b) Theoretical Description

The normal equations are typically stated for the estimator of the p vector $B^{*\prime} = (\beta_1, \ldots, \beta_p)$ since $\hat{\boldsymbol{\beta}}_0 = \bar{\mathbf{y}} - \sum_{i=i}^{p} \hat{\boldsymbol{\beta}}_i \bar{x}_i$. They are of the form $R_{xx}B^* = \mathbf{R}_{xy}$, where R_{xx} is the $p \times p$ matrix with (i, j)th element $\sum_{k=1}^{n} (x_{ik} - \bar{x}_k)(x_{jk} - \bar{x}_j)$ and $\mathbf{R}_{xy}$ is the p vector with ith element $\sum_{k=1}^{n} (x_{ik} - \bar{x}_i)(\mathbf{y}_k - \bar{\mathbf{y}})$. The solution of the normal equations is given by $\hat{\mathbf{B}}^* = R_{xx}^{-1}\mathbf{R}_{xy}$. Now suppose that we calculate the solution by the numerical analysis method of Gaussian elimination. The very nature of this calculation process produces as a by-product the wherewithal to go from estimates of a regression with p_1 independent variables to estimates of the regression with either one more variable added to or one variable deleted from the set of p_1 variables.

The essence of the method is that of successively applying the so-called "pivot" operation on a partitioned matrix of the form

$$A = \begin{bmatrix} R_{xx} & \mathbf{R}_{xy} & I \\ \mathbf{R}_{xy}' & \mathbf{s}_{yy} & D \\ -I & B & C \end{bmatrix},$$

where I is a $p \times p$ identity matrix, $\mathbf{s}_{yy} = \sum_{k=1}^{n} (\mathbf{y}_k - \bar{\mathbf{y}})^2$, and B, C, and D are as-yet-unspecified matrices which are initially set at 0. The "pivot" operation when variable k is selected as the "pivoting" variable ($1 \le k \le p$) produces from A a new matrix in which the new a_{ij}'s are computed from the elements of the previous A matrix via the relation:

$$a_{ij} = \begin{cases} (a_{ij}a_{kk} - a_{ik}a_{kj})/a_{kk}, & i \neq k, \quad j \neq k, \\ a_{kj}/a_{kk}, & i = k, \quad j \neq k, \\ -a_{ik}/a_{kk}, & i \neq k, \quad j = k, \\ 1/a_{kk}, & i = j = k. \end{cases}$$

This procedure has the flexibility, when applied to regression parameter estimation, to enable the statistician at each step of the process to select the variable k to be used as a pivot. (Without intervention by the statistician, the Gaussian elimination method will automatically successively take k to be 1, 2, 3, ..., p.) Details of this procedure are given in Efroymson [1962] and in Draper and Smith [1966] (first edition only). (A general set of statistical operators derived from this procedure and usable in many other statistical contexts is set forth in Beaton [1964] (see also Goodnight [1979]).)

To select the appropriate variable on which to pivot, one calculates, for each independent variable not currently used in the regression, the reduction in variance if that variable is added to the set used. One then selects the variable yielding the largest reduction in variance, performs some "test" to see if the contribution to variance reduction is mterial, and if so, pivots on that variable. These variance reductions are merely the changes in the a_{ik} corresponding to the elements of the R_{xy} vector, and so are a by-product of the computation.

To begin, one selects at the first step that independent variable most correlated with y, and pivots on that variable. Suppose x_{k_1} is selected. Then one tests to see if this variable significantly contributes to the prediction of the dependent variable. The standard approach would be to test whether the regression coefficient of x_{k_1} is sufficiently different from 0, via the usual t test on the estimated regression coefficient. Equivalently, and in conformity with the procedures to the used after subsequent pivots, one can calculate the F statistic for testing the significance of the regression based on the analysis of variance table presented in Section 1 above. (That these procedures are equivalent is a consequence of the fact that the F statistic is the square of the t statistic in this case.)

Suppose that this F statistic indicates that x_{k_1} has made a significant contribution to the regression. One then pivots on this variable and produces an updated version of the A matrix. The determination of which variable is next for inclusion is based on a comparison of the partial correlation of each x_i, $i \neq k_1$, with y, given x_{k_1}; these quantities are given by $a_{i,p+1}^2/a_{ii}$, $i = 1, \ldots, p$, $i \neq k_1$, where the a's are elements of the current A matrix.

Now suppose x_{k_2} is selected. We must next test to see if this independent variable makes a sigificant additional contribution in predicting the dependent variable's value. To do this one calculates the "partial F value for x_{k_2} given

x_{k_1}", which is given by

$$\frac{(n-2)a_{k_2,p+1}^2}{a_{p+1,p+1}a_{p+1+k_2,p+1+k_2}}.$$

This value is compared with a critical value or cutoff to determine whether or not the inclusion of x_{k_2} is worthwhile. (We will discuss the nature of this critical value later.) If x_{k_2} is not included, the stepwise process terminates.

Suppose x_{k_2} is included. Before proceeding further, one should determine whether, given the presence of x_{k_2}, the presence of x_{k_1} is warranted. To do this we calculate the "partial F value of x_{k_1} given x_{k_2}" and compare it with an appropriate cutoff value. The partial F value is given by

$$\frac{(n-2)a_{k_1,p+1}^2}{a_{p+1,p+1}a_{p+1+k_1,p+1+k_1}}.$$

If this value is smaller than the cutoff value, x_{k_1} is deleted from the set of independent variables.

If x_{k_1} is deleted, the appropriate next pivot row is row $k_1 + p + 1$. If x_{k_1} is not delted one must determine the next row on which to pivot, i.e., the next variable for inclusion. Once more one considers the quantities $a_{i,p+1}^2/a_{ii}$, but this time for $i = 1, \ldots, p$, $i \neq k_1$, $i \neq k_2$. The pivot row k_3 is that value of i which maximizes this quantity.

The next pivot step, step 3, produces a new A matrix. At this point there are r_3 independent variables in the regression, where

$$r_3 = \begin{cases} 1 & \text{if } x_{k_1} \text{ was deleted in step 2, } \textit{or} \\ 1 & \text{if the selection process terminated with only } x_{k_1} \text{ selected,} \end{cases}$$

$$r_3 = 3 \quad \text{if } x_{k_1} \text{ and } x_{k_2} \text{ are included and } x_{k_3} \text{ is being considered.}$$

In general, if the stepwise process has not terminated, then at step l, letting r_l be the number of variables being considered, the partial F value calculated for the decision to include variable x_j is

$$\frac{(n-r_l)a_{j,p+1+j}^2}{a_{p+1,p+1}a_{p+1+j,p+1+j}},$$

and that calculated for the decision to delete the previously included variable x_i is

$$\frac{(n-r_l)a_{i,p+1}^2}{a_{p+1,p+1}a_{p+1+i,p+1+i}}.$$

As in the case of step 2, the general step consists of:

(1) a pivot step;

(2) a determination of whether the currently introduced variable makes a significant contribution or whether we should terminate;

(3) a determination of whether, if we do not terminate, to delete a previously included variable; and finally,
(4) given that no variable is to be deleted, a determination of the next pivot variable.

(c) Cutoff Rules

The values of the appropriate cutoff for inclusion and exclusion of variables are moot subjects. Even at step 1, when we test β_{k_1} for significance, since k_1 is determined by

$$k_1 = \max \mathbf{r}_{i,p+1}, \qquad i = 1, \ldots, p,$$

k_1 is a random variable. Our hypothesis testing structure requires that k_1 be given prior to perusal of the data, and not selected on the basis of the data. To test that $\beta_{k_1} = 0$ when $k_1 = \max_i \mathbf{r}_{i,p+1}$ is a different problem from testing that $\beta_{k_1} = 0$ for a given k_1. For the latter the usual F ratio is appropriate, and has an $F(1, n-2)$ distribution. For the former we should use as a cutoff the 95% point (say) of the distribution of the maximum of p correlated F ratios.

The joint distribution of these F ratios (or equivalently their square roots, t statistics with $n-2$ degrees of freedom) can be deduced from the work of Dunnett and Sobel [1954]. The appropriate critical value is that value of c such that $\Pr\{\mathbf{f}_1 \leq c, \mathbf{f}_2 \leq c, \ldots, \mathbf{f}_p \leq c\} = 0.95$, where $\mathbf{f}_i$ is the F ratio based on selection of x_i as the one independent variable in the regression. Tables of c for very special cases are given by Krishnaiah and Armitage [1965] and Pope [1969] (see also Pope and Webster [1972]).

The problem of cutoff value is even more difficult at step 2 and later, for now our "F ratio" is dependent on the outcome of the previous steps as well, and no distribution theory has been developed for this situation. Even if this problem were solved, we would still be in the situation of performing multiple tests, each at level α, on the same data set. Thus if k steps are taken, then k tests will be performed, each at level α and the type I error rate for the set of tests will be $1 - (1 - \alpha)^k$. An approach to take to circumvent the effect of this is to set α^* as the level of significance of each test, where, if α is the desired overall level, α^* satisfies $\alpha = 1 - (1 - \alpha^*)^k$ or $\alpha^* = 1 - (1 - \alpha)^{1/k}$.

With all this as background, it is no wonder that practitioners choose to pretend that each of the various "F ratios" has an F distribution with the appropriate degrees of freedom ... or, as Draper and Smith [1966] suggest, use a single cutoff value, loosely based on the F distribution, for all steps. As long as they do not claim to be performing a formal test or affixing a level of significance to their procedure, but rather view their use of the cutoff as merely a data analysis convenience, their use of such cutoff values is legitimate.

What is illegitimate is the use and interpretation of the so-called "probability levels" of the "F ratios" as probabilities rather than merely figures of merit. Viewed as providing a figure of merit, and not as a statistical test procedure, the calculation of "F ratios" may indeed be quite good. In their study, Bendel and Afifi [1977] compared this figure of merit with some others, including C_p and $\bar{r}^2$, and found that it "generally performed well", with "best" choice of α between 0.15 and 0.25, depending on the degrees of freedom of the "F ratio". Their general recommendation is $\alpha = 0.15$.

(d) Example

Let us illustrate the stepwise procedure on the data of Table 1. The A matrix begins as

$$A = \begin{bmatrix} 1 & 0.01698 & -0.57951 & 0.66534 & 1 & 0 & 0 \\ 0.01698 & 1 & -0.70451 & 0.74751 & 0 & 1 & 0 \\ -0.57951 & -0.70451 & 1 & -0.91682 & 0 & 0 & 1 \\ 0.66534 & 0.74751 & -0.91682 & 1 & 0 & 0 & 0 \\ -1 & 0 & 0 & 0 & 0 & 0 & 0 \\ 0 & -1 & 0 & 0 & 0 & 0 & 0 \\ 0 & 0 & -1 & 0 & 0 & 0 & 0 \end{bmatrix}.$$

Since r_{34}^2 is larger than r_{14}^2 and r_{24}^2, $k_1 = 3$. The F statistic for the regression of y on x_3 is given by

$$13r_{34}^2/(r_{44}^2 - r_{34}^2) = 13 \times (0.91682)^2/(1 - 0.91682^2) = 68.53,$$

which is highly significant.

We now pivot on row 3, producing the following A matrix:

$$A = \begin{bmatrix} 0.66417 & -0.39129 & 0 & 0.13403 & 1 & 0 & 0.57951 \\ -0.39129 & 0.50307 & 0 & 0.10160 & 0 & 1 & 0.70451 \\ -0.57951 & -0.70451 & 1 & -0.91682 & 0 & 0 & 1 \\ 0.13403 & 0.10160 & 0 & 0.15944 & 0 & 0 & 0.91682 \\ -1 & 0 & 0 & 0 & 0 & 0 & 0 \\ 0 & -1 & 0 & 0 & 0 & 0 & 0 \\ -0.57951 & -0.70451 & 0 & -0.91682 & 0 & 0 & 1 \end{bmatrix}.$$

The partial correlation of x_1 with y given x_3 is $a_{14}^2/a_{11} = (0.13403)^2/0.66417 = 0.02705$; that of x_2 with y given x_3 is $a_{24}^2/a_{22} = (0.10160)^2/0.50367 = 0.02049$. Thus x_1 is selected for inclusion, and $k_2 = 1$.

We now pivot on x_1 to produce the A matrix

$$A = \begin{bmatrix} 1 & -0.58914 & 0 & 0.20181 & 1.50564 & 0 & 0.87254 \\ 0 & 0.27314 & 0 & 0.18057 & 0.58914 & 1 & 1.04592 \\ 0 & -1.04592 & 1 & -0.79987 & 0.87254 & 0 & 1.50564 \\ 0 & 0.18057 & 0 & 0.13239 & -0.20181 & 0 & 0.79987 \\ 0 & -0.58914 & 0 & 0.20181 & 1.50564 & 0 & 0.87254 \\ 0 & -1 & 0 & 0 & 0 & 0 & 0 \\ 0 & -1.04592 & 0 & -0.79987 & 0.87254 & 0 & 1.50564 \end{bmatrix},$$

and check whether the inclusion of x_1 made a significant contribution to the regression. To do this we calculate the partial F value for x_1 given x_3, namely

$$\frac{13a_{14}^2}{a_{44}a_{55}} = \frac{13 \times (0.20181)^2}{(0.13239)(1.50564)} = 2.656.$$

If we use as our cutoff the 85% pivot of the $F(1, 13)$ distribution, namely 2.341, we would include x_1 as well. If, however, we used the 95% point, 4.665, we would terminate.

We also check whether x_3 can be eliminated, by calculating the partial F value for x_3 given x_1, namely

$$\frac{13a_{34}^2}{a_{44}a_{77}} = \frac{13 \times (0.79987)^2}{(0.13239)(1.50564)} = 41.726.$$

Since this value exceeds the cutoff, we retain x_3.

The only remaining variable is x_2, and so we take $k_3 = 2$, pivot, and find that the A matrix is

$$A = \begin{bmatrix} 1 & 0 & 0 & 0.59128 & 2.77639 & 2.15694 & 3.12853 \\ 0 & 1 & 0 & 0.66108 & 2.15694 & 3.66114 & 3.82927 \\ 0 & 0 & 1 & -0.10843 & 3.12853 & 3.82927 & 5.51078 \\ 0 & 0 & 0 & 0.01302 & -0.59128 & -0.66108 & 0.10843 \\ 0 & 0 & 0 & 0.59128 & 2.77639 & 2.15694 & 3.12853 \\ 0 & 0 & 0 & 0.66108 & 2.15694 & 3.66114 & 3.82927 \\ 0 & 0 & 0 & -0.10843 & 3.12853 & 3.82927 & 5.51078 \end{bmatrix}.$$

The standardized regression coefficients of x_1, x_2, and x_3 are given by a_{14}, a_{24}, and a_{34}, respectively. The regression coefficients themselves can be obtained by multiplying these standardized coefficients by the ratio of the standard deviations of y and the appropriate x_i. For example, since $s_y = 5.77342$ and $s_x = 3.85202$, the regression coefficient of x_1 is $0.59128 \times 5.77342/3.85202 = 0.88621$.

The analysis of variance table for regression is given by the following table:

Source of variation	Sum of squares	df
Regression	$s_y^2 a_{p+1,p+1}$	r_l
Residual	$s_y^2(1 - a_{p+1,p+1})$	$n - r_l - 1$
Total	1.0	$n - 1$

which in this case is, in standardized units,

Source of variation	Sum of squares	df
Regression	0.1302	3
Residual	0.98698	12
Total	1.0	14

To convert to the original units, all sums of squares should be multiplied by $s_y^2 = 33.33238$.

Finally, the standard error of the ith regression coefficient is given by

$$\text{STD. DEV. OF RESIDUALS} \times \sqrt{\frac{a_{p+1+i,p+1+i}}{s_{x_i}}}$$

$$= \frac{s_y\sqrt{1 - a_{p+1,p+1}}}{\sqrt{n - r_l - 1}} \sqrt{\frac{a_{p+1+i,p+1+i}}{s_{x_i}}}.$$

For $i = 1$, this quantity is

$$5.77342 \times \frac{\sqrt{0.98698}}{\sqrt{12}} \sqrt{\frac{2.77639}{3.85202}} = 1.4057.$$

3. Multicollinearity

(a) Description of the Phenomenon

The $n \times p + 1$ matrix of observations on the $p + 1$ independent variables X is of rank at most $p + 1$ (assuming that $n \geq p + 1$, i.e., we have at least as many observations as parameters to be estimated). The usual estimator of B, given by $\hat{\mathbf{B}} = (X'X)^{-1}X'\mathbf{Y}$, makes the tacit assumption that the rank of X is equal to $p + 1$, for otherwise the inverse of $X'X$ would not exist. When the rank of X is less than $p + 1$, we say that the independent variables are *multicollinear*. The phenomenon arises because, by virtue of the rank of X being less than $p + 1$, at least one column of X can be reproduced by a suitable linear combination of the other columns of X. (More precisely, let the rank of X be $s < p + 1$. Then there are exactly s columns of X such that all other columns of X are expressible as linear combinations of these s columns.)

What does all this linear algebra jargon mean operationally? Let us look at two simple examples. First, ket $p = 2$ and our model be

$$\mathbf{y}_i = \beta_0 + \beta_1 x_{1i} + \beta_2 x_{2i} + \mathbf{u}_i.$$

Saying that x_{2i} is a linear combination of the other two independent variables

(1 and x_{1i}) means that there are constants γ_1 and γ_2 such that, for all i, $x_{2i} = \gamma_0 + \gamma_1 x_{1i}$. Substituting this expression into the above model yields the revised model

$$\begin{aligned}\mathbf{y}_i &= \beta_0 + \beta_1 x_{1i} + \beta_2(\gamma_0 + \gamma_1 x_{1i}) + \mathbf{u}_i \\ &= (\beta_0 + \beta_2\gamma_0) + (\beta_1 + \beta_2\gamma)x_{1i} + \mathbf{u}_i.\end{aligned}$$

That is, the true model, based on this re-expression, is a regression with $p = 1$, namely

$$\mathbf{y}_i = \beta_0^* + \beta_1^* x_{1i} + \mathbf{u}_i,$$

with parameters β_0^* (expressible in terms of the parameters of the old model as $\beta_0 + \beta_2\gamma_0$) and β_1^* (expressible in terms of the parameters of the old model as $\beta_1 + \beta_2\gamma_1$). There are only two underlying parameters and not the three of the original model statement.

Note that saying that an independent variable cannot be expressed as a linear combination of other independent variables (i.e., an independent variable of a regression model is linearly independent [in the linear algebra sense of the terms] of the other independent variables) does not mean that the variable cannot be functionally dependent on the other independent variables. If, for example, x_{2i} is expressible as $\gamma_0 + \gamma_1 x_{1i}^2$, say, then x_{2i} is no longer *linearly* related to the other independent variables $(1, x_{1i})$. The model with the three independent variables $(1, x_{1i}, x_{2i})$ would *not* be multicollinear. (In this example, an equivalent model would involve the three independent variables $(1, x_{1i}, x_{1i}^2)$.)

As our second example, let $p = 3$ and our model be

$$\mathbf{y}_i = \beta_0 + \beta_1 x_{1i} + \beta_2 x_{2i} + \beta_3 x_{3i} + \mathbf{u}_i,$$

and let $x_{3i} = \gamma_0 + \gamma_1 x_{1i} + \gamma_2 x_{2i}$ be the linear combination of $(1, x_{1i}, x_{2i})$ which determines x_{3i} for all i. A bit of algebra produces the true relationship, namely

$$\mathbf{y}_i = (\beta_0 + \beta_3\gamma_0) + (\beta_1 + \beta_3\gamma_1)x_{1i} + (\beta_2 + \beta_3\gamma_2)x_{2i} + \mathbf{u}_i,$$

a regression with $p = 2$, $\beta_0^* = \beta_0 + \beta_3\gamma_0$, $\beta_1^* = \beta_1 + \beta_3\gamma_1$, and $\beta_2^* = \beta_2 + \beta_3\gamma_2$.

The reason for presenting these two algebraically quite similar examples is because the two are illustrative of different forms of the underlying cause of the multicollinearity. With each form there is an associated method for multicollinearity detection. Let our model be the form

$$\mathbf{y}_i = \beta_0 + \beta_1 x_{1i} + \beta_2 x_{2i} + \cdots + \beta_p x_{pi} + \mathbf{u}_i.$$

The first example of multicollinearity is one in which an independent variable (variable x_2) is a linear function of a *single* other independent variable (variable x_1).The second example is one in which an independent variable (in this case, variable x_3) is a linear function of a *number* of other independent variables (in this case, variables x_1 and x_2).

We want to detect the presence of multicollinearly *ex post*, i.e., after performing the regression and either having the matrix inversion of $X'X$ blow up or, worse yet, produce spurious estimates of B based not on the data but on the rounding error generated by the computer program used in "successfully inverting" the noninvertible matrix $X'X$. The basic concept underlying methods for *ex ante* multicollinearity detection is that if a perfect linear relationship exists among a set of variables then the squared multiple correlation coefficient of one of these variables on the remainder of that set of variables will be equal to 1. Since multicollinearity is defined as the existence of such a perfect linear relationship among the regression's independent variables, it can be checked for by a series of calculations of r^2's.

The detection of multicollinearity of the type exemplified by the first example merely entails an examination of the matrix of correlations amongst the independent variables.[1] If a pair of independent variables is perfectly correlated, i.e., $r = \pm 1$, then multicollinearity of this type exists. But what if an r is very, very large, say $r = 0.98$? Strictly speaking, multicollinearity does not exist. But unlike pregnancy, one *can* be "a little bit multicollinear". For perhaps r was only 0.98 and not 1.0 because of a data error. Or, unbeknownst to the modeler, he has included two essentially redundant independent variables into his model.

What then is an appropriate cutoff value for r to rule out pairwise multicollinearity? The answer to this question cannot be found by significance testing, for the null hypothesis that the true correlation coefficient equals 1 will surely be rejected by any set of data that does not display perfect correlation. One approach is based on an analysis of the determinant of the correlation matrix R as a function of a single element r_{ij}. Let $d_R(r_{ij})$ denote this determinant, and expand it in a first-order Taylor series around r_{ij}^0. Then

$$d_R(r_{ij}) \approx d_R(r_{ij}^0) + d_R'(r_{ij}^0)(r_{ij} - r_{ij}^0).$$

If the true determinant is 0 but our calculated determinant, $d_R(r_{ij}^0)$, is nonzero, then

$$|r_{ij} - r_{ij}^0| \simeq -d_R(r_{ij}^0)/d_R'(r_{ij}^0).$$

But, since the derivative of the determinant of a symmetric matrix with respect to a nondiagonal element is twice the adjoint of that element, we see that

$$|r_{ij} - r_{ij}^0| \simeq -1/2r_0^{ij},$$

where r_0^{ij} is the (i, j)th element of R_0^{-1}. Thus, if the true correlation is 1, then

$$r_{ij}^0 \simeq 1 \pm 1/2r_0^{ij},$$

depending on the sign of r_0^{ij}.

[1] It almost goes without saying that one should also check the standard deviations of the independent variables. An independent variable with 0 standard deviation is a constant, hence linearly related to the "independent variable" associated with β_0, namely 1.

Table 4. Imports, Production, Stock Formation and Consumption in France (1949–66) in Milliards of New Francs at 1966 Prices.

IMPORT	DOPROD	STOCK	CONSUM
159	1493	42	1081
164	1612	41	1148
190	1715	31	1232
191	1755	31	1269
188	1808	11	1321
204	1907	22	1377
227	2021	21	1460
265	2124	56	1541
281	2261	50	1623
276	2319	51	1643
263	2390	7	1676
311	2580	56	1768
333	2698	39	1866
370	2884	31	1997
433	3045	46	2139
490	3234	70	2238
503	3368	12	2320
566	3539	45	2429

Glossary: IMPORT—Imports. DOPROD—Gross domestic production. STOCK—Stock-formation. CONSUM—Consumption.

SOURCE: Malinvaud [1966]. Reprinted from *Statistical Methods of Econometrics* with permission from Rand–McNally.

To illustrate this procedure, consider the data set given in Table 4, taken from Malinvaud [1966], p. 17. The correlation matrix of the three independent variables, DOPROD, STOCK, and CONSUM, is

$$\begin{bmatrix} 1.0 & 0.21545 & 0.99893 \\ 0.21545 & 1.0 & 0.21369 \\ 0.99893 & 0.21369 & 1.0 \end{bmatrix}$$

Letting $r_{ij}^0 = 0.99893$, we find that $d_R'(r_{ij}^0) = 1.90867$, $d_R(r_{ij}^0) = 0.0020372$, so that $r_{ij}^0 \approx 1 - 0.0020372/1.90867 = 1 - 0.0010674 = 0.9989326$. Since the observed value is at the cutoff value (to five places), we say that this data set exhibits pairwise multicollinearity.

What about the more complex example of multicollinearity? Examination of the correlation matrix of independent variables is insufficient to detect the presence of multicollinearity, as can be seen by the following example based on the data of Table 5, taken from Roberts and Ling ([1982], p. A-10). I first

Table 5. Monthly Sales Quotations and Sales (Thousands of Dollars).

Sales	Quotes	Sales	Quotes	Sales	Quotes
1451	6411	2423	8632	1358	6678
2095	7013	2434	7713	1355	6624
1380	6423	2721	8697	2152	6901
2811	7027	971	6805	1844	4648
915	4971	1645	8880	1885	8407
2890	9339	3662	7973	2208	7219
2069	10846	1314	8554	916	4641
2499	8736	2621	7381	868	5865
2479	13660	1764	8272	851	11499
2320	12134	1357	8665	1256	8568
3011	12198	1453	9151	1662	8891
2547	8748	1857	6406	2276	6855

SOURCE: Roberts and Ling [1982]. Reprinted from *Conversational Statistics with IDA* with permission from McGraw-Hill.

created the variable $x_{2i} = x_{1,i-1}$, $i = 2, \ldots, 36$. Next I created the variable $x_{3i} = x_{1i} - x_{2i}$. Thus for $i = 2, \ldots, 36$ the (putative) three independent variables are linearly related by the equation $x_{3i} = x_{1i} - x_{2i}$. Now the correlation matrix of (x_1, x_2, x_3) for these data is

$$\begin{bmatrix} 1.0 & & \\ 0.44160 & 1.0 & \\ 0.52529 & -0.53149 & 1.0 \end{bmatrix}$$

so that pairwise multicollinearity is not indicated.

It bespeaks well of IDA that a regression of y on the triple (x_1, x_2, x_3) produced the following:

VARIABLE	B(STD. V)	B	STD. ERROR(B)	T
%SQRT: negative arg; result = SQRT(ABS(arg)) at IDA34 + 244 (PC 513761)				
X1	−0.3448	−1.1691E − 01	4.3325E + 02	0.000
%SQRT: negative arg; result = SQRT(ABS(arg)) at IDA34 + 244 (PC 513761)				
X2	0.8091	2.7311E − 01	4.3325E + 02	0.001
X3	0.6250	2.0007E − 01	4.3325E + 02	0.000
CONSTANT	0	6.6944E + 02	1.8869E + 02	3.548

But when I augmented the set of independent variables with the new variable $x_{4i} = i$ and regressed y onto (x_1, x_2, x_3, x_4) IDA produced the following:

VARIABLE	B(STD. V)	B	STD. ERROR(B)	T
X1	−0.4041	$-1.3701E-01$	$4.1271E+02$	0.000
X2	0.7740	$2.6124E-01$	$4.1271E+02$	0.001
X3	0.6146	$1.9675E-01$	$4.1271E+02$	0.000
X4	−0.3317	$-2.2763E+01$	$1.1158E+01$	−2.040
CONSTANT	0	$1.3622E+03$	$3.3892E+02$	4.030

with no indication that anything was amiss. Thus one cannot rely on the regression package in all cases to signal the presence of multicollinearity.

A strategy for *ex ante* multicollinearity detection is to perform p regressions, where the jth regression has x_j as the dependent variable and the remaining $p-1$ x_k's as the set of independent variables. When $s = p$, i.e., one of the independent variables is a linear combination of the others, all of these regressions will produce an associated $r^2 = 1$. But when $s < p$ all but $p - s$ regressions will produce $r^2 = 1$; the remaining $p - s$ regressions will blow up due to multicollinearity in the set of $p - 1$ independent variables.

Our emphasis so far has been on finding the source of the multicollinearity and exorcising it from the model. This approach is consistent with the first dictum of applied statistics, "Know thy data". But many don't want to go through the trouble of performing p multiple regressions on the independent variable set preparatory to the one desired multiple regression of y on the set (or a subset) of independent variables. Others may argue that they want to use "all" the data, that is, do not want to discard any independent variable or be forced to choose which member of the redundant set to discard. For them we recommend an approach to estimation of B which has the virtues of:

(a) using all the independent variables,
(b) reducing to the usual estimator of B when there is no multicollinearity, and
(c) producing a least squares estimator of B when multicollinearity exists.

Its drawbacks are:

(a) the procedure is not part of any standard statistical package that I know of; and
(b) the procedure involves computation of eigenvectors and eigenvalues of symmetric matrices rather than matrix inversion, and so is more computationally complex.

The procedure centers around the concept of the generalized inverse of a matrix, and the theory underlying the procedure is described in, for example, Madansky [1976]. Briefly, the general least squares estimator of B is given by $\hat{B} = X^+ Y$, where X^+ is the generalized inverse of X. Now X is an $n \times p + 1$ matrix, assumed to be of rank $s \le p + 1$. The symmetric matrix $X'X$ can be expressed as $X'X = QDQ'$, where Q is the $p + 1 \times p + 1$ orthogonal matrix of eigenvectors and D is the diagonal matrix of eigenvalues of $X'X$. Since D has s nonzero diagonal elements, we cannot define D^{-1}, but can define D^+ to be a diagonal matrix whose (i, i)th element is 0 if the (i, i)th element of D is 0 and $1/d_{ii}$ otherwise (where d_{ii} is (i, i)th element of D). Then X^+ is given by

$$X^+ = QD^+Q'X'.$$

When $s = p + 1$, $X^+ = (X'X)^{-1}X'$ and so $\hat{B}$ reduces to the usual least squares estimate of B.

(b) Diagnostics

Old-fashioned algorithms for matrix inversion require first a calculation of the matrix's determinant. If the determinant is "zero" then the matrix is judged to be singular. But consider for example a 50×50 diagonal matrix each of whose diagnonal elements is 0.1. The determinant of that matrix is 10^{-50}, which is typically smaller than the smallest allowable nonzero number in a computer, and so a perfectly nonsingular matrix is dubbed "singular" and not inverted. Yet all we need do to circumvent this problem is to multiply the matrix by 10, and we have the identity matrix to invert, after which we can readjust one inverse by dividing it by 10. That is, in general it may be possible by suitable scaling of the matrix to have it "pass" the nonzero determinant pretest of these inversion routines. So it is not the "zeroness" of the determinant that is useful as an index of the multicollinearity problem.

Numerical analysts have studied the problem of the stability of numeric procedures for matrix inversion and developed a way of classifying the stability via a construct called the "condition number". Briefly, the "norm" of an $m \times n$ matrix A, written $\|A\|$, is defined as

$$\|A\| = \sqrt{\sum_{i=1}^{m} \sum_{j=1}^{n} a_{ij}^2}.$$

If A is an $n \times n$ matrix whose inverse exists, the condition number of A with respect to inversion is defined as

$$\eta(A) = \|A\| \, \|A^{-1}\|.$$

If $\eta(A)$ is larger, the inverse of A is sensitive to small perturbations in A and the problem of computing the inverse of A is said to be ill-conditioned. An algebraically equivalent expression for $\eta(A)$ is the ratio of the largest to the smallest eigenvalues of A.

Since $\mathscr{V}\hat{\mathbf{B}} = \sigma^2(X'X)^{-1}$, the variance of $\hat{\boldsymbol{\beta}}_i$ is σ^2 times the (i, i)th element of $(X'X)^{-1}$. If the independent variables are "standardized", i.e., have zero arithmetic mean and sum of squares equal to 1, then $X'X$ is the sample correlation matrix of the independent variables. In this case, the (i, i)th element of $(X'X)^{-1}$ is just the (i, i)th element of the inverse of the correlation matrix of the independent variables. This element, which we shall refer to as v_{ii}, is sometimes called the VIF (Variance Inflation Factor).[2] It can be shown that the VIF is

[2] MINITAB's regression command has a VIF subcommand to produce these values routinely.

given by $v_{ii} = 1/(1 - r_{xi}^2)$, where r_{xi} is the multiple correlation coefficient from the regression of x_i on all the other independnt variables. As $r_{xi}^2 \to 1$, i.e., as we approach multicollinearity, $v_{ii} \to \infty$. The VIF is thus an index of the "degree of multicollinearity" in the data. One rule of thumb is that a VIF in excess of 10 is an indication that multicollinearity will cause difficulties in estimation. An overall measure of degree of multicollinearity suggested is the measure $r_L = \sum_{i=1}^{p} v_{ii}/p$.

We illustrate some of these concepts on the highly multicollinear data of Table 4. If we tried to estimate the regression coefficients directly, we would be facing the matrix inversion of the matrix

$$X'X = \begin{bmatrix} 18 & 42{,}753 & 662 & 30{,}128 \\ & 188{,}403{,}941 & 1{,}612{,}871 & 76{,}044{,}089 \\ & & 29{,}502 & 11{,}134{,}345 \\ & & & 53{,}366{,}850 \end{bmatrix}.$$

It is clear at a glance that the diversity of magnitudes of the matrix entries will lead to numeric instability. To alleviate this, one notes that, since

$$\hat{\boldsymbol{\beta}}_0 = \bar{\mathbf{y}} - \hat{\mathbf{B}}^*\bar{X},$$

we can reduce the inversion problem to that of estimating B^* by

$$\hat{\mathbf{B}}^* = (X^{*\prime}X^*)^{-1}X^{*\prime}(\mathbf{Y} - \bar{\mathbf{y}}E),$$

where X^* is the $n \times p$ matrix

$$X^* = X^+ - \bar{X}E,$$

$\bar{X}$ is the $p \times 1$ vector of column means of X^+, the last p columns of the X matrix, and E is a p vector of 1's. Equivalently, we can express $\hat{\mathbf{B}}^*$ as

$$\hat{\mathbf{B}}^* = \left(\frac{1}{n}X^{*\prime}X^*\right)^{-1} X^{*\prime}\frac{1}{n}(\mathbf{Y} - \bar{\mathbf{y}}E)$$

(and note parenthetically that the matrix $S = (1/n)X^{*\prime}X^*$ would be the maximum likelihood estimate of the covariance matrix of the p vector of independent variables if they were randomly drawn from a multivariate normal distribution). In our example, the S matrix is given by

$$S = \begin{bmatrix} 381{,}024.33 & 2{,}250.5912 & 249{,}170.39 \\ & 286.39503 & 1{,}461.3394 \\ & & 163{,}292.97 \end{bmatrix}.$$

The eigenvalues[3] of this matrix are 544087.0, 275.980, and 241.095, so that the condition number of this matrix is 2256.7.

[3] As a bit of numerical analysis advice, let me offer the observation that the usual IMSL eigenvalue program for symmetric matrices breaks down and produces negative eigenvalues for both S and R (defined below). One must resort to IMSL's singular value decomposition program to produce correct eigenvalues.

One might rescale further by noting that, letting s_i^2 be the ith diagonal element of S, the sample correlation matrix is given by

$$R = D_s^{-1} S D_s^{-1},$$

where D_s is a diagonal matrix with the s_i as diagonal elements. Thus $S^{-1} = D_s^{-1} R^{-1} D_s^{-1}$, so that finding R^{-1} would simplify the determination of S^{-1}. For this example,

$$D_s = \begin{bmatrix} 617.2717 & & \\ & 16.92321 & \\ & & 404.0952 \end{bmatrix},$$

and

$$R = \begin{bmatrix} 1 & 0.21544555 & 0.99893295 \\ & 1 & 0.21369020 \\ & & 1 \end{bmatrix},$$

so that the multicollinearity is apparent. The eigenvalues of R are 2.08389, 0.915049, and 0.00106543, so that the condition number of R is 1956.

The matrix of eigenvectors of R is given by

$$Q = \begin{bmatrix} -0.681039 & -0.189707 & -0.707246 \\ -0.269595 & 0.962973 & -0.00130264 \\ -0.680812 & -0.191557 & 0.706966 \end{bmatrix},$$

so that $R = QDQ'$, where D is the diagonal matrix of eigenvalues of R. Thus

$$R^{-1} = \begin{bmatrix} 469.740 & -0.976244 & -469.030 \\ & 1.04988 & 0.750854 \\ & & 469.369 \end{bmatrix},$$

and

$$S^{-1} = D_s^{-1} R^{-1} D_s^{-1} = D_s^{-1} Q D^{-1} Q' D_s^{-1}$$

$$= \begin{bmatrix} 0.00123283 & -0.0000934543 & -0.00188036 \\ & 0.00366584 & 0.000109797 \\ & & 0.0028744 \end{bmatrix}.$$

The diagonal elements of R^{-1} are the variance inflation factors, so that for this data set we find that $r_L = 313.4$, far greater than 10.

Since

$$X^{*\prime} \frac{1}{n} (\mathbf{Y} - \bar{\mathbf{y}} E) = \begin{bmatrix} 73688.29 \\ 545.84 \\ 48268.57 \end{bmatrix},$$

we see that

$$\hat{\mathbf{B}}^* = \begin{bmatrix} 0.03184 \\ 0.41422 \\ 0.24260 \end{bmatrix}.$$

(c) Ridge Regression

Prescription

Another approach to combatting the "multicollinearity problem", i.e., to cater to the reluctance to relinquish some independent variables from the data set, is to use a regression procedure called *ridge regression.* Succinctly, ridge regression recommends replacing the matrix $X'X$ with the matrix $X'X + kI$, for "suitably" chosen k, in the expression for the estimator of B. Thus the ridge regression estimator is

$$\tilde{\mathbf{B}} = (X'X + kI)^{-1}X'\mathbf{Y}.$$

The rule most often adduced for "suitably choosing" k is one based on a graphical display called the "ridge trace". The ridge trace is a plot of each estimator $\tilde{\mathbf{b}}_i(k)$ as a function of k. It has been observed in a number of examples that at a certain value k^* the estimator $\tilde{\mathbf{b}}_i(k)$ will "stabilize". i.e., not vary too much as k varies around k^*. However, as shown by Vinod and Ullah ([1981], p. 181), the ridge trace may appear to be more stable for large k even for data in which $X'X = I$, i.e., the case, if ever there was one, in which $k = 0$ should be best. It has been suggested that, in place of plotting $\tilde{\mathbf{b}}_i(k)$ as a function of k, the ridge trace should plot $\tilde{\mathbf{b}}_i(k)$ as a function of m, where

$$m = p - \sum_{i=1}^{p} \frac{d_i}{d_i + k}.$$

Since $0 \le m \le p$, this suggestion also leads to a plot over a finite range, in contrast to the semi-infinite range of k (though Hoerl and Kennard recommend the range $0 \le k \le 1$, for most practical situations).

The calculation of the ridge trace is quite tedious, involving estimating the regression coefficients for each selected value of k. A computation which bypasses this is a criterion called the ISRM (Index of Stability of Relative Magnitudes) due to Vinod [1976]. The ISRM is given by

$$\text{ISRM} = \sum_{i=1}^{p} \left[\frac{(d_i + k)^2}{\displaystyle\sum_{j=1}^{p} \frac{d_j}{(d_j + k)^2}} - 1 \right]^2.$$

Once the eigenvalues d_i of $X'X$ are calculated, one can easily calculate ISRM for each choice of k. Vinod suggests selecting the k corresponding to the smallest local minimum of ISRM for use in the ridge regression.

To return to our earlier example, the eigenvalues of R are $d_1 = 2.08389$, $d_2 = 0.915049$, and $d_3 = 0.00106543$, so that we can calculate the ISRM as a function of k handily from these values. (We base our determination on R rather than on S because our "baseline" case is one where $X'X = I$, i.e., the X matrix is standardized, and this, as noted earlier, leads to an analysis of R.) Following are values of ISRM and m as a function of some values of k for this matrix:

k	m	ISRM
0.015	0.957	1.767
0.016	0.962	1.716
0.017	0.967	1.676
0.018	0.972	1.647
0.019	0.976	1.629
0.020	0.980	1.620
0.021	0.984	1.621
0.022	0.988	1.631
0.023	0.991	1.649
0.024	0.994	1.675
0.025	0.998	1.707

Thus $k = 0.02$ is indicated as the appropriate value for ridge regression.

Following also are the values of $\tilde{\mathbf{B}}$ for these values of k. Note that the deviation of $\tilde{\mathbf{B}}$ from $\hat{\mathbf{B}} = [0.0318356, 0.414219, 0.242596]$ is minimal, despite the high degree of multicollinearity of the data.

		$\tilde{\mathbf{B}}$	
k			
0.015	0.0318424	0.414195	0.242585
0.016	0.0318428	0.414194	0.242585
0.017	0.0318433	0.414192	0.242584
0.018	0.0318438	0.414191	0.242583
0.019	0.0318442	0.414189	0.242583
0.020	0.0318447	0.414188	0.242582
0.021	0.0318451	0.414186	0.242581
0.022	0.0318456	0.414185	0.242581
0.023	0.0318460	0.414183	0.242580
0.024	0.0318465	0.414182	0.242579
0.025	0.0318469	0.414180	0.242479

Theoretical Background

What properties does this estimator have, other than its built-in ability to use all the data and produce a nonsingular matrix $X'X + kI$ to be inverted? First, we can express $\tilde{\mathbf{B}}$ as

$$\tilde{\mathbf{B}} = (I + k(X'X)^{-1})^{-1}(X'X)^{-1}X'\mathbf{Y}$$
$$= (I + k(X'X)^{-1})^{-1}\hat{\mathbf{B}},$$

whenever $X'X$ is invertible. Thus $\mathscr{E}\tilde{\mathbf{B}} = (I + k(X'X)^{-1})^{-1}B$ in that case, so that $\tilde{\mathbf{B}}$ is a biased estimate of B. But

$$\mathscr{V}\tilde{\mathbf{B}} = (I + k(X'X)^{-1})^{-1}\mathscr{V}\hat{\mathbf{B}}(I + k(X'X)^{-1})^{-1}$$
$$= (I + k(X'X)^{-1})^{-1}\sigma^2(X'X)^{-1}(I + k(X'X)^{-1})^{-1}$$
$$= \sigma^2(X'X + kI)^{-1}(X'X)(X'X)(X'X + kI)^{-1},$$

so that the mean squared error matrix of $\tilde{\mathbf{B}}$, $\mathscr{M}\tilde{\mathbf{B}}$, is given by

$$\mathscr{M}\tilde{\mathbf{B}} = \mathscr{V}\tilde{\mathbf{B}} + (\mathscr{E}\tilde{\mathbf{B}} - B)(\mathscr{E}\tilde{\mathbf{B}} - B)'.$$

The most general optimality property of the least squares estimate $\tilde{\mathbf{B}}$ is that, among all unbiased estimators of B which are linear functions of $\mathbf{y}$, it has "smallest" covariance matrix in the following sense. Let A be any p vector, and let Σ be the covariance matrix of any other unbiased estimator of B which is a linear function of $\mathbf{y}$. Then

$$A'\mathscr{V}\tilde{\mathbf{B}}A \le A'\Sigma A$$

for all choices of A. In particular, this result implies that the variance of the least squares estimator of each of the elements of B is no larger than that of any other linear (in $\mathbf{y}$) unbiased estimator of that element, and hence that the sum of the variances of the elements of $\tilde{\mathbf{B}}$ is no larger than the sum of the variances of any other linear estimator of B. This latter property is expressible matricially as

$$\operatorname{tr} \mathscr{V}\hat{\mathbf{B}} \le \operatorname{tr} \Sigma$$

for covariance matrices Σ of all linear estimators of B.

The ridge regression estimator is usually adjudged relative to the least squares estimator by involving a variant of this latter measure to assess optimality. More precisely, one compares $\operatorname{tr} \mathscr{M}\tilde{\mathbf{B}}$ with $\operatorname{tr} \mathscr{M}\hat{\mathbf{B}}$. Since $\tilde{\mathbf{B}} = \hat{\mathbf{B}}$ when $k = 0$, it suffices to study $\operatorname{tr} \mathscr{M}\tilde{\mathbf{B}}$ as a function of k. Now

$$\operatorname{tr} \mathscr{M}\tilde{\mathbf{B}} = \sigma^2 \operatorname{tr}[(I + k(X'X)^{-1})^{-1}(X'X)^{-1}(I + k(X'X)^{-1})^{-1}]$$
$$+ B'(I + k(X'X)^{-1})^{-1}(I + k(X'X)^{-1})^{-1}B$$
$$= \sigma^2 \operatorname{tr}[Q(k)(X'X)^{-1}Q(k)] + B'Q(k)Q(k)B,$$

where $Q(k) = [I + k(X'X)^{-1}]^{-1}$. Hoerl and Kennard [1970] show that the derivative of $\mathscr{M}\tilde{\mathbf{B}}$ with respect to k is negative near $k = 0$, and hence that positive values of k exist which made the trace of the mean square error matrix of the ridge regression estimator smaller than that of the least squares estimator. Vinod and Ullah [1981] summarize a number of conditions under which such an improvement occurs. One sufficient condition is $k \leq 2\sigma^2/B'B$. Another cited by Vinod and Ullah and due to Swindel and Chapman is the bound

$$k < \frac{-2}{\min(0, \xi)},$$

where

$$\xi = \frac{1}{d_1} - \frac{\sqrt{B'B}}{\sigma^2},$$

and d_1 is the largest eigenvector of $X'X$. Of course, these bounds depend on knowledge of B and σ^2, so to replace these theoretical bounds one needs a more empirically based rule for selection of k. Those rules are given in the "prescription" section above.

Ridge regression, as described herein, appears to be an *ad hoc* procedure, motivated entirely by the artifact that relative to one criterion, the trace of the mean squared error matrix, a ridge regression will perform better than ordinary least squares regression. To dissipate (though not dispel) this impression, I note that if in fact it is known as *priori* that $B'B \leq c^2$, then the ridge regression minimizes the sum of squares of residuals, i.e., is the least squares solution, subject to this constraint, when k satisfies

$$Y'X(X'X + kI)^{-2}X'Y = c^2.$$

(There is also a Bayesian interpretation of the ridge regression estimator based on a prior distribution on B, which gives higher probability to values of the β_i close to 0 than larger values of the B_i. See Goldstein and Smith [1974].)

Another motivation of ridge regression is that it is designed to overcome the multicollinearity problem (which is why I include it in a section on multicollinearity rather than in a section of its own). The phenomenon to which ridge regression advocates point most often as the "triumph" of the procedure is that least squares estimates of some β_i have signs opposite of what theory would dictate, whereas the signs of the ridge regression estimates are consonant with theory. They attribute this typically to the fact that $X'X$ is nearly singular, causing instability in least squares estimates. Since, as mentioned earlier, they wish to retain all the independent variables in the relation, they switch to the ridge regression solution. Indeed, they advocate the ridge trace as an indicator of multicollinearity. If k^*, the point of stabilization of the ridge trace, is close to 0, multicollinearity is absent. Its presence is reflected in the distance of k^* from 0.

References

Abt, K. 1967. On the identification of the significant independent variables in linear models. *Metrika* **12**: 2–15.

Akaike, H. 1972. Information theory and an extension of the maximum likelihood principle. *Proceedings, Second International Symposium on Information Theory,* 267–81.

Allen, D. M. 1971. The prediction sum of squares as a criterion for selecting prediction variables. Technical Report 23. Department of Statistics. University of Kentucky.

Allen, D. M. 1974. The relationship between variable selection and data augmentation and a method for prediction. *Technometrics* **16** (February): 125–27.

Amemiya, T. 1976. Selection of regressors. Technical Report 225. Institute for Mathematical Studies in the Social Sciences. Stanford University.

Beaton, A. E. 1964. The use of special matrix operators in statistical calculus. *Research Bulletin* RB 64 51 (October). Princeton: Educational Testing Service.

Bendel, R. B. and Afifi, A. A. 1977. Comparison of stopping rules in forward "Stepwise" regression. *Journal of the American Statistical Association* **72** (March): 46–53.

Daniel, C. and Wood, F. S. 1980. *Fitting Equations to Data.* New York: Wiley.

Draper, N. R. and Smith, H. 1966. *Applied Regression Analysis.* New York: Wiley.

Dunnett, C. W. and Sobel, M. 1954. A bivariate generalization of Student's distribution, with tables for certain special cases. *Biometrika* **41** (April): 153–69.

Efroymson, M. A. 1962. Multiple regression analysis. In *Mathematical Methods for Digital Computers,* ed. A. Ralston, and H. S. Wilf. New York: Wiley.

Fisher, R. A. 1934. *Statistical Methods for Research Workers.* New York: Hafner.

Goldstein, M. and Smith, A. F. M. 1974. Ridge type estimators for regression analysis. *Journal of the Royal Statistical Society, Series B* **36** (December): 284–91.

Goodnight, J. H. 1979. A tutorial on the SWEEP operator. *American Statistician* **33** (August): 149–58.

Gorman, J. W. and Toman, R. J. 1966. Selection of variables for fitting equations to data. *Technometrics* **8** (February): 27–51.

Hocking, R. R. 1972. Criteria for selection of a subset regression: Which one should be used? *Technometrics* **14** (November): 967–70.

Hocking, R. R. 1976. The analysis and selection of variables in linear regression. *Biometrics* **32** (March): 1–49.

Hoerl, A. E. and Kennard, R. W. 1970. Ridge regression: Biased estimation of non-orthogonal problems. *Technometrics* **12** (February): 55–67.

Krishnaiah, P. R. and Armitage, J. V. 1. 1965. *Probability Integrals of the Multivariate F Distribution, with Tables and Applications,* ARL 65-236. Ohio: Wright–Patterson AFB.

Kullback, S. and Leibler, R. A. 1951. On information and sufficiency. *Annals of Mathematical Statistics* **22** (March): 79–66.

Madansky, A. 1976. *Foundations of Econometrics.* Amsterdam: North-Holland.

Malinvaud, E. 1966. *Statistical Methods of Econometrics.* Chicago: Rand–McNally.

Mallows, C. L. 1967. Choosing a subset regression. Bell Telephone Laboratories, unpublished report.

Pope, P. T. 1969. On the stepwise construction of a prediction equation. Tech. Report 37. THEMIS, Statistics Department, Southern Methodist University, Dallas, Texas.

Pope, P. T. and Webster, J. T. 1972. The use of an *F*-statistic *n* stepwise regression procedures. *Technometrics* **14** (May): 327–40.

Roberts, H. V. and Ling, R. F. 1982. *Conversational Statistics with IDA.* New York: Scientific Press and McGraw-Hill.

Theil, H. 1961. *Economic Forecasts and Policy.* Amsterdam: North-Holland.

Thompson, M. L. 1978a. Selection of variables in multiple regression: Part I. A review and evaluation. *International Statistical Review* **46** (April): 1–19.

Thompson, M. L. 1978b. Selection of variables in multiple regression: Part II. Chosen procedures, computations, and examples. *International Statistical Review* **46** (April): 129–46.

Vinod, H. D. 1976. Application of new ridge regression methods to a study of Bell system scale economies. *Journal of the American Statistical Association* **71** (December): 929–33.

Vinod, H. D. and Ullah, A. 1981. *Recent Advances in Regression Analysis.* New York: Marcel Dekker.

CHAPTER 7

Categorical Variables in Regression

"I know the Kings of England, and I quote the fights historical,
From Marathon to Waterloo, in order categorical."

Gilbert [1879]

0. Introduction

In this chapter we will, for the most part, treat regression problems in which some of the variables, both independent and dependent, are categorical. We begin, though, in Section 1 with a brief treatment of two sample tests, including a description of nonparametric tests of differences in location. This section provides a classical background for the problem treated in Section 2, the analysis of variance. Our treatment is unusual in that we view the problem as a special case of multiple regression in which a set of independent variables, each taking on a value which is either 0 or 1, is constructed, so that the composite test of significance of at least one of the regression coefficients is equivalent to a test of equality of k population means.

The analysis of variance framework of Section 2 is sometimes dubbed "model I", to contrast it with an alternative framework, the components of variance model, sometimes dubbed "model II". This latter model is described in Section 3. Finally, Section 4 treats the (unrelated) problem in which the dependent variable is categorical.

1. Two Sample Tests

Throughout this section we will interrupt the technical description of procedures, to illustrate each with an example. All of the examples are based on the data of Table 1, operating costs per mile driven for each car in a fleet of 18 Chevrolets, 17 Fords, and 18 Plymouths. Table 1 also contains the sample means and standard deviations of these three samples.

Table 1. Operating Costs per Mile for a Fleet of Automobiles.

	Chevrolet	Ford	Plymouth
	3.926	4.700	2.430
	3.450	4.150	2.980
	2.000	4.550	3.040
	2.280	3.310	4.940
	3.494	2.130	3.150
	4.250	4.686	2.460
	2.382	2.680	3.340
	3.020	2.360	2.384
	3.260	3.934	2.270
	4.080	1.560	2.524
	3.670	4.290	3.100
	2.940	1.740	3.530
	5.900	2.170	3.060
	2.180	1.970	2.570
	5.390	4.689	3.480
	2.740	2.870	5.940
	3.492	3.170	2.516
	2.700	1.610	
Mean	3.39744	3.23288	3.07356
Standard Deviation	1.05174	1.13871	1.00239
n	18	17	18

(a) Parametric Procedures

Let $\mathbf{x}_{11}, \ldots, \mathbf{x}_{1n_1}$ be a random sample from a $N(\mu_1, \sigma_1^2)$ distribution, and $\mathbf{x}_{21}, \ldots, \mathbf{x}_{2n_2}$ be an independent random sample from a $N(\mu_2, \sigma_2^2)$ distribution. In testing the hypothesis $\mu_1 = \mu_2$ we distinguish a number of cases.

Case I. σ_1^2, σ_2^2 *known.*
In this case, the statistic

$$\mathbf{z} = \frac{\bar{\mathbf{x}}_1 - \bar{\mathbf{x}}_2}{\sqrt{\dfrac{\sigma_1^2}{n_1} + \dfrac{\sigma_2^2}{n_2}}}$$

has a $N(0, 1)$ distribution. The test procedure of accepting the null hypothesis if $-k_\alpha < \mathbf{z} < k_\alpha$, where k_α is the upper $100\alpha\%$ point of the $N(0, 1)$ distribution, is the uniformly most powerful test of this hypothesis (see Lehmann [1959], p. 117).

Thus if we were comparing Chevrolets with Fords and know that $\sigma_1^2 = 1.1$

and $\sigma_2^2 = 1.3$, then z would be

$$\frac{3.39744 - 3.23288}{\sqrt{\dfrac{1.1}{18} + \dfrac{1.3}{17}}} = 0.4436$$

and we would conclude that the means were not different.

Case II. σ_1^2, σ_2^2 *unknown, but* $\sigma_1^2 = \sigma_2^2$.
In this case we note that if σ^2, the common value of σ_1^2 and σ_2^2, were known, then

$$\mathbf{z} = \frac{\bar{\mathbf{x}}_1 - \bar{\mathbf{x}}_2}{\sigma\sqrt{\dfrac{1}{n_1} + \dfrac{1}{n_2}}}$$

has a $N(0, 1)$ distribution and is used as in Case I to test the null hypothesis.

When σ^2 is unknown, one uses the test statistic

$$\mathbf{u} = \frac{\bar{\mathbf{x}}_1 - \bar{\mathbf{x}}_2}{\sqrt{\dfrac{\sum_{i=1}^{n_1}(\mathbf{x}_{i1} - \bar{\mathbf{x}}_1)^2 + \sum_{i=1}^{n_2}(\mathbf{x}_{i2} - \bar{\mathbf{x}}_2)^2}{n_1 + n_2 - 2}\left(\dfrac{1}{n_1} + \dfrac{1}{n_2}\right)}},$$

which has a t distribution with $n_1 + n_2 - 2$ degrees of freedom. Here in our example:

$$u = \frac{3.39744 - 3.23288}{\sqrt{\dfrac{17 \times (1.05174)^2 + 16 \times (1.13871)^2}{18 + 17 - 2}\left(\dfrac{1}{18} + \dfrac{1}{17}\right)}} = 0.4445,$$

and again we would conclude that the means were not different.

Case III. σ_1^2, σ_2^2 *unknown,* $\sigma_1^2 \neq \sigma_2^2$.
Here the obvious statistic is the sample correlative of the $\mathbf{z}$ of Case I, namely

$$\mathbf{w} = \frac{\bar{\mathbf{x}}_1 - \bar{\mathbf{x}}_2}{\sqrt{\dfrac{\mathbf{s}_1^2}{n_1} + \dfrac{\mathbf{s}_2^2}{n_2}}}.$$

One problem with this statistic is that its sampling distribution is not functionally independent of the parameters σ_1^2 and σ_2^2. Thus a test of the hypothesis $\mu_1 = \mu_2$ based on this statistic will not be a "similar" test, i.e., will not have a constant significance level independent of the values of σ_1^2 and σ_2^2. (This is by contrast to the situation in Case II, wherein the sampling distribution of $\mathbf{u}$ is independent of σ^2.) Moreover, the exact sampling distribution of $\mathbf{w}$ has not been tabulated.

Fortunately, a reasonably good approximation to the sampling distribution of $\mathbf{w}$ has been developed by Welch [1947], and this approximation is independent of σ_1^2 and σ_2^2. This approximation uses the student t dsitribution with **f** degrees of freedom as the surrogate for the exact distribution of **w**, where **f** is given by

$$\mathbf{f} = \frac{\left(\dfrac{\mathbf{s}_1^2}{n_1} + \dfrac{\mathbf{s}_2^2}{n_2}\right)^2}{\dfrac{\left(\dfrac{\mathbf{s}_1^2}{n_1}\right)^2}{n_1 + 1} + \dfrac{\left(\dfrac{\mathbf{s}_2^2}{n_2}\right)^2}{n_2 + 1}} - 2.$$

Here in our example:

$$w = \frac{3.39744 - 3.23288}{\sqrt{\dfrac{1.10616}{18} + \dfrac{1.29666}{17}}} = 0.4434,$$

and f is computed to be 34.341. We thus compare 0.4434 with the 95% point of a t distribution with 34 degrees of freedom and accept the hypothesis that the means are equal.

Case IV. *Matched samples.*

A special two-sample comparison-of-means situation is one wherein we observe n pairs $(\mathbf{x}_1, \mathbf{y}_1), \ldots, (\mathbf{x}_n, \mathbf{y}_n)$ drawn at random from a bivariate normal distribution with mean vector $\boldsymbol{\mu}' = (\mu_1, \mu_2)$ and covariance matrix

$$\Sigma = \begin{bmatrix} \sigma_1^2 & \rho\sigma_1\sigma_2 \\ \rho\sigma_1\sigma_2 & \sigma_2^2 \end{bmatrix}.$$

If we create new random variables $\mathbf{z}_i = \mathbf{x}_i - \mathbf{y}_i$, $i = 1, \ldots, n$, then the $\mathbf{z}$'s are distributed as $N(\mu_1 - \mu_2, \sigma_1^2 + \sigma_2^2 - 2\rho\sigma_1\sigma_2)$. With respect to our null hypothesis that $\mu_1 = \mu_2$, it can be viewed as the null hypothesis that $\mathscr{E}\mathbf{z} = 0$ which is tested by the one-sample t test based on

$$\mathbf{t} = \frac{\bar{\mathbf{z}} - 0}{\sqrt{\dfrac{\sum_{i=1}^{n} (\mathbf{z}_i - \bar{\mathbf{z}})^2}{n(n-1)}}}$$

$$= \frac{\bar{\mathbf{x}} - \bar{\mathbf{y}}}{\sqrt{\dfrac{\sum_{i=1}^{n} (\mathbf{x}_i - \mathbf{y}_i - \bar{\mathbf{x}} + \bar{\mathbf{y}})^2}{n(n-1)}}}$$

$$= \frac{\bar{\mathbf{x}} - \bar{\mathbf{y}}}{\sqrt{\dfrac{\mathbf{s}_1^2 + \mathbf{s}_2^2 - 2\sum_{i=1}^{n} (\mathbf{x}_i - \bar{\mathbf{x}})(\mathbf{y}_i - \bar{\mathbf{y}})/(n-1)}{n}}}.$$

(Note that the denominator is an estimator of the standard deviation of $\bar{\mathbf{z}}$.)

Table 2. Paired Comparison of Operating Costs for Same Driver.

Chevrolet	Plymouth	Differential
3.92600	2.43000	1.49600
3.45000	2.98000	0.47000
2.00000	3.04000	−1.04000
2.28000	4.94000	−2.66000
3.49400	3.15000	0.34400
4.25000	2.46000	1.79000
2.38200	3.34000	−0.95800
3.02000	2.38400	0.63600
3.26000	2.27000	0.99000
4.08000	2.52400	1.55600
3.67000	3.10000	0.57000
2.94000	3.53000	−0.59000
5.90000	3.06000	2.84000
2.18000	2.57000	−0.39000
5.39000	3.48000	1.91000
2.74000	5.94000	−3.20000
3.49200	2.51600	0.97600
2.70000	1.61000	1.09000

This statistic has a t distribution with $n - 1$ degrees of freedom, in contrast to the two sample t test based on independent samples which has **f** (on the order of magnitude of $2n$) degrees of freedom. But, for example, if **x** and **y** are positive correlated, the denominator of **t** will be smaller than that of **w**, so that **t** will be larger than **w**. Thus if one used the two-sample test based on **w** and rejected the null hypothesis, one could pretty well say that the one-sample sample test based on **t** would have rejected the null hypothesis.

Moreover, as in Case I, this test is uniformly most powerful among unbiased procedures (Lehmann [1959], p. 206).

Here as an example we compare the Chevrolet data with the Plymouth data, assuming that the underlying basis for the matching of the pairs of observations in Table 2 is that the same driver drove both cars in the matched pair. Here $\bar{z} = 0.32388$ and the standard deviation of the z's is 1.56606, so that

$$t = \frac{0.32388}{\sqrt{(1.56606)^2/18}} = 0.8774.$$

Again the hypothesis of equality of means is accepted. By contrast, if we had ignored the correlation between the Chevrolet and Plymouth data ($r = -0.162$), pooled the two standard deviations 1.05174 and 1.00239 to obtain 1.05547, and calculated

$$u = \frac{0.32388}{1.05547 \times \sqrt{\frac{1}{18} + \frac{1}{18}}} = 0.9206,$$

we would have obtained (incorrectly) a larger critical value.

Theoretical Background

When σ^2 is unknown, one is tempted to use a statistic of the same form, but with an estimate of σ replacing the σ in the denominator of $\mathbf{z}$. One genre of estimators is based on a weighted average of the separate estimators of σ_1^2 and σ_2^2, i.e.,

$$\mathbf{s}_\lambda^2 = \lambda \frac{\sum_{i=1}^{n_1} (\mathbf{x}_{i1} - \overline{\mathbf{x}}_1)^2}{n_1 - 1} + (1 - \lambda) \frac{\sum_{i=1}^{n_2} (\mathbf{x}_{i2} - \overline{\mathbf{x}}_2)^2}{n_2 - 1} = \lambda \mathbf{s}_1^2 + (1 - \lambda)\mathbf{s}_2^2,$$

where $0 \le \lambda \le 1$. Another potential estimator, correct under the null hypothesis, is to pool both samples and calculate a pooled estimate of σ^2, namely

$$\mathbf{s}_P^2 = \frac{\sum_{i=1}^{2} \sum_{j=1}^{n_i} (\mathbf{x}_{ij} - \overline{\mathbf{x}})^2}{n_1 + n_2 - 1},$$

where $\overline{\mathbf{x}} = (n_1 \overline{\mathbf{x}}_1 + n_2 \overline{\mathbf{x}}_2)/(n_1 + n_2)$.

One can of course determine the sampling distribution of the "$\mathbf{z}$" statistic for each choice of denominator. But a few observations are in order. First, $\overline{\mathbf{x}}_1 - \overline{\mathbf{x}}_2$ is not independent of $\mathbf{s}_P^2$, so that the "$\mathbf{z}$" using $\mathbf{s}_P^2$ does not have the t distribution. Second, although $\mathbf{s}_\lambda^2$ is independent of $\overline{\mathbf{x}}_1 - \overline{\mathbf{x}}_2$ for all λ, only for $\lambda = (n_1 - 1)/(n_1 + n_2 - 2)$ does $\mathbf{s}_\lambda^2$ have a $\sigma^2\chi^2(f)/f$ distribution, with $f = n_1 + n_2 - 2$. Thus the statistic $\mathbf{u}$ given above has a t distribution with $n_1 + n_2 - 2$ degrees of freedom. Moreover, this test is uniformly most powerful among unbiased test procedures (Lehmann [1959], pp. 172–73) and uniformly most powerful invariant with respect to the group of scale and translation transformations of x and y (Lehmann [1959], p. 251).

(b) Nonparametric Procedures

Exact procedures for comparisons of means of two nonnormal populations have been developed for a few distributions. (For example, for comparing means of exponential distributions, see Epstein and Tsao [1953], and for comparing uniform distributions, see Barr [1966]. The comparison of two binomials is equivalent to the analysis of the 2×2 contingency table treated elsewhere in this book, and reviewed in Paulson and Wallis [1947].) Approximate procedures rely on the asymptotic normality of the sample means, and so rely on the procedures given in Section 1(a) above and are merely approximate. An alternative approach to the comparison of means is to eschew any distributional assumptions and to develop procedures which work for samples from all (or almost all) distributions. Strictly speaking, these procedures compare population medians, since for some populations (e.g., those governed by the Cauchy distribution), the population mean does not exist.

We consider herein only procedures for testing differences in location, i.e., that there is a difference in medians for two populations whose population densities are otherwise identical. (Amongst the cases considered earlier, this corresponds to the generalization of Case II.) Moreover, we only consider procedures based on the ranks of the observations from the two populations. We rely on the theorem of Hájek and Sĭdák ([1967], p. 67) which characterizes the locally most powerful rank test for shift in location, i.e., the test which is uniformly most powerful for small changes in location. Their theorem states that the optimal test is based on the statistic

$$\mathbf{s} = \sum_{i=1}^{n_1} a_{\mathbf{r}_{1i}}(f),$$

where the $\mathbf{r}_{1i}$ are the ranks of the $\mathbf{x}_{1i}$ in the combined sample and

$$a_i(f) = \mathscr{E}\left[\frac{-f'(\mathbf{x}_i)}{f(\mathbf{x}_i)}\right].$$

Nonparametric methods based on ranks for testing the hypothesis of equality of location have been invented without recourse to the optimality theory just referenced, simply because each procedure "works" for *all* possible underlying distributions. This result, though, allows us to look carefully at the adduced nonparametric procedures and discover for which distribution the procedure is optimal in the sense defined above. We will thus, as we present the four most prominent nonparametric procedures, also point out the underlying distribution for which the procedure is optimal.

Moreover, it has been shown that each of the statistics has an asymptotically normal sampling distribution. In addition to a presentation of tables of percentage points of the exact sampling distribution of each of these statistics for small samples, we shall therefore also cite the parameters of the asymptotic normal distribution of these statistics.

If the **x**'s come from a normal distribution, the optimal statistic reduces to

$$\mathbf{s} = \sum_{i=1}^{n_1} \Phi^{-1}(u_{\mathbf{r}_{1i}}),$$

where u_j is the jth order statistic from a uniform (0, 1) distribution. This test statistic is called the *normal score* statistic and was first proposed by Fisher and Yates [1938]. The most extensive set of tables of $\Phi^{-1}(u_j)$ are those of Harter [1969]. Tables of critical value of **s** for $n \leq 20$ are given by Klotz [1964] and reproduced as Table 3. The statistic **s** has an asymptotic normal distribution with mean 0 and variance

$$\frac{n_1 n_2}{(n_1 + n_2)(n_1 + n_2 + 1)} \sum_{i=1}^{n_1+n_2} [\Phi^{-1}(u_i)]^2.$$

One can approximate $a_i(f)$ by

$$\begin{aligned} a_i(f) &\simeq \Phi(\mathscr{E}\mathbf{u}_i, f) \\ &= \Phi\left(\frac{i}{n+1}, f\right). \end{aligned}$$

Table 3. Critical Values of the Normal Score Statistic.

s $P\{S \geq s\}$	†		s $P\{S \geq s\}$	†		s $P\{S \geq s\}$	†		s $P\{S \geq s\}$	†	
6	3	2.11051 0.05000	12	4	2.54800 0.06051	16	3	2.48073 0.05000	19	2	2.10822 0.05263
7	2	2.10965 0.04762	12	5	2.65507 0.06061	16	4	2.75626 0.06000	19	3	2.53302 0.04954
7	3	2.10955 0.05714	12	6	2.74496 0.06195	16	5	2.93686 0.04991	19	4	2.82201 0.05005
8	2		13	2	2.05632 0.05128	16	6	3.06730 0.05007	19	5	3.06191 0.04997
8	3	2.12331 0.05367	13	3	2.44374 0.04896	16	7	3.15249 0.04991	19	6	3.22606 0.04968
8	4	2.27583 0.05714	13	4	2.61754 0.06035	16	8	3.15989 0.06004	19	7	3.34991 0.04999
9	2	2.05695 0.05556	13	5	2.77561 0.04973	17	2	2.12617 0.06147	19	8	3.42021 0.06000
9	3	2.33151 0.04762	13	6	2.83207 0.05128	17	3	2.49423 0.05000	19	9	3.45902 0.06000
9	4	2.41731 0.04762	14	2	2.10903 0.05496	17	4	2.78845 0.06000	20	2	2.18241 0.04737
10	2	2.19481 0.04444	14	3	2.45330 0.04946	17	5	2.96939 0.06026	20	3	2.55009 0.05000
10	3	2.31748 0.05000	14	4	2.67192 0.04995	17	6	3.11859 0.06018	20	4	2.86646 0.04096
10	4	2.44791 0.05238	14	6	2.82312 0.04996	17	7	3.21996 0.06003	20	5	3.07968 0.04999
10	5	2.57068 0.04762	14	6	2.91128 0.05162	17	8	3.26458 0.05006	20	6	3.26636 0.06000
11	2	2.04841 0.05455	14	7	2.94835 0.05012	18	2	2.17087 0.06229	20	7	3.39273 0.06003
11	3	2.31528 0.04848	15	2	2.19562 0.04782	18	3	2.62936 0.06026	20	8	3.48683 0.06000
11	4	2.54017 0.04848	15	3	2.46816 0.06065	18	4	2.80709 0.06000	20	9	3.54228 0.04999
11	5	2.60838 0.04978	15	4	2.69622 0.04982	18	5	3.01899 0.06007	20	10	3.55706 0.04997
12	2	2.16607 0.04545	15	5	2.88136 0.04995	18	6	3.17544 0.04977			
12	3	2.42271 0.05000	15	6	2.98502 0.04975	18	7	3.27683 0.04999			
			15	7	3.04666 0.04988	18	8	3.34539 0.04989			
			16	2	2.16221 0.05000	18	9	3.37553 0.06004			

SOURCE: Klotz [1964]. Reprinted from the *Journal of the American Statistical Association*, Vol. 59, with the permission of the ASA.

Using this approximation, van der Waerden [1953] proposed the statistic

$$\mathbf{s}' = \sum_{i=1}^{n_1} \Phi^{-1}\left(\frac{\mathbf{r}_{1i}}{n+1}\right).$$

Tables of critical values for $n \leq 50$, $|n_1 - n_2| \leq 5$, are given in van der Waerden and Nievergelt [1956] and are reproduced as Table 4. The statistic $\mathbf{s}'$ has an asymptotic normal distribution mean 0 and variance

$$\frac{n_1 n_2}{(n_1 + n_2)(n_1 + n_2 + 1)} \sum_{i=1}^{n_1+n_2} \left[\Phi^{-1}\left(\frac{i}{n_1 + n_2 + 1}\right)\right]^2.$$

If f is of the logistic type, then the statistic $\mathbf{s}$ reduces to

$$\mathbf{s} = \sum_{i=1}^{n_1} \mathbf{r}_{1i},$$

which is the well-known Wilcoxon [1945] statistic for testing for location. (An equivalent procedure is based on the Mann–Whitney [1947] statistic $\mathbf{t} = n_1 n_2 + n_2(n_2 + 1)/2 - \mathbf{s}$. The most extensive tables of the Mann–Whitney statistic are those of Milton [1964] for $n_1 \leq 20$, $n_2 \leq 40$, reproduced as Table 5. Thus critical values of $\mathbf{s}$ are equal to $n_1 n_2 + n_2(n_2 + 1)/2$ minus the critical values of $\mathbf{t}$.) The statistic $\mathbf{s}$ has an asymptotic normal distribution with mean $n_1(n_1 + n_2 + 1)/2$ and variance $n_1 n_2(n_1 + n_2 + 1)/12$.

If f is the double-exponential density, then the approximation to $\mathbf{s}$ produces the statistic

$$\mathbf{s}' = \sum_{i=1}^{n_1} \operatorname{sign}\left[\mathbf{r}_{1i} - \frac{(n+1)}{2}\right].$$

Tables of the distribution of $\mathbf{s}'$ are given by Westenberg [1952] and reproduced as Table 6. A related statistic, $\mathbf{s} = (\mathbf{s}' + n_1)/2$, is called the median test statistic, mentioned in Mood [1950]. The statistic $\mathbf{s}$ has an asymptotic normal distribution with mean $n_1/2$ and variance given by

$$\frac{n_1 n_2}{4(n_1 + n_2 - 1)} \quad \text{if } n_1 + n_2 \text{ is even,}$$

$$\frac{n_1 n_2}{4(n_1 + n_2)} \quad \text{if } n_1 + n_2 \text{ is odd.}$$

Once again we consider the Chevrolet and Ford data to illustrate the nonparametric techniques. Table 7 presents the basic ingredients of the nonparametric procedures. The first column (OPCOST) provides the ranked operating costs for the 35 automobiles. This is followed by a column (SOURCE) with a 1 if the auto is a Chevrolet, 2 if it is a Ford. Column 3 (RANK) records the rank of each observation. NSCORE in column 4 is the set of normal scores $\Phi^{-1}(u_j)$ for a sample size 35. Column 5 (VDW) contains $r_i/36$, which is used to determine VANDW, $\Phi^{-1}(r_i/36)$, which is recorded in column 6. Column 7 (MED) contains $r_i = 18$, and column 8 contains SRANK, the sign of the entry in column 7.

Table 4. Critical Values of the van der Waerden Statistic.

	One-sided 2.5%				One-sided 1%				One-sided 0.5%		
n	$g - k =$ 0 or 1	$g - k =$ 2 or 3	$g - k =$ 4 or 5	n	$g - k =$ 0 or 2	$g - k =$ 2 or 3	$g - k =$ 4 or 5	n	$g - k =$ 0 or 1	$g - k =$ 2 or 3	$g - k =$ 4 or 5
6	∞	∞	∞	6	∞	∞	∞	6	∞	∞	∞
7	∞	∞	∞	7	∞	∞	∞	7	∞	∞	∞
8	2.40	2.30	∞	8	∞	∞	∞	8	∞	∞	∞
9	2.38	2.20	∞	9	2.80	∞	∞	9	∞	∞	∞
10	2.60	2.49	2.30	10	3.00	2.90	2.80	10	3.20	3.10	∞
11	2.72	2.58	2.40	11	3.20	3.0	2.90	11	3.40	3.40	∞
12	2.86	2.79	2.68	12	3.29	3.30	3.20	12	3.60	3.58	3.40
13	2.96	2.91	2.78	13	3.50	3.36	3.18	13	3.71	3.68	3.50
14	3.11	3.06	3.00	14	3.62	3.55	3.46	14	3.94	3.88	3.76
15	3.24	3.19	3.06	15	3.74	3.68	3.57	15	4.07	4.05	3.88
16	3.39	3.36	3.28	16	3.92	3.90	3.80	16	4.26	4.25	4.12
17	3.49	3.44	3.36	17	4.06	4.01	3.90	17	4.44	4.37	4.23
18	3.63	3.60	3.53	18	4.23	4.21	4.14	18	4.60	4.58	4.50
19	3.73	3.69	3.61	19	4.37	4.32	4.23	19	4.77	4.71	4.62
20	3.86	3.84	3.78	20	4.52	4.50	4.44	20	4.94	4.92	4.85
21	3.96	3.92	3.85	21	4.66	4.62	4.53	21	5.10	5.05	4.96
22	4.08	4.06	4.01	22	4.80	4.78	4.72	22	5.26	5.24	5.17
23	4.18	4.15	4.08	23	4.92	4.89	4.81	23	5.40	5.36	5.27
24	4.29	4.27	4.23	24	5.06	5.04	4.99	24	5.55	5.53	5.48
25	4.39	4.36	4.30	25	5.18	5.14	5.08	25	5.68	5.65	5.58
26	4.50	4.48	4.44	26	5.30	5.29	5.24	26	5.83	5.81	5.76
27	4.59	4.56	4.51	27	5.42	5.39	5.33	27	5.95	5.92	5.85
28	4.69	4.68	4.64	28	5.54	5.52	5.48	28	6.09	6.07	6.03
29	4.78	4.76	4.72	29	5.65	5.62	5.57	29	6.22	6.19	6.13
30	4.88	4.87	4.84	30	5.77	5.75	5.72	30	6.35	6.34	6.30
31	4.97	4.95	4.91	31	5.87	5.85	5.80	31	6.47	6.44	6.39
32	5.07	5.06	5.03	32	5.99	5.97	5.94	32	6.60	6.58	6.55
33	5.15	5.13	5.10	33	6.09	6.07	6.02	33	6.71	6.69	6.64
34	5.25	5.24	5.21	34	6.20	6.19	6.16	34	6.84	6.82	6.79
35	5.33	5.31	5.28	35	6.30	6.28	6.24	35	6.95	6.92	6.88
36	5.42	5.41	5.38	36	6.40	6.39	6.37	36	7.06	7.05	7.02
37	5.50	5.48	5.45	37	6.50	6.48	6.45	37	7.17	7.15	7.11
38	5.59	5.58	5.55	38	6.60	6.59	6.57	38	7.28	7.27	7.25
39	5.67	5.65	5.62	39	6.70	6.68	6.65	39	7.39	7.37	7.33
40	5.75	5.74	5.72	40	6.80	6.79	6.77	40	7.50	7.49	7.47
41	5.83	5.81	5.79	41	6.89	6.88	6.85	41	7.62	7.60	7.56
42	5.91	5.90	5.88	42	6.99	6.98	6.96	42	7.72	7.71	7.69
43	5.99	5.97	5.95	43	7.08	7.07	7.04	43	7.82	7.81	7.77
44	6.06	6.06	6.04	44	7.17	7.17	7.14	44	7.93	7.92	7.90
45	6.14	6.12	6.10	45	7.26	7.25	7.22	45	8.02	8.01	7.98
46	6.21	6.21	6.19	46	7.35	7.35	7.32	46	8.13	8.12	8.10
47	6.29	6.27	6.25	47	7.44	7.43	7.40	47	8.22	8.21	8.18
48	6.36	6.35	6.34	48	7.53	7.52	7.50	48	8.32	8.31	8.29
49	6.43	6.42	6.39	49	7.61	7.60	7.57	49	8.41	8.40	8.37
50	6.50	6.50	6.48	50	7.70	7.69	7.68	50	8.51	8.50	8.48
	Two-sided 5%				Two-sided 2%				Two-sided 1%		

SOURCE: van der Waerden and Nievergelt [1956]. Reprinted from *Tafeln zum Vergleich zweier Stichprobem mittels X test and Zeichentest* with permission from Springer-Verlag. (Their g is our n_2, their k is our n_1.)

Table 5. Critical Value of the Wilcoxon Statistic.
Table of the greatest integer u such that $\Pr(U_{y<s} \leq u) \leq 0.05$.

m	n = 1	2	3	4	5	6	7	8	9	10	11	12	13	14	15	16	17	18	19	20
1	—																			
2	—	—																		
3	—	—	0																	
4	—	—	0	1																
5	—	0	1	2	4															
6	—	0	2	3	5	7														
7	—	0	2	4	6	8	11													
8	—	1	3	5	8	10	13	15												
9	—	1	4	6	9	12	15	18	21											
10	—	1	4	7	11	14	17	20	24	27										
11	—	1	5	8	12	16	19	23	27	31	34									
12	—	2	5	9	13	17	21	26	30	34	38	42								
13	—	2	6	10	15	19	24	28	33	37	42	47	51							
14	—	3	7	11	16	21	26	31	36	41	46	51	56	61						
15	—	3	7	12	18	23	28	33	39	44	50	55	61	66	72					
16	—	3	8	14	19	25	30	36	42	48	54	60	65	71	77	83				
17	—	3	9	15	20	26	33	39	45	51	57	64	70	77	83	89	96			
18	—	4	9	16	22	28	35	41	48	55	61	68	75	82	88	95	102	109		
19	0	4	10	17	23	30	37	44	51	58	65	72	80	87	94	101	109	116	123	
20	0	4	11	18	25	32	39	47	54	62	69	77	84	92	100	107	115	123	130	138
21	0	5	11	19	26	34	41	49	57	65	73	81	89	97	105	113	121	130	138	146
22	0	5	12	20	28	36	44	52	60	68	77	85	94	102	111	119	128	136	145	154
23	0	5	13	21	29	37	46	54	63	72	81	90	98	107	116	125	134	143	152	161
24	0	6	13	22	30	39	48	57	66	75	85	94	103	113	122	131	141	150	160	169
25	0	6	14	23	32	41	50	60	69	79	89	98	108	118	128	137	147	157	167	177
26	0	6	15	24	33	43	53	62	72	82	92	103	113	123	133	143	154	164	174	185
27	0	7	15	25	35	45	55	65	75	86	96	107	117	128	139	149	160	171	182	192
28	0	7	16	26	36	46	57	68	78	89	100	111	122	133	144	156	167	178	189	200
29	0	7	17	27	38	48	59	70	82	93	104	116	127	138	150	162	173	185	196	208
30	0	7	17	28	39	50	61	73	85	96	108	120	132	144	156	168	180	192	204	216
31	0	8	18	29	40	52	64	76	88	100	112	124	136	149	161	174	186	199	211	224
32	0	8	19	30	42	54	66	78	91	103	116	128	141	154	167	180	193	206	218	231
33	0	8	19	31	43	56	68	81	94	107	120	133	146	159	172	186	199	212	226	239
34	0	9	20	32	45	57	70	84	97	110	124	137	151	164	178	192	206	219	233	247
35	0	9	21	33	46	59	73	86	100	114	128	141	156	170	184	198	212	226	241	255
36	0	9	21	34	48	61	75	89	103	117	131	146	160	175	189	204	219	233	248	263
37	0	10	22	35	49	63	77	91	106	121	135	150	165	180	195	210	225	240	255	271
38	0	10	23	36	50	65	79	94	109	124	139	154	170	185	201	216	232	247	263	278
39	1	10	23	38	52	67	82	97	112	128	143	159	175	190	206	222	238	254	270	286•
40	1	11	24	39	53	68	84	99	115	131	147	163	179	196	212	228	245	261	278	294•

SOURCE: Milton [1964]. Reprinted from the *Journal of the American Statistical Association*, Vol. 59, with the permission of the ASA.

Table 8 reproduces the relevant variables from Table 7 strictly for the Ford sample. Thus the normal score statistic sums the entries in NSCORE and produces $s = -1.93$. Since $\sum_{i=1}^{35} [\Phi^{-1}(n_i)]^2 = 32.515$, the asymptotic standard deviation of **s** is

$$\sqrt{\frac{17 \times 18}{35 \times 36} \times 32.515} = \sqrt{7.897} = 2.81,$$

so that s is less than one standard deviation from its asymptotic mean of 0, and so we accept the hypothesis of no difference in location.

Table 6. Critical Values of the Median Test Statistic.

			$N_1 = 6$	10	20	50	100	200	500	1000	2000	
											31.5	
											33.0	
											34.8	
2000											37.2	2000
											41.2	
											44.9	
											52.5	
										22.4	25.8	
										23.4^5	27.0	
										24.8	28.4	
1000										26.5	30.7	1000
										29.3	33.7	
										31.9	36.8	
										37.2	43.1	
									16.0	18.4	20.1	
									16.7	19.3	21.0	
									17.6	20.3	22.2	
500									18.8	21.8	23.8	500
									20.8	24.0	26.2	
									22.7	26.1	28.5	
									26.4	30.5^5	33.3	
								10.3	12.2	13.1	13.7	
								10.7	12.8	13.7	14.3	
								11.3	13.4	14.5	15.1	
200								12.1	14.4	15.5^5	16.2	200
								13.3^5	15.9	17.1	17.8	
								14.5	17.2	18.6	19.4	
								16.9	20.1	21.7	22.6	
							7.4	8.5	9.4	9.8	10.0^5	
							7.7	8.9	9.9	10.3	10.5	
							8.1^5	9.4	10.4	10.8	11.1	
100							8.6^5	10.0	11.1	11.5	11.8	100
							9.5	11.0	12.2	12.7	13.0	
							10.4	11.9	13.2	13.8	14.1^5	
							12.1	13.9	15.4	16.1	16.4	
						5.4	6.2	6.7	7.1	7.2	7.3	
						5.6	6.4	7.0	7.4	7.5	7.6	
						5.9	6.7	7.3	7.8	7.9^5	8.0^5	
50						6.3	7.1	7.8	8.3	8.5	8.6	
						6.9	7.9	8.6	9.1	9.3	9.4	50
						7.5	8.5	9.3	9.9	10.1	10.1^5	
						8.6	9.9	10.8	11.5	11.7	11.8	
					3.5^5	4.2	4.4^5	4.6^5	4.8	4.8	4.8	
					3.7	4.4	4.6	4.8	4.9^5	5.0	5.0	
					3.9	4.6	4.8^5	5.1	5.2	5.2^5	5.2^5	
20					4.2	4.9	5.2	5.3	5.5	5.6	5.6	20
					4.6	5.3	5.6	5.9	6.0	6.1	6.1	
					4.9	5.7	6.1	6.3	6.4^5	6.5^5	6.5^5	
					5.6	6.5	7.0	7.3	7.4	7.5	7.5	
				2.6	3.0	3.3	3.3^5	3.3^5	3.5	3.5	3.5	
				2.7	3.1	3.4	3.5	3.5	3.6^5	3.6^5	3.6^5	
				2.8^5	3.2	3.5^5	3.7	3.7	3.8	3.8	3.8	
10				3.1	3.4	3.7	3.9	3.9^5	4.0	4.0	4.0	10
				3.4	3.7	4.1	4.3	4.3	4.3	4.3	4.3	
				3.6	4.0	4.3^5	4.5	4.5^5	4.5^5	4.6	4.6	
				4.0^5	4.6	4.9	—	—	—	—	—	
	$2\frac{1}{2}\%$	5%	$\lvert\delta\rvert = 2.1^5$	2.3^5	2.5	2.6^5	2.7^5	2.8	2.8	2.8	2.8	
	2%	4%	2.2^5	2.4	2.6	2.7^5	2.8^5	2.9	2.9	2.9	2.9	
	$1\frac{1}{2}\%$	3%	2.4	2.5	2.7	2.9	2.9^5	3.0	—	—	—	
$N_2 = 6$	1%	2%	2.5	2.6^5	2.9	—	—	—	—	—	—	$N_2 = 6$
	$\frac{1}{2}\%$	1%	2.7	2.9	—	—	—	—	—	—	—	
	$\frac{1}{4}\%$	$\frac{1}{2}\%$	2.8^5	—	—	—	—	—	—	—	—	
	$\frac{1}{20}\%$	$\frac{1}{10}\%$	—	—	—	—	—	—	—	—	—	
	Unilateral tailerror	Bilateral tailerror	$N_1 = 6$	10	20	50	100	200	500	1000	2000	

SOURCE: Westenberg [1952]. Reprinted from the *Proceedings Koninklijke Nederlandse Akademic van Wetenschappen*, Vol. 55, with the permission of North-Holland Publishing Company.

Table 7. Computations Underlying Various Nonparametric Tests of Automobile Operating Cost Data.

UPCOST	SOURCE	RANK	NSCORE	VDW	VANDW	MED	SRANK
1.56000	2.00000	1.00000	−2.10661	0.02778	−1.91493	−17.00000	−1.00000
1.74000	2.00000	2.00000	−1.69023	0.05556	−1.59356	−16.00000	−1.00000
1.97000	2.00000	3.00000	−1.44762	0.08333	−1.38323	−15.00000	−1.00000
2.00000	1.00000	4.00000	−1.26860	0.11111	−1.22078	−14.00000	−1.00000
2.13000	2.00000	5.00000	−1.12295	0.13889	−1.08536	−13.00000	−1.00000
2.17000	2.00000	6.00000	−0.99790	0.16667	−0.96736	−12.00000	−1.00000
2.18000	1.00000	7.00000	−0.88681	0.19444	−0.86149	−11.00000	−1.00000
2.28000	1.00000	8.00000	−0.78574	0.22222	−0.76448	−10.00000	−1.00000
2.36000	2.00000	9.00000	−0.69214	0.25000	−0.67419	−9.00000	−1.00000
2.38200	1.00000	10.00000	−0.60427	0.27778	−0.58910	−8.00000	−1.00000
2.68000	2.00000	11.00000	−0.52084	0.30556	−0.50808	−7.00000	−1.00000
2.70000	1.00000	12.00000	−0.44091	0.33333	−0.43029	−6.00000	−1.00000
2.74000	1.00000	13.00000	−0.36371	0.36111	−0.35505	−5.00000	−1.00000
2.87000	2.00000	14.00000	−0.28863	0.38889	−0.28179	−4.00000	−1.00000
2.94000	1.00000	15.00000	−0.21515	0.41667	−0.21005	−3.00000	−1.00000
3.02000	1.00000	16.00000	−0.14282	0.44444	−0.13941	−2.00000	−1.00000
3.17000	2.00000	17.00000	−0.07123	0.47222	−0.06951	−1.00000	−1.00000
3.26000	1.00000	18.00000	0.00000	0.50000	0.00000	0.00000	−1.00000
3.31000	2.00000	19.00000	0.07123	0.52778	0.06951	1.00000	1.00000
3.45000	1.00000	20.00000	0.14282	0.55556	0.13941	2.00000	1.00000
3.49200	1.00000	21.00000	0.21515	0.58333	0.21005	3.00000	1.00000
3.49400	1.00000	22.00000	0.28863	0.61111	0.28179	4.00000	1.00000
3.67000	1.00000	23.00000	0.36371	0.63889	0.35505	5.00000	1.00000
3.92600	1.00000	24.00000	0.44091	0.66667	0.43029	6.00000	1.00000
3.93400	2.00000	25.00000	0.52084	0.69444	0.50808	7.00000	1.00000
4.08000	1.00000	26.00000	0.60427	0.72222	0.58910	8.00000	1.00000
4.15000	2.00000	27.00000	0.69214	0.75000	0.67419	9.00000	1.00000
4.25000	1.00000	28.00000	0.78574	0.77778	0.76448	10.00000	1.00000
4.29000	2.00000	29.00000	0.88681	0.80556	0.86149	11.00000	1.00000
4.55000	2.00000	30.00000	0.99790	0.83333	0.96736	12.00000	1.00000
4.68600	2.00000	31.00000	1.12295	0.86111	1.08536	13.00000	1.00000
4.68900	2.00000	32.00000	1.26860	0.88889	1.22078	14.00000	1.00000
4.70000	2.00000	33.00000	1.44762	0.91667	1.38323	15.00000	1.00000
5.39000	1.00000	34.00000	1.69023	0.94444	1.59356	16.00000	1.00000
5.90000	1.00000	35.00000	2.10661	0.97222	1.91493	17.00000	1.00000

The van der Waerden statistic is the sum of entries in VANDW, and so $s' = -1.708$. Since $\sum_{i=1}^{35} [\Phi^{-1}(i/36)]^2 = 29.138235$, the asymptotic standard deviation of $\mathbf{s}'$ is less than one standard deviation from its asymptotic mean of 0, and so we accept the null hypothesis. To use the exact tables, we compare s' with the value associated with $n = 35$, $n_2 - n_1 = 1$, of Table 4, namely -5.33, and accept the null hypothesis.

The Wilcoxon statistic is the sum of RANK, namely $s = 294$, and its asymptotic mean and standard deviation are $17 \times 36/2 = 306$ and $\sqrt{17 \times 18 \times 36/12} = \sqrt{918} = 30.299$. Since s is less than one standard deviation from its asymptotic mean, we accept the null hypothesis. Alternatively, we can use Table 5 and compare 294 with the critical value associated with $n_1 = 17$, $n_2 = 18$, namely $17 \times 18 + 18 \times 19/2 - 102 = 375$, and conclude that the null hypothesis is accepted.

The median test statistic is based on the sum of SRANK, namely $s' = 1$, so that $s = (1 + 17)/2 = 9$ and the asymptotic mean and standard deviation

Table 8. Computations Underlying Various Nonparametric Tests of Ford Operating Cost Data.

OPCOST	RANK	NSCORE	VDW	VANDW	MED	SRANK
1.56000	1.00000	−2.10661	0.02778	−1.91493	−17.00000	−1.00000
1.74000	2.00000	−1.69023	0.05556	−1.59356	−16.00000	−1.00000
1.97000	3.00000	−1.44762	0.08333	−1.38323	−15.00000	−1.00000
2.13000	5.00000	−1.12295	0.13889	−1.08536	−13.00000	−1.00000
2.17000	6.00000	−0.99790	0.16667	−0.96736	−12.00000	−1.00000
2.36000	9.00000	−0.69214	0.25000	−0.67419	−9.00000	−1.00000
2.68000	11.00000	−0.52084	0.30556	−0.50808	−7.00000	−1.00000
2.87000	14.00000	−0.28863	0.38889	−0.28179	−4.00000	−1.00000
3.17000	17.00000	−0.07123	0.47222	−0.06951	−1.00000	−1.00000
3.31000	19.00000	0.07123	0.52778	0.06951	1.00000	1.00000
3.93400	25.00000	0.52084	0.69444	0.50808	7.00000	1.00000
4.15000	27.00000	0.69214	0.75000	0.67419	9.00000	1.00000
4.29000	29.00000	0.88681	0.80556	0.86149	11.00000	1.00000
4.55000	30.00000	0.99790	0.83333	0.96736	12.00000	1.00000
4.68600	31.00000	1.12295	0.86111	1.08536	13.00000	1.00000
4.68900	32.00000	1.26860	0.88889	1.22078	14.00000	1.00000
4.70000	33.00000	1.44762	0.91667	1.38323	15.00000	1.00000

of s are $17/2 = 8.5\sqrt{17 \times 18/4 \times 35} = \sqrt{2.1857} = 1.478$. Again the statistic is within one standard deviation from its mean, and so we accept the null hypothesis. Using Table 6, we compare $s' = 1$ with the 95% value for $n_1 = n_2 = 20$, namely 3.5, and see that the null hypothesis is accepted.

Theoretical Background

Let $\mathbf{u}_1 \leq \cdots \leq \mathbf{u}_n$ denote n ordered random variables from a uniform distribution on the interval (0, 1). Let f denote the common density of all the x's under the null hypothesis and F the common cumulative distribution function. Define $F^{-1}(u) = \inf\{x | F(x) \geq u\}$, and

$$\phi(u, f) = \frac{f'(F^{-1}(u))}{f(F^{-1}(u))}, \qquad 0 < u < 1.$$

Let

$$\begin{aligned} a_i(f) &= \mathscr{E}\phi(\mathbf{u}_i, f) \\ &= n\binom{n-1}{i-1}\int_0^1 \phi(u, f)u^i(1-u)^{n-i}\,du \\ &= n\binom{n-1}{i-1}\int_{-\infty}^{\infty} f'(x)[F(x)]^{i-1}[1-F(x)]^{n-i}\,dx \\ &= \mathscr{E}\left[\frac{-f'(\mathbf{x}_i)}{f(\mathbf{x}_i)}\right]. \end{aligned}$$

Now let $n = n_1 + n_2$, and suppose we combined the two samples into a single set of n observations, preserving the identity of the population from which the sample was drawn, and ranked the n observations. Let $\mathbf{r}_{11}, \ldots, \mathbf{r}_{1n_1}$ denote the rank of $\mathbf{x}_{11}, \ldots, \mathbf{x}_{1n_1}$ and $\mathbf{r}_{21}, \ldots, \mathbf{r}_{2n_2}$ denote the ranks of $\mathbf{x}_{21}, \ldots, \mathbf{x}_{2n_2}$ in this combined sample. Hájek and Sĭdák have shown that the locally optimal rank test of equality of locations of two populations is based on the statistic

$$\mathbf{s} = \sum_{i=1}^{n_1} a_{\mathbf{r}_{1i}}(f).$$

2. Analysis of Variance via Regression (Model I)

(a) One-Way Anova

We consider in this section the generalization of Case II above, wherein $\mathbf{x}_{i1}, \ldots, \mathbf{x}_{in_i}$ is a random sample from a $N(\mu_i, \sigma^2)$ population, $i = 1, \ldots, k$, and wherein the samples are independent of each other. The procedure for testing the hypothesis that $\mu_1 = \mu_2 = \cdots = \mu_k$, the one-way analysis of variance, is usually presented in the following prototypic table:

Source of variation	Sum of squares	df	Mean square
Between populations	$\sum_{i=1}^{k} n_i(\overline{\mathbf{x}}_{i\cdot} - \overline{\mathbf{x}}_{\cdot\cdot})^2$	$k - 1$	$\sum_{i=1}^{k} n_i(\overline{\mathbf{x}}_i - \overline{\mathbf{x}}_{\cdot\cdot})^2/(k-1)$
Within populations	$\sum_{i=1}^{k} \sum_{j=1}^{k_i} (\mathbf{x}_{ij} - \overline{\mathbf{x}}_{i\cdot})^2$	$n - k$	$\sum_{i=1}^{k} \sum_{j=1}^{n_i} (\mathbf{x}_{ij} - \overline{\mathbf{x}}_{i\cdot})^2/(n-k)$
Total	$\sum_{i=1}^{k} \sum_{j=1}^{n_i} (\mathbf{x}_{ij} - \overline{\mathbf{x}}_{\cdot\cdot})^2$	$n - 1$	

where

$$\overline{\mathbf{x}}_{i\cdot} = \sum_{j=1}^{n_i} \mathbf{x}_{ij}/n_i, \qquad \overline{\mathbf{x}}_{\cdot\cdot} = \sum_{i=1}^{k} \sum_{j=1}^{n_i} \mathbf{x}_{ij}/n, \qquad n = \sum_{i=1}^{k} n_i.$$

The test statistic is given by

$$\mathbf{v} = \frac{\sum_{i=1}^{k} n_i(\overline{\mathbf{x}}_{i\cdot} - \overline{\mathbf{x}}_{\cdot\cdot})^2/(k-1)}{\sum_{i=1}^{k} \sum_{j=1}^{n_i} (\mathbf{x}_{ij} - \overline{\mathbf{x}}_i)^2/(n-k)},$$

and has an $F(k-1, n-k)$ distribution. The ratio of the expected values of the numerator and denominator $\mathbf{v}$ is

$$\frac{\sigma^2 + \sum_{i=1}^{k} n_i(\mu_i - \bar{\mu})^2/(k-1)}{\sigma^2},$$

where $\bar{\mu} = \sum_{i=1}^{k} n_i\mu_i/n$. If all the μ_i's are equal, the ratio of expected values of numerator and denominator would equal 1. Large values of $\mathbf{v}$ are indicative of some difference among the values of the μ_i's.

The computation of this statistic can be accomplished via a regression approach. Let us combine all the observations into a single n vector, $\mathbf{Y}' = (\mathbf{x}_{11}, \ldots, \mathbf{x}_{1n}, \mathbf{x}_{21}, \ldots, \mathbf{x}_{2n_2}, \ldots, \mathbf{x}_{k1}, \ldots, \mathbf{x}_{kn_k})$. Let us define $k - 1$ n vectors of independent variables $Z_1, \ldots, Z_{k-1}$, where Z_1 has 1's as its first n_1 elements and 0's elsewhere, Z_2 has 1's as its $(n_1 + 1)$th through n_2th elements and 0's elsewhere, and Z_{k-1} has 1's as its $(n_1 + \cdots + n_{k-2} + 1)$th through $(n_1 + \cdots + n_{k-1})$th elements and 0's elsewhere. The regression model

$$\mathbf{Y} = \alpha E + \beta_1 Z_1 + \cdots + \beta_{k-1} Z_{k-1} + \mathbf{U},$$

where $E' = (1, \ldots, 1)$ an $\mathbf{U}$ is the vector of residuals, bears the following relationship to the original model:

$$\begin{aligned}
\mu_1 &= \alpha + \beta_1, \\
\mu_2 &= \alpha + \beta_2, \\
&\vdots \\
\mu_{k-1} &= \alpha + \beta_{k-1}, \\
\mu_k &= \alpha.
\end{aligned}$$

Thus the hypothesis that $\mu_1 = \cdots = \mu_k$ is equivalent to the hypothesis that $\beta_1 = \cdots = \beta_{k-1} = 0$. That hypothesis is tested traditionally by the analysis of variance associated with the regression analysis, which, for example, in IDA is produced by the ANOV command.

Let us consider once again our example data. Table 9 arrays the dependent variable along with two independent variables, z_1 having a value of 1 if the data is Chevrolet data and 0 otherwise, and z_2 having a value of 1 if the data is Ford data and 0 otherwise. The regression results in estimates of α, β_1, and β_2 as follows:

$$\begin{aligned}
\hat{\beta}_1 &= 0.32389, \\
\hat{\beta}_2 &= 0.15933, \\
\hat{\alpha} &= 3.0736.
\end{aligned}$$

Note that $\hat{\alpha}$ is the average of the Plymouth data, $\hat{\alpha} + \hat{\beta}_1 = 3.39749$ is the mean of the Chevrolet data, and $\hat{\alpha} + \hat{\beta}_2 = 3.23293$ is the mean of the Ford data (with differences in the fifth decimal place due to rounding error).

Table 9. Automobile Operating Cost Data and Dummy Variables for the One-Way Anova.

y	z_1	z_2	y	z_1	z_2
3.530	0	0	4.700	0	1
2.000	1	0	2.460	0	0
2.382	1	0	2.700	1	0
1.970	0	1	4.080	1	0
2.360	0	1	3.450	1	0
2.516	0	0	3.260	1	0
4.940	0	0	3.020	1	0
2.570	0	0	3.494	1	0
3.670	1	0	2.280	1	0
3.492	1	0	4.686	0	1
5.390	1	0	2.130	0	1
4.250	1	0	2.524	0	0
4.689	0	1	2.270	0	0
3.934	0	1	3.480	0	0
1.740	0	1	1.610	0	0
1.560	0	1	2.940	1	0
2.870	0	1	2.740	1	0
5.940	0	0	3.170	0	1
3.340	0	0	4.550	0	1
3.040	0	0	3.310	0	1
2.980	0	0	3.060	0	0
2.430	0	0	3.150	0	0
5.900	1	0	3.100	0	0
2.180	1	0	2.384	0	0
2.170	0	1	3.926	1	0
4.150	0	1	4.290	0	1
2.680	0	1			

$z_{1i} = 1$ if Chevrolet
$z_{2i} = 1$ if Ford

The analysis of variance table for these data is:

Source of variation	Sum of squares	df	Mean square
Between populations	0.944215	2	0.472108
Within populations	56.6327	50	1.13265
Total	57.5769	52	

and so $v = 0.472108/1.13265 = 0.42$ is to be compared with the 95% point of the $F(2, 50)$ distribution. This comparison results in the conclusion that the three population means are equal.

What is the relation between the regression results and the standard analysis of variance table? In the regression setup, the $r - 1$ parameters are estimated to be

$$\hat{\beta}_i = \bar{x}_{i\cdot} - \bar{x}_{r\cdot}, \qquad i = 1, \ldots, r - 1.$$

Now

$$\sum_{i=1}^{r-1} n_{i\cdot}\hat{\beta}_i^2 = \sum_{i=1}^{r-1} n_{i\cdot}\bar{x}_{i\cdot}^2 + (n - n_{r\cdot})\bar{x}_{r\cdot}^2 - 2\bar{x}_{r\cdot}\sum_{i=1}^{r-1} n_{i\cdot}\bar{x}_{i\cdot}.$$

Since $n\bar{x}_{\cdot\cdot} = \sum_{i=1}^{r} n_{i\cdot}\bar{x}_{i\cdot}$, we see that

$$\sum_{i=1}^{r-1} n_{i\cdot}\hat{\beta}_i^2 = \sum_{i=1}^{r-1} n_{i\cdot}\bar{x}_{i\cdot}^2 + (n - n_{r\cdot})\bar{x}_{r\cdot}^2 - 2\bar{x}_{r\cdot}(n\bar{x}_{\cdot\cdot} - n_{r\cdot}\bar{x}_{r\cdot})$$

$$= \sum_{i=1}^{r} n_{i\cdot}\bar{x}_{i\cdot}^2 + n\bar{x}_{r\cdot}(\bar{x}_{r\cdot} - 2\bar{x}_{\cdot\cdot}).$$

Also

$$\sum_{i=1}^{r-1} n_{i\cdot}\hat{\beta}_i = \sum_{i=1}^{r-1} n_{i\cdot}\bar{x}_{i\cdot} - (n - n_{r\cdot})\bar{x}_{r\cdot}$$

$$= \sum_{i=1}^{r} n_{i\cdot}\bar{x}_{i\cdot} - n\bar{x}_{r\cdot}$$

$$= n(\bar{x}_{\cdot\cdot} - \bar{x}_{r\cdot}),$$

so

$$\sum_{i=1}^{r-1} n_{i\cdot}\hat{\beta}_i^2 - \frac{1}{n}\left(\sum_{i=1}^{r-1} n_{i\cdot}\hat{\beta}_i\right)^2$$

$$= \sum_{i=1}^{r} n_{i\cdot}\bar{x}_{i\cdot}^2 + n\bar{x}_{r\cdot}(\bar{x}_{r\cdot} - 2\bar{x}_{\cdot\cdot}) - n(\bar{x}_{\cdot\cdot}^2 + \bar{x}_{r\cdot}^2 - 2\bar{x}_{\cdot\cdot}\bar{x}_{r\cdot})$$

$$= \sum_{i=1}^{r} n_{i\cdot}\bar{x}_{i\cdot}^2 - n\bar{x}_{\cdot\cdot}^2.$$

But the between populations sum of squares is

$$\sum_{i=1}^{r} n_{i\cdot}(\bar{x}_{i\cdot} - \bar{x}_{\cdot\cdot})^2 = \sum_{i=1}^{r} n_{i\cdot}\bar{x}_{i\cdot}^2 + n\bar{x}_{\cdot\cdot}^2 - 2\bar{x}_{\cdot\cdot}\sum_{i=1}^{r} n_{i\cdot}\bar{x}_{i\cdot}$$

$$= \sum_{i=1}^{r} n_{i\cdot}\bar{x}_{i\cdot}^2 + n\bar{x}_{\cdot\cdot}^2 - 2n\bar{x}_{\cdot\cdot}^2$$

$$= \sum_{i=1}^{r} n_{i\cdot}\bar{x}_{i\cdot}^2 - n\bar{x}_{\cdot\cdot}^2.$$

Therefore the between populations sum of squares is given by

$$\sum_{i=1}^{r-1} n_{i\cdot}\hat{\beta}_i^2 - \frac{1}{n}\left(\sum_{i=1}^{r-1} n_{i\cdot}\hat{\beta}_i\right)^2.$$

Since this derivation does not depend on the number of main effects, we have shown that one can obtain the between populations sum of squares from an appropriate manipulation of the regresssion coefficients of the dummy variables pertaining to that main effect.

(b) Unbalanced Two-Way Anova with Only Main Effects

We have just studied the model in which $\mathscr{E}\mathbf{x}_{ij} = \mu_i$, for each $j = 1, \ldots, n_i$, $i = 1, \ldots, k$. To use regression methods, we rewrote the model as $\mathscr{E}\mathbf{x}_{ij} = \alpha + \beta_i$, $i = 1, \ldots, k-1$, $\mathscr{E}\mathbf{x}_{kj} = \alpha$. Yet another rewrite of the model is to set $\mathscr{E}\mathbf{x}_{ij} = \alpha^* + \beta_i^*$, $i = 1, \ldots, k$, but with $\sum_{i=1}^k \beta_i^* = 0$. This reparametrization may not be particularly useful for estimation purposes, but it does lead to a series of generalizations for multiway classified data. We illustrate this by considering the simplest such classification, the two-way classification. Here we have $k = rc$ populations, indexed by the pair (i, j), $i = 1, \ldots, r$, $j = 1, \ldots, c$, and n_{ij} observations on population (i, j), namely $\mathbf{x}_{ijl}$, $l = 1, \ldots, n_{ij}$. Our model is

$$\mathscr{E}\mathbf{x}_{ijl} = \alpha^* + \beta_i^* + \gamma_j^*,$$

with $\sum_{i=1}^r \beta_i^* = \sum_{j=1}^c \gamma_j^* = 0$. The statistical hypothesis testing problem we consider here is that $\beta_1^* = \cdots = \beta_r^*$ (or $\gamma_1^* = \cdots = \gamma_c^*$). Given the constraint, we see that our hypothesis is that the common value is 0.

Once more we structure a regression to test our null hypothesis. We construct two sets of independent variables, $z_1, \ldots, z_{r-1}$, where $z_{i'} = 1$ if $i = i'$ and $z_{i'} = 0$ otherwise, and $w_1, \ldots, w_{c-1}$, where $w_{j'} = 1$ if $j = j'$ and $w_{j'} = 0$ otherwise. Our regression is of the form

$$\mathbf{x}_{ijl} = \alpha + \beta_1 z_1 + \cdots + \beta_{r-1} z_{r-1} + \gamma_1 w_1 + \cdots + \gamma_{c-1} w_{c-1} + \mathbf{u}_{ijl}.$$

Now, parametrizing as in the one-way anova, we have

$$\begin{aligned}
\mathscr{E}\mathbf{x}_{ijl} &= \alpha + \beta_i + \gamma_j, && i = 1, \ldots, r-1, \quad j = 1, \ldots, c-1,\\
\mathscr{E}\mathbf{x}_{rjl} &= \alpha + \gamma_j, && j = 1, \ldots, c-1,\\
\mathscr{E}\mathbf{x}_{icl} &= \alpha + \beta_i, && i = 1, \ldots, r-1,\\
\mathscr{E}\mathbf{x}_{rcl} &= \alpha.
\end{aligned}$$

Thus

$$\begin{aligned}
\alpha &= \alpha^* + \beta_r^* + \gamma_c^*,\\
\alpha + \beta_i &= \alpha^* + \beta_i^* + \gamma_c^*, && i = 1, \ldots, r-1,\\
\alpha + \gamma_j &= \alpha^* + \beta_r^* + \gamma_j^*, && j = 1, \ldots, c-1,\\
\alpha + \beta_i + \gamma_j &= \alpha^* + \beta_i^* + \gamma_j^*, && i = 1, \ldots, r-1, \quad j = 1, \ldots, c-1.
\end{aligned}$$

The general approach for analyzing this two-way classification (see Snedecor and Cochran [1967], §16.7 or Yates [1934]) begins by first estimating α, the β_i, and the γ_j by a regression of the x_{ijl}'s onto the set of dummy independent variables as described above. The sum of squares of residuals derived from this regression is used as the "error sum of squares" or "within populations sum of squares" in the anova table.

Before proceeding with the remainder of the procedure, let us illustrate this step. We use an augmented version of our previous example, one in which we

Table 10. Driver Designation for Automobile Operating Cost Data.

3.530	0	3.340	2	4.686	3
2.000	0	3.040	2	2.130	3
2.382	0	2.980	2	2.524	4
1.970	0	2.430	2	2.270	4
2.360	0	5.900	2	3.480	4
2.516	1	2.180	2	1.610	4
4.940	1	2.170	2	2.940	4
2.570	1	4.150	2	2.740	4
3.670	1	2.680	2	3.170	4
3.492	1	4.700	2	4.550	4
5.390	1	2.460	3	3.310	4
4.250	1	2.700	3	3.060	5
4.689	1	4.080	3	3.150	5
3.934	1	3.450	3	3.100	5
1.740	1	3.260	3	2.384	5
1.560	1	3.020	3	3.926	5
2.870	1	3.494	3	4.290	5
5.940	2	2.280	3		

introduce the fact that there were six different drivers, with the driver designation given in Table 10. The dummy variables for this regression are given in Table 11, and following are the resulting estimators:

$$\hat{\alpha} = 3.4328,$$
$$\hat{\beta}_1 = -0.24120,$$
$$\hat{\beta}_2 = 0.2781,$$
$$\hat{\gamma}_1 = -1.0474,$$
$$\hat{\gamma}_2 = 0.0032318,$$
$$\hat{\gamma}_3 = 0.21811,$$
$$\hat{\gamma}_4 = -0.44734,$$
$$\hat{\gamma}_5 = -0.43250.$$

Finally, the ANOV command of IDA produced the following analysis of variance table:

Source of variation	Sum of squares	df	Mean square
Regression	7.93875	7	1.13411
Residuals	49.6382	45	1.10307
Total	57.5769	52	

Table 11. Automobile Operating Cost Data and Dummy Variables for the Two-Way Anova Without Interaction.

y	z_1	z_2	w_1	w_2	w_3	w_4	w_5	y	z_1	z_2	w_1	w_2	w_3	w_4	w_5
3.530	0	0	1	0	0	0	0	4.700	0	1	0	0	1	0	0
2.000	1	0	1	0	0	0	0	2.460	0	0	0	0	0	1	0
2.382	1	0	1	0	0	0	0	2.700	1	0	0	0	0	1	0
1.970	0	1	1	0	0	0	0	4.080	1	0	0	0	0	1	0
2.360	0	1	1	0	0	0	0	3.450	1	0	0	0	0	1	0
2.516	0	0	0	1	0	0	0	3.260	1	0	0	0	0	1	0
4.940	0	0	0	1	0	0	0	3.020	1	0	0	0	0	1	0
2.570	0	0	0	1	0	0	0	3.494	1	0	0	0	0	1	0
3.670	1	0	0	1	0	0	0	2.280	1	0	0	0	0	1	0
3.492	1	0	0	1	0	0	0	4.686	0	1	0	0	0	1	0
5.390	1	0	0	1	0	0	0	2.130	0	1	0	0	0	1	0
4.250	1	0	0	1	0	0	0	2.524	0	0	0	0	0	0	1
4.689	0	1	0	1	0	0	0	2.270	0	0	0	0	0	0	1
3.934	0	1	0	1	0	0	0	3.480	0	0	0	0	0	0	1
1.740	0	1	0	1	0	0	0	1.610	0	0	0	0	0	0	1
1.560	0	1	0	1	0	0	0	2.940	1	0	0	0	0	0	1
2.870	0	1	0	1	0	0	0	2.740	1	0	0	0	0	0	1
5.940	0	0	0	0	1	0	0	3.170	0	1	0	0	0	0	1
3.340	0	0	0	0	1	0	0	4.550	0	1	0	0	0	0	1
3.040	0	0	0	0	1	0	0	3.310	0	1	0	0	0	0	1
2.980	0	0	0	0	1	0	0	3.060	0	0	0	0	0	0	0
2.430	0	0	0	0	1	0	0	3.150	0	0	0	0	0	0	0
5.900	1	0	0	0	1	0	0	3.100	0	0	0	0	0	0	0
2.180	1	0	0	0	1	0	0	2.384	0	0	0	0	0	0	0
2.170	0	1	0	0	1	0	0	3.926	1	0	0	0	0	0	0
4.150	0	1	0	0	1	0	0	4.290	0	1	0	0	0	0	0
2.680	0	1	0	0	1	0	0								

The test for significance of the row effect involves developing a decomposition of the regression sum of squares into two components, that due to the rows, called the "adjusted row sum of squares", and that due to columns, called the "unadjusted column sum of squares". Similarly, the test for significance of the column effect involves decomposing the regression sum of squares into an "adjusted column sum of squares" and an "unadjusted row sum of squares". The unadjusted row sum of squares is calculable directly as

$$\sum_{i=1}^{r} n_{i\cdot}(\bar{x}_{i\cdot\cdot} - \bar{x}_{\cdot\cdot\cdot})^2, \quad \text{where} \quad \bar{x}_{i\cdot\cdot} = \sum_{j=1}^{c} n_{ij}\bar{x}_{ij\cdot}/n_{i\cdot} \quad \text{and} \quad \bar{x}_{\cdot\cdot\cdot} = \sum_{i=1}^{r}\sum_{j=1}^{c} n_{ij}\bar{x}_{ij\cdot}/n.$$

The unadjusted column sum of squares is given by

$$\sum_{j=1}^{c} n_{\cdot j}(\bar{x}_{\cdot j\cdot} - \bar{x}_{\cdot\cdot\cdot})^2, \quad \text{where} \quad \bar{x}_{\cdot j\cdot} = \sum_{i=1}^{r} n_{ij}\bar{x}_{ij\cdot}/n_{\cdot j}.$$

The adjusted row and column sum of squares can then be obtained by subtraction of the unadjusted column and row sum of squares, respectively, from the regression sum of squares.

In our example, this decomposition would yield the tables:

Source of variation	Sum of squares	df	Mean square
Rows (adjusted)	1.983584	2	0.991792
Columns (unadjusted)	5.955166	5	—
Residuals	49.6382	45	1.10307
Total	57.5769	52	

Source of variation	Sum of squares	df	Mean square
Row (unadjusted)	0.942781	2	—
Columns (adjusted)	6.995969	5	1.39919
Residuals	49.6382	45	1.10307
Total	57.5769	52	

The appropriate F statistic for testing a significance of difference of row effects is the ratio of the adjusted row mean square to the residual mean square; in our example, we compare $0.991792/1.10307 = 0.899$ with the 95% point of an $F(2, 45)$ variable. Similarly, the ratio of the adjusted column mean square to the residual mean square is the test statistic for equality of column effects; in our sample, we compare $1.39919/1.10307 = 1.268$ with the 95% point of an $F(5, 45)$ variable. In both cases we accept the hypothesis of no difference attributable to either automobile or driver on operating costs.

(c) Unbalanced Two-Way Anova with Interaction

Because we have replications in each cell, we can enrich the model to also include an interaction term δ_{ij}^*, so that $\mathscr{E}\mathbf{x}_{ijl} = a^* + \beta_i^* + \gamma_j^* + \delta_{ij}^*$, where $\sum_{i=1}^r \delta_{ij}^* = \sum_{j=1}^c \delta_{ij}^* = 0$. To analyze this model one introduces additional dummy independent variables

$$v_{11}, \ldots, v_{1,c-1}, v_{21}, \ldots, v_{2,c-1}, \ldots, v_{r-1,1}, \ldots, v_{r-1,c-1},$$

where $v_{i'j'} = 1$ if $i = i'$ and $j = j'$ and $v_{i'j'} = 0$ otherwise. The model is now

$$\mathbf{x}_{ijl} = \alpha + \sum_{i=1}^{r-1} \beta_i z_i + \sum_{j=1}^{c-1} \gamma_j w_j + \sum_{i=1}^{r-1} \sum_{j=1}^{c-1} \delta_{ij} v_{ij} + \mathbf{u}_{ijl},$$

and algebra similar to that used in the no-interaction case can be used to relate the "unstarred" parameters to the "starred" parameters.

Table 12 presents the data for the regression approach to the model

Table 12. Automobile Operating Cost Data and Dummy Variables for the Two-Way Anova with Interaction.

y	z_1	z_2	w_1	w_2	w_3	w_4	w_5	v_{11}	v_{12}	v_{13}	v_{14}	v_{15}	v_{21}	v_{22}	v_{23}	v_{24}	v_{25}
3.530	0	0	1	0	0	0	0	0	0	0	0	0	0	0	0	0	0
2.000	1	0	1	0	0	0	0	1	0	0	0	0	0	0	0	0	0
2.382	1	0	1	0	0	0	0	1	0	0	0	0	0	0	0	0	0
1.970	0	1	1	0	0	0	0	0	0	0	0	0	1	0	0	0	0
2.360	0	1	1	0	0	0	0	0	0	0	0	0	1	0	0	0	0
2.516	0	0	0	1	0	0	0	0	0	0	0	0	0	0	0	0	0
4.940	0	0	0	1	0	0	0	0	0	0	0	0	0	0	0	0	0
2.570	0	0	0	1	0	0	0	0	0	0	0	0	0	0	0	0	0
3.670	1	0	0	1	0	0	0	0	1	0	0	0	0	0	0	0	0
3.492	1	0	0	1	0	0	0	0	1	0	0	0	0	0	0	0	0
5.390	1	0	0	1	0	0	0	0	1	0	0	0	0	0	0	0	0
4.250	1	0	0	1	0	0	0	0	1	0	0	0	0	0	0	0	0
4.689	0	1	0	1	0	0	0	0	0	0	0	0	0	1	0	0	0
3.934	0	1	0	1	0	0	0	0	0	0	0	0	0	1	0	0	0
1.740	0	1	0	1	0	0	0	0	0	0	0	0	0	1	0	0	0
1.560	0	1	0	1	0	0	0	0	0	0	0	0	0	1	0	0	0
2.870	0	1	0	1	0	0	0	0	0	0	0	0	0	1	0	0	0
5.940	0	0	0	0	1	0	0	0	0	0	0	0	0	0	0	0	0
3.340	0	0	0	0	1	0	0	0	0	0	0	0	0	0	0	0	0
3.040	0	0	0	0	1	0	0	0	0	0	0	0	0	0	0	0	0
2.980	0	0	0	0	1	0	0	0	0	0	0	0	0	0	0	0	0
2.430	0	0	0	0	1	0	0	0	0	0	0	0	0	0	0	0	0
5.900	1	0	0	0	1	0	0	0	0	1	0	0	0	0	0	0	0
2.180	1	0	0	0	1	0	0	0	0	1	0	0	0	0	0	0	0
2.170	0	1	0	0	1	0	0	0	0	0	0	0	0	0	1	0	0
4.150	0	1	0	0	1	0	0	0	0	0	0	0	0	0	1	0	0
2.680	0	1	0	0	1	0	0	0	0	0	0	0	0	0	1	0	0
4.700	0	1	0	0	1	0	0	0	0	0	0	0	0	0	1	0	0
2.460	0	0	0	0	0	1	0	0	0	0	0	0	0	0	0	0	0
2.700	1	0	0	0	0	1	0	0	0	0	1	0	0	0	0	0	0
4.080	1	0	0	0	0	1	0	0	0	0	1	0	0	0	0	0	0
3.450	1	0	0	0	0	1	0	0	0	0	1	0	0	0	0	0	0
3.260	1	0	0	0	0	1	0	0	0	0	1	0	0	0	0	0	0
3.020	1	0	0	0	0	1	0	0	0	0	1	0	0	0	0	0	0
3.494	1	0	0	0	0	1	0	0	0	0	1	0	0	0	0	0	0
2.280	1	0	0	0	0	1	0	0	0	0	1	0	0	0	0	0	0
4.686	0	1	0	0	0	1	0	0	0	0	0	0	0	0	0	1	0
2.130	0	1	0	0	0	1	0	0	0	0	0	0	0	0	0	1	0
2.524	0	0	0	0	0	0	1	0	0	0	0	0	0	0	0	0	0
2.270	0	0	0	0	0	0	1	0	0	0	0	0	0	0	0	0	0
3.480	0	0	0	0	0	0	1	0	0	0	0	0	0	0	0	0	0
1.610	0	0	0	0	0	0	1	0	0	0	0	0	0	0	0	0	0
2.940	1	0	0	0	0	0	1	0	0	0	0	1	0	0	0	0	0
2.740	1	0	0	0	0	0	1	0	0	0	0	1	0	0	0	0	0
3.170	0	1	0	0	0	0	1	0	0	0	0	0	0	0	0	0	1
4.550	0	1	0	0	0	0	1	0	0	0	0	0	0	0	0	0	1
3.310	0	1	0	0	0	0	1	0	0	0	0	0	0	0	0	0	1
3.060	0	0	0	0	0	0	0	0	0	0	0	0	0	0	0	0	0
3.150	0	0	0	0	0	0	0	0	0	0	0	0	0	0	0	0	0
3.100	0	0	0	0	0	0	0	0	0	0	0	0	0	0	0	0	0
2.384	0	0	0	0	0	0	0	0	0	0	0	0	0	0	0	0	0
3.926	1	0	0	0	0	0	0	0	0	0	0	0	0	0	0	0	0
4.290	0	1	0	0	0	0	0	0	0	0	0	0	0	0	0	0	0

involving interaction. The resulting parameter estimates are:

$$\hat{\alpha} = 2.9235,$$
$$\hat{\beta}_1 = 1.0025,$$
$$\hat{\beta}_2 = 1.3665,$$
$$\hat{\gamma}_1 = 0.6065,$$
$$\hat{\gamma}_2 = 0.4185,$$
$$\hat{\gamma}_3 = 0.6225,$$
$$\hat{\gamma}_4 = -0.4635,$$
$$\hat{\gamma}_5 = -0.4525,$$
$$\hat{\delta}_{11} = -2.3415,$$
$$\hat{\delta}_{12} = -0.1440,$$
$$\hat{\delta}_{13} = -0.5085,$$
$$\hat{\delta}_{14} = -0.27907,$$
$$\hat{\delta}_{15} = -0.6335,$$
$$\hat{\delta}_{21} = -2.7315,$$
$$\hat{\delta}_{22} = -1.7499,$$
$$\hat{\delta}_{23} = -1.4875,$$
$$\hat{\delta}_{24} = -0.4185,$$
$$\hat{\delta}_{25} = -0.16084,$$

and the ANOV table is

Source of variation	Sum of squares	df	Mean square
Regression	16.5156	17	0.971507
Residuals	41.0613	35	1.17318
Total	57.5769	52	

The sum of squares due to regression when interaction terms are included can be partitioned into

regression sum of squares in noninteraction model

+ interaction sum of squares

= regression sum of squares in interaction model.

(In our example, then, the interaction sum of squares is found by subtraction to be 16.5156 − 7.9388 = 8.5768.)

One can bypass the entire construction of dummy variables and estimation of the regression coefficients and calculate the interaction regression sum of squares, sometimes called the "sum of squares between subclasses", directly as

$$\sum_{i=1}^{r}\sum_{j=1}^{c} n_{ij}(\bar{x}_{ij.} - \bar{x}_{...})^2.$$

Thus even though one is interested in the interaction effects, one need only calculate one regression, without interaction terms, and perform a few additional calculations to develop the full anova table for this case.

The test for significant interaction coefficients is based on the statistic

$$w = \frac{\text{interaction sum of squares}/(r-1)(c-1)}{\text{noninteraction residual sum of squares}/(n-r-c+1)},$$

which has an F distribution with $(r-1)(c-1)$ and $n-r-c+1$ degrees of freedom. (In our example, $w = [8.5768/10] \div 1.10307 = 0.778$, leading us to conclude that there is no significant interaction.)

If the interaction were significant, then the test for row effect and column effect given earlier is altered so that the interaction mean square is used as the denominator of the test statistic in place of the residual mean square, and an F table with denominator degrees of freedom equal to $(r-1)(c-1)$ is used to check for significance of the test statistic.

To summarize, following is the complete anova table for the unbalanced two-way analysis of variance.

Source of variation	df		Sum of squares
Regression without interaction	$r+c-2$	(1)	(9) − (8)
Rows (unadjusted)	—	(2)	$\sum_{i=1}^{r} n_{i.}(\bar{x}_{i..} - \bar{x}_{...})^2$
Columns (adjusted)	$c-1$	(3)	(1) − (2)
Rows (adjusted)	$r-1$	(4)	(1) − (5)
Columns (unadjusted)	—	(5)	$\sum_{j=1}^{c} n_{.j}(\bar{x}_{.j.} - \bar{x}_{...})^2$
Regression with interaction	$rc-1$	(6)	$\sum_{i=1}^{r}\sum_{j=1}^{c} n_{ij}(\bar{x}_{ij.} - \bar{x}_{...})^2$
Interaction	$(r-1)(c-1)$	(7)	(6) − (1)
Residuals	$n-r-c+1$	(8)	$\sum_{i=1}^{r}\sum_{j=1}^{c}\sum_{k=1}^{n_{ij}} (x_{ijk} - \bar{x}_{i..} - \bar{x}_{.j.} + \bar{x}_{...})^2$
Total	$n-1$	(9)	$\sum_{i=1}^{r}\sum_{j=1}^{c}\sum_{k=1}^{n_{ij}} (x_{ijk} - \bar{x}_{...})^2$

For an alternate presentation of this material, the reader is referred to §16.7 of Snedecor and Cochran [1967] cited above or to Chapter 20 of Kleinbaum and Kupper [1978].

This procedure is incorporated as the standard output into the ANOVA of the SPSS statistical package (Nie *et. al* [1975]). The BMD statistical package (Dixon and Brown [1979]), however, views the hypothesis testing problem differently, following a formulation of Kutner [1974]. Kutner distinguishes between three different null hypotheses about the row effects:

(1) $\mathscr{E}\mathbf{y}_{1..} = \cdots = \mathscr{E}\mathbf{y}_{r..}$;
(2) $\sum_{j=1}^{c} n_{1j}\mathscr{E}\mathbf{y}_{1j.}/n_{1.} = \cdots = \sum_{j=1}^{c} n_{rj}\mathscr{E}\mathbf{y}_{rj.}/n_{r.}$;
(3) hypothesis (1) assuming no interaction.

Hypothesis (1) is equivalent to the hypothesis $\beta_1^* = \cdots = \beta_r^* = 0$. Hypothesis (2) is appropriate if we view the $r \times c$ table as containing a set of observations drawn at random from a population, that is, if we model the unbalancedness of the table as the result of the random sampling, and wherein the $\mathbf{n}_{ij}/n_{..}$ are unbiased estimates of the population proportions in each of the cells.

The unadjusted mean square is the appropriate numerator of the F test if one wishes to test hypothesis (2). The adjusted mean square is the appropriate numerator to test hypothesis (3). Thus the procedure we have described (and that tested in SPSS) is really a "conditional" test, namely a testing of the equality of the β's given no interaction. The BMD package tests hypothesis (1), as does SPSS if OPTION 9 is specified. Francis [1973] states that "these are also the kinds of tests that most people would really want to do".

The SAS statistical package (SAS Institute [1985]) produces three different analyses of variance for this situation. Their Type I analysis contains the unadjusted row sum of squares and the adjusted column sum of squares. (This also is the output provided by OPTION 10 of SPSS.) Note that this analysis depends on the definition of row and column variables. Their Type II analysis is the one we have described herein, containing both adjusted row and column sums of squares. Their Type III and Type IV analyses are equivalent as long as all rc cells contain at least one observation, and this analysis is identical with that of BMD.

Following is a summary of the various anova tables produced for our data:

Source of variation	Sum of squares SPSS/SAS II	BMD/SPSS 9/SAS III	SPSS 10/SAS I
Car	1.981	0.827	0.944
Driver	6.995	5.675	6.995
Interaction	8.577	8.577	8.577
Residuals	41.061	41.061	41.061
Total	57.577	57.577	57.577

When $n_{ij} = n$ for all i, j, this procedure simplifies greatly and reduces to the following anova table:

Source of variation	df	Sum of squares
Rows	$r - 1$	$\sum_{i=1}^{r} nc(\bar{x}_{i..} - \bar{x}_{...})^2$
Columns	$c - 1$	$\sum_{j=1}^{c} nr(\bar{x}_{.j.} - \bar{x}_{...})^2$
Interaction	$(r - 1)(c - 1)$	$\sum_{i=1}^{r} \sum_{j=1}^{c} n(\bar{x}_{ij.} - \bar{x}_{i..} - \bar{x}_{.j.} + \bar{x}_{...})^2$
Residuals	$rc(n - 1)$	$\sum_{i=1}^{r} \sum_{j=1}^{c} \sum_{k=1}^{n} (x_{ijk} - \bar{x}_{ij.})^2$
Total	$rcn - 1$	$\sum_{i=1}^{r} \sum_{j=1}^{c} \sum_{k=1}^{n} (x_{ijk} - \bar{x}_{...})^2$

It is impossible to test for the existence of interaction parameters δ_{ij} in this way when one has only a single observation for each i, j. Tukey [1949] derived the following procedure for teasing out of the residual sum of squares a sum of squares due to "nonadditivity", i.e., due to the existence of some interaction in the model. Let

$$\mathbf{v} = \frac{\left[\sum_{i=1}^{r} \sum_{j=1}^{c} \mathbf{x}_{ij}(\bar{\mathbf{x}}_{i.} - \bar{\mathbf{x}}_{..})(\bar{\mathbf{x}}_{.j} - \bar{\mathbf{x}}_{..})\right]^2}{\left[\sum_{i=1}^{r} (\bar{\mathbf{x}}_{i.} - \bar{\mathbf{x}}_{..})^2 \sum_{j=1}^{c} (\bar{\mathbf{x}}_{.j} - \bar{\mathbf{x}}_{..})^2\right]},$$

and

$$\mathbf{w} = \sum_{i=1}^{r} \sum_{j=1}^{c} (\mathbf{x}_{ij} - \bar{\mathbf{x}}_{i.} - \bar{\mathbf{x}}_{.j} + \bar{\mathbf{x}}_{..})^2 - \mathbf{v}.$$

Then $\mathbf{v}$ and $\mathbf{w}$ are independent, the statistic

$$\mathbf{u} = \frac{\mathbf{v}((r - 1)(c - 1) - 1)}{\mathbf{w}}$$

has an $F(1, (r - 1)(c - 1) - 1)$ distribution, and large values of $\mathbf{u}$ are indicative of the existence of nonadditivity.

This procedure was derived without reference to a specific functional form for the interaction. Studies have indicated, though, that the procedure is quite powerful when the interaction is a product of row and column effects. For the likelihood ratio test under these circumstances, as well as a good set of references to research in this area, see Johnson and Graybill [1972].

(d) Median Polish

The two-way analysis of variance model postulates that $\mathscr{E}\mathbf{x}_{ij} = \alpha^* + \beta_i^* + \gamma_j^*$, and the regression analysis produces estimates of these parameters, based on the estimates of the α^*, β_i^*, and γ_j^* parameters. But suppose $\mathbf{x}_{ij}$ does not satisfy the normality assumption. One might model a comparable relationship, even for the case in which $\mathscr{E}\mathbf{x}_{ij}$ does not exist, as a decomposition of the population median of $\mathbf{x}_{ij}$, $\mathscr{N}\mathbf{x}_{ij}$, namely

$$\mathscr{N}\mathbf{x}_{ij} = \alpha + \beta_i + \gamma_j.$$

A nonparametric statistical procedure which has been devised to estimate these parameters is the data analysis procedure called "median polish" (Tukey [1977]). We describe the procedure algorithmically, interpolating at each step of the algorithm an example to illustrate the mechanics of each step.

We use as our example the 3×6 table of data consisting of the averages of the operating costs for each automobile/driver combination in our original data set. Our data then is the table:

						Row median
3.530	3.340	3.546	2.460	2.471	2.924	3.132
2.191	4.201	4.040	3.183	2.840	3.926	3.5545
2.165	2.959	3.425	3.408	3.677	4.290	3.4165

Step 1. Subtract the row medians from each row element (e.g., 0.398 = 3.530 − 3.132).

							Row median
	0.398	0.208	0.414	−0.672	−0.661	−0.208	3.132
	−1.3635	0.6465	0.4855	−0.3715	−0.7145	0.3715	3.5545
	−1.2515	−0.4575	0.0085	−0.0085	0.2605	0.8735	3.4165
Column median	−1.2515	0.208	0.414	−0.3715	−0.661	0.3715	

Median of column medians = −0.08175,
Median of row medians = 3.4165.

Step 2. Subtract the column medians from each column element (e.g., 1.6495 = 0.398 − (−1.2515)). Subtract median of row medians from each row median to obtain adjusted row median (e.g., −0.2845 = 3.132 − 3.4165).

						Row median	Adjusted row median
1.6495	0	0	−0.3005	0	−0.5795	0	−0.2845
−0.112	0.4385	0.0715	0	−0.0535	0	0	0.138
0	−0.6655	−0.4055	0.363	0.9215	0.502	0.1815	0

At this point we have a first-round estimate of our parameters, from the column medians of Step 1, the adjusted row medians of Step 2, and the median of row medians of Step 1, namely:

Parameter	Estimate
α^*	3.4165
β_1^*	−0.2845
β_2^*	0.1380
β_3^*	0.0000
γ_1^*	−1.2515
γ_2^*	0.2080
γ_3^*	0.4140
γ_4^*	−0.3715
γ_5^*	−0.6610
γ_6^*	0.3715

But another pair of steps like the above two will improve the estimates.

Step 3. Subtract the row medians from each row element (e.g., $-0.847 = -0.6655 - 0.1815$). Subtract the median of column medians from each column median to obtain adjusted column median (e.g., $-1.16975 = -1.2515 - (-0.08175)$). Add a new median to adjusted row median (e.g., $0.1815 = 0 + 0.1815$).

							Adjusted row median
	1.6495	0	0	−0.3005	0	−0.5795	−0.2845
	−0.112	0.4385	0.0715	0	−0.0535	0	0.138
	−0.1815	−0.847	−0.587	0.1815	0.74	0.3205	0.1815
Column median	−0.112	0	0	0	0	0	
Adjusted column median	−1.16975	0.28975	0.49575	−0.28975	−0.57925	0.45325	

Median of row medians = 0.138

Step 4. Subtract the column medians from each column element (e.g., 1.7615 = 1.6495 − (−0.112)). Subtract median of row medians from each row median to obtain adjusted row median (e.g., −0.4225 = −0.2845 − 0.138). Add column median to adjusted column median (e.g., −1.28175 = −1.16976 − 0.112)

							Row median	Adjusted row median
	1.7615	0	0	−0.3005	0	−0.5795	0	−0.4225
	0	0.4385	0.0715	0	−0.0535	0	0	0
	−0.0695	−0.847	−0.587	0.1815	0.74	0.3205	−0.0585	−0.0125
Adjusted column median	−1.28175	0.28975	0.49575	−0.28975	−0.57925	0.45325	3.4165	

Thus in our terminology the median polish procedure produces the following estimates of our parameters:

Parameter	Estimate
α^*	3.4165
β_1^*	−0.4225
β_2^*	0.0000
β_3^*	−0.0125
γ_1^*	−1.2818
γ_2^*	0.2898
γ_3^*	0.4958
γ_4^*	−0.2898
γ_5^*	−0.5793
γ_6^*	0.4533

If instead we had performed a regression of these x_{ij} on appropriate indicator variables, we would have obtained the following coefficients:

$$\hat{\alpha} = 3.7796,$$

$$\hat{\beta}_1 = -0.27546,$$

$$\hat{\beta}_2 = 0.076277,$$

$$\hat{\gamma}_1 = -1.0845,$$

$$\hat{\gamma}_2 = 0.21347,$$

$$\hat{\gamma}_3 = -0.042833,$$

$$\hat{\gamma}_4 = -0.69602,$$

$$\hat{\gamma}_5 = -0.71728.$$

From this we can calculate estimates of the expected values of the x_{ij}, viz:

							Row mean
	2.4196	3.7176	3.4613	2.8081	2.7869	3.5041	3.1163
	2.7714	4.0693	3.8130	3.1599	3.1386	3.8559	3.4680
	2.6951	3.9931	3.7368	3.0836	3.0623	3.7796	3.3918
Column mean	2.6287	3.9267	3.6704	3.0172	2.9959	3.7132	3.3253

The analysis of variance would thus produce the following estimates of our parameters:

Parameter	Estimate
α^*	3.3253
β_1^*	−0.2090
β_2^*	0.1427
β_3^*	0.0665
γ_1^*	−0.6966
γ_2^*	0.6014
γ_3^*	0.3451
γ_4^*	−0.3081
γ_5^*	−0.3294
γ_6^*	0.3879

3. Components of Variance (Model II)

One potential difficulty of applying the analysis of variance model described above to the analysis of data is that the underlying structure of the model is not realistic. The k "populations" may in reality be k independent samples from some superpopulation. For example, if the $\mathbf{x}_{ij}$ are diameters of bolts punched out by machines, $i = 1, \ldots, k$, we may not be interested so much in whether those k particular machines have equal μ_i's, but rather in whether the factory producing these machines, of which these k are a random sample, produce machines whose average bolt diameters do not vary much from each other. That is, we want to extend the inference beyond the particular set of k machines we happen to be analyzing to the set of all possible machines from which this random sample was selected.

To do this let us look more closely at the analysis of variance table given at the beginning of Section 2. In particular, let us append a column to the table, a column containing the expected value of each of the mean squares.

As noted earlier, under Model I this column would contain the following:

Source of variation	Expected mean square
Between populations	$\sigma^2 + \sum_{i=1}^{k} n_i(\mu_i - \bar{\mu})^2/(k-1)$
Within populations	σ^2

In Model II this column is interpreted as containing the *conditional* expected mean square, where the expectation is taken conditional on the k values $\mu_1, \ldots, \mu_k$ drawn from our superpopulation of μ's. To obtain the unconditional expected mean square in a simple case, suppose $n_i = n$ for all i. We note that $\mathscr{E} n \sum_{i=1}^{k} (\boldsymbol{\mu}_i - \bar{\boldsymbol{\mu}})^2 = n(k-1)\sigma_\mu^2$, where σ_μ^2 is the variance of the superpopulation of μ's. Thus in Model II this last column would read as follows:

Source of variation	Expected mean square
Between populations	$\sigma^2 + n\sigma_\mu^2$
Within populations	σ^2

Since the Within Populations Mean Square (WPMS) is an unbiased estimate of σ^2 and since the Between Populations Mean Square (BPMS) is an unbiased estimate of $\sigma^2 + n\sigma_\mu^2$, an unbiased estimate of σ_μ^2 is given by

$$\hat{\boldsymbol{\sigma}}_\mu^2 = \frac{\text{BPMS} - \text{WPMS}}{n}.$$

Unfortunately, sometimes this "traditional" estimator can lead to a negative estimate of σ_μ^2. Herbach [1959] considered the maximum likelihood approach to the problem and obtained the following estimates:

$$\hat{\sigma}^2 = \text{WPMS}, \qquad \hat{\sigma}_\mu^2 = \frac{[(k-1)/k]\,\text{BPMS} - \text{WPMS}}{n}$$

$$\text{when} \quad (k-1)/k\ \text{BPMS} > \text{WPMS},$$

$$\hat{\sigma}^2 = 0, \qquad \hat{\sigma}_\mu^2 = \frac{[(k-1)\,\text{BPMS} + k(n-1)\,\text{WPMS}]}{nk}, \quad \hat{\sigma}^2 = 0,$$

$$\text{when} \quad (k-1)/k\ \text{BPMS} < \text{WPMS}.$$

Thompson [1962], adopting the restricted maximum likelihood approach, i.e., starting with the joint likelihood of WPMS and BPMS instead of with the joint likelihood of all the observations, found that the traditional estimator, when positive, is the restricted maximum likelihood estimator of σ^2 and σ_μ^2. When, however, BPMS < WPMS, the estimators are as follows:

$$\hat{\sigma}^2 = 0, \qquad \hat{\sigma}_\mu^2 = [(k-1)\,\text{BPMS} + k(n-1)\,\text{WPMS}]/(nk-1), \qquad \hat{\sigma}_\mu^2 = 0.$$

For this case (though not in general) the maximum likelihood estimator has smaller mean square error than does the restricted maximum likelihood estimator (see Harville [1977]). Klotz, Milton, and Zacks [1969] have shown that the mean square error of the maximum likelihood estimator of σ_μ^2 is uniformly smaller than that of the unbiased estimator $\hat{\sigma}_\mu^2$. Moreover, they introduced the estimator

$$\max\left(0, \frac{[(k-1)/(k+1)]\text{ BPMS} - \text{WPMS}}{n}\right),$$

and proved that its mean square error is uniformly smaller than that of the maximum likelihood estimator.

The model can be generalized to any analysis of variance situation. Moreover, one can have a mix of effects, some being of the fixed variety of Model I and others being the random effects of Model II. Such a model is called a "mixed" model analysis of variance. The general statement of such a model is that

$$\mathbf{Y} = XA + Z\mathbf{B} + \mathbf{U},$$

where $\mathbf{Y}$ is an n vector of observations, X and Z are, respectively, the matrices of independent variables associated with the p fixed and q random effects, A is the p vector of fixed effect coefficients, $\mathbf{B}$ is the q vector of random effect coefficients, and $\mathbf{U}$ is an error term. In this model, we let $\Delta = \mathscr{V}\mathbf{B}$, $\Sigma = \mathscr{V}\mathbf{U}$, and so

$$\mathscr{V}\mathbf{y} = \Sigma + Z\Delta Z'.$$

The general problem is then to estimate A, Δ, and Σ. A review of numeric procedures for maximum likelihood and restricted maximum likelihood estimation is given in Harville [1977]. A review of other procedures, along with an extensive catalog of the resulting estimators for a variety of models, can be found in Searle [1971].

Recent literature has focused on another criterion, that of finding estimates of the variance components that have minimum mean squared error at some point in parameter space. Olson, Seeley, and Birkes [1976] have found the minimal complete class of quadratic unbiased estimators of the variance components for the one-way anova problem treated above, and found that the "traditional" estimator is not a member of the class.

4. Dichotomous Dependent Variables

Suppose our dependent variables $\mathbf{y}_1, \ldots, \mathbf{y}_n$ are outcomes of Bernoulli trials, i.e., only take on values of 0 or 1. Let us further assume that these observations can be grouped into r subsamples such that the n_i observations in subsample i are Bernoulli trials wherein the probability that $\mathbf{y} = 1$ is given by θ_i. Let us reorganize our data so that the y's are in the order $\mathbf{y}_{11}, \ldots, \mathbf{y}_{1n_1}, \mathbf{y}_{21}, \ldots, \mathbf{y}_{2n_2}, \ldots, \mathbf{y}_{r1}, \ldots, \mathbf{y}_{rn_r}$, where $\mathbf{y}_{ij}$ is the jth observation ($j = 1, \ldots, n_i$) from the Bernoulli trial generator whose parameter is θ_i.

Consider now the regression model in which, for the observations from population i,

$$\mathscr{E}\mathbf{y}_{ij} = X'_{ij}B, \qquad j = 1, \ldots, n_i,$$

where X_{ij} is a p vector of independent variable values. Since $\mathscr{E}\mathbf{y}_{ij} = \theta_i$, the relation $\theta_i = X'_{ij}B$ cannot hold for all j, i.e., unless the value of the vector of ent variables is constant for all Bernoulli trials from group i. Also, since the $\mathbf{y}_{ij}$ are the result of Bernoulli trials, we see that $\mathscr{V}\mathbf{y}_{ij} = \theta_i(1 - \theta_i)$, and so the regression model $\mathbf{y}_{ij} = X'_{ij}B + \mathbf{u}_{ij}$ does not satisfy the usual homoscedasticity assumption about the residuals $\mathbf{u}_{ij}$, except in the trivial case $\theta_1 = \theta_2 = \cdots = \theta_r$. Moreover, the $\mathbf{u}_{ij}$ are not normally distributed. Finally, since $0 \leq \theta_i \leq 1$, this restriction induces constraints on the components of B in this regression model.

For these reasons the modeling of the relationship between a dichotomous dependent variable $\mathbf{y}_{ij}$ and a constant independent variable vector X_i are not done by linear regression. The most commonly used model is the linear logistic regression model, wherein it is assumed that

$$\theta_i = \mathscr{E}\mathbf{y}_{ij} = \frac{e^{X'_iB}}{1 + e^{X'_iB}},$$

so that

$$1 - \theta_i = \frac{1}{1 + e^{X'_iB}},$$

and

$$\lambda_i = \log\left(\frac{\theta_i}{1 - \theta_i}\right) = X'_iB.$$

The statistical inferential procedure for estimating B from the $\mathbf{y}_{ij}$'s is based on the asymptotic normality of the random variables

$$\mathbf{z}_i = \log\left(\frac{\mathbf{y}_{i\cdot}}{n_i - \mathbf{y}_{i\cdot}}\right),$$

where $\mathbf{y}_{i\cdot} = \sum_{j=1}^{n_i} \mathbf{y}_{ij}$. For large values of n_i, $\mathbf{z}_i$ is approximately normally distributed with mean λ_i and with variance consistently estimated by

$$\mathbf{v}_{i\cdot}^2 = \frac{n_i}{\mathbf{y}_{i\cdot}(n_i - \mathbf{y}_{i\cdot})}.$$

Thus

$$\frac{\mathbf{z}_i}{\mathbf{v}_i} \simeq \frac{1}{\mathbf{v}_i} X'_iB,$$

and

$$\frac{\mathbf{z}_i}{\mathbf{v}_i} \simeq 1,$$

so that, if we ignore the fact that the $\mathbf{v}_i$ on the right-hand side of the regression relation are random variables, we can use $\mathbf{z}_i/\mathbf{v}_i$ as the dependent variables, $X_i/\mathbf{v}_i$ as the independent variables, and estimate B using standard linear regression methods.

Unfortunately, when $\mathbf{y}_{i\cdot} = 0$ or $\mathbf{y}_{i\cdot} = n_i$ the variable $\mathbf{z}_i$ is undefined. A modified definition of $\mathbf{z}_i$ eliminating this difficulty is to use

$$\mathbf{z}_i^* = \log\left(\frac{\mathbf{y}_{i\cdot} + 0.5}{n_i - \mathbf{y}_{i\cdot} + 0.5}\right).$$

This adjustment makes z_i^* as close as possible to λ_i (see Haldane [1955] and Anscombe [1956]). With this modification to $\mathbf{z}_i$, one must also reconsider the definition of $\mathbf{v}_i^2$. Gart and Zweifel [1967] have suggested the variables

$$\mathbf{v}_i^{*2} = \frac{(n_i + 1)(n_i + 2)}{n_i(\mathbf{y}_{i\cdot} + 1)(n_i - \mathbf{y}_{i\cdot} + 1)}$$

as a close approximation to $\mathbf{z}_i^*$.

More generally, one can view the relationship between θ_i and $X_i'B$ as $\theta_i = F(X_i'B)$, where F is some cumulative distribution function. For example, if F is the cumulative of the logistic distribution

$$F(t) = \frac{1}{1 + e^{-t}},$$

we obtain the model analyzed earlier. Other models include taking $F(t)$ to be the cumulative unit normal distribution (probit analysis, see Finney [1947]).

To illustrate the procedure, we consider the data of Table 5.3 of Cox [1977], reproduced as Table 13. Clearly, with the preponderance of 0 values of $y_{i\cdot}$, the adjustment procedure described above is called for. Table 13 also contains the values of the z_i^* and v_i^*. From this we can regress the z_i^*/v_i^* onto the values of x_{1i}/v_i^* and x_{2i}/v_i^* to obtain the relationship

$$\frac{z_i^*}{v_i^*} = -2.5577 - 0.28882\frac{x_{1i}}{v_i^*} + 0.008783\frac{x_{2i}}{v_i^*}.$$

An alternative approach is to attack the estimation problem directly using the method of maximum likelihood. The log likelihood function is

$$l(B) = B' \sum_{i=1}^{r} X_i y_{i\cdot} - \sum_{i=1}^{r} n_i \log(1 + e^{X_i'B}),$$

which is to be maximized with respect to B. Various iterative procedures have been suggested for solving the equations $\partial l(B)/\partial B = 0$ for the elements of B. The advantage of using maximum likelihood estimation is that there is no

Table 13. Regression Variables for the Cox Ingot Data.

i	$y_{i.}$	n_i	x_{1i}	x_{2i}	z_i^*	v_i	z_i^*/v_i^*	x_{1i}/v_i^*	x_{2i}/v_i^*
1	0	10	1.0	7	-3.04452	1.09545	-2.77926	0.91287	6.39010
2	0	31	1.0	14	-4.14313	1.03175	-4.01562	0.96922	13.56913
3	1	56	1.0	27	-3.61092	0.72602	-4.97358	1.37737	37.18907
4	3	13	1.0	51	-1.09861	0.60591	-1.81315	1.65040	84.17023
5	0	17	1.7	7	-3.55535	1.05719	-3.36302	1.60804	6.62134
6	0	43	1.7	14	-4.46591	1.02299	-4.36554	1.66179	13.68535
7	4	44	1.7	27	-2.19722	0.47905	-4.58662	3.54868	56.36141
8	0	1	1.7	51	-1.09861	1.73205	-0.63428	0.98150	29.44486
9	0	7	2.2	7	-2.70805	1.13389	-2.38828	1.94022	6.17342
10	2	33	2.2	14	-2.53370	0.61289	-4.13403	3.58956	22.84268
11	0	21	2.2	27	-3.76120	1.04654	-3.59395	2.10217	25.79939
12	0	1	2.2	51	-1.09861	1.73205	-0.63428	1.27017	29.44486
13	0	12	2.8	7	-3.21888	1.08012	-2.98010	2.59230	6.48074
14	0	31	2.8	14	-4.14313	1.03175	-4.01562	2.71383	13.56913
15	1	22	2.8	27	-2.66259	0.75515	-3.52592	3.70788	35.75460
16	0	9	4.0	7	-2.94444	1.10554	-2.66335	3.61814	6.33174
17	0	19	4.0	14	-3.66356	1.05131	-3.48474	3.80476	13.31666
18	1	16	4.0	27	-2.33537	0.77308	-3.02086	5.17409	34.92513
19	0	1	4.0	51	-1.09861	1.73205	-0.63428	2.30940	29.44486

$y_{i.}$ = number of ingots not ready for rolling.

n_i = number of ingots tested.

x_{1i} = soaking time.

x_{2i} = heating time.

$$z_i^* = \log\left[\frac{y_{i.} + 0.5}{n_i - y_{i.} + 0.5}\right].$$

$$v_i^* = \sqrt{\frac{(n_i + 1)(n_i + 2)}{n_i(y_{i.} + 1)(n_i - y_{i.} + 1)}}.$$

adjustment necessary for extreme values of $y_{i.}$. The disadvantage is that special computer programs must be used, in contrast to the simple regression procedure described above.

References

Anscombe, F. J. 1956. On estimation of binomial response relations. *Biometrika* **43** (December): 641–4.

Barr, D. R. 1966. On testing the equality of uniform and related distributions. *Journal of the American Statistical Association* **61** (September): 856–64.

Cox, D. R. 1977. *Analysis of Binary Data.* London: Chapman and Hall.

Dixon, W. J. and Brown, M. B. 1979. *BMDP*-79. Berkeley, CA: University of California Press.

Epstein, B. and Tsao, C. K. 1953. Some tests based on ordered observations from two exponential populations. *Annals of Mathematical Statistics* **24** (September): 458–66.

Finney, D. J. 1947. *Probit Analysis.* Cambridge: Cambridge University Press.

Fisher, R. A. and Yates, F. 1938. *Statistical Tables for Biological, Agricultural, and Medical Research.* New York: Hafner.

Francis, I. 1973. Comparison of several analysis of variance programs. *Journal of the American Statistical Association* **68** (December): 860–65.
Gart, J. J. and Zweifel, J. R. 1967. On the bias of various estimators of the logit and its variance with application to quantal bioassay. *Biometrika* **54** (June): 181–87.
Gilbert, W. S. 1879. *The Pirates of Penzance or The Slave of Duty*. London: Chappell.
Hájek, J. and Sĭdák, Z. 1967. *Theory of Rank Tests*. New York: Academic Press.
Haldane, J. B. S. 1955. The estimation and significance of the logarithm of a ratio of frequencies. *Annals of Human Genetics* **20**: 309–11.
Harter, H. L. 1969. *Order Statistics and Their Use in Testing and Estimation*, Vol. 2. Washington, DC: U.S. Government Printing Office.
Harville, D. A. 1977. Maximum likelihood approaches to variance component estimation and to related problems. *Journal of the American Statistical Association* **72** (June): 320–37.
Herbach, L. H. 1959. Properties of model II-type analysis of variance tests. *Annals of Mathematical Statistics* **30** (December): 939–59.
Johnson, D. E. and Graybill, F. A. 1972. An analysis of a two-way model with interaction and no replication. *Journal of the American Statistical Association* **67** (December): 862–68.
Kleinbaum, D. G. and Kupper, L. L. 1978. *Applied Regression Analysis and Other Multivariable Methods*. North Scituate, MA: Duxbury Press.
Klotz, J. H. 1964. On the normal scores two-sample rank test. *Journal of the American Statistical Association* **59** (September): 652–64.
Klotz, J. H., Milton, R. C., and Zacks, S. 1969. Mean square efficiency of estimators of variance components. *Journal of the American Statistical Association* **64** (December): 1383–1402.
Kutner, M. H. 1974. Hypothesis testing in linear models (Eisenhart model I). *The American Statistician* **28** (August): 98–100.
Lehmann, E. L. 1959. *Testing Statistical Hypotheses*. New York: Wiley.
Mann, H. B. and Whitney, D. R. 1947. On a test of whether one of two random variables is stochastically larger than the other. *Annals of Mathematical Statistics* **18** (March): 50–60.
Milton, R. C. 1964. An extended table of critical values for the Mann–Whitney (Wilcoxon) two-sample statistic. *Journal of the American Statistical Association* **59** (September): 925–34.
Mood, A. M. 1950. *Introduction to the Theory of Statistics*. New York: McGraw-Hill.
Nie, N. H., Hill, C. H., Jenkins, J. G., Steinbrenner, K., and Bent, D. H. 1975. *Statistical Package for the Social Sciences*. New York: McGraw-Hill.
Olsen, A., Seely, J., and Birkes, D. 1976. Invariant quadratic unbiased estimation for two variance components. *Annals of Statistics* **4** (September): 878–90.
Paulson, E. and Wallis, W. A. 1947. Planning and analyzing experiments for comparing two percentages. In *Techniques of Statistical Analysis*, 247–65, Chapter 7. ed. C. Eisenhart, M. W. Hastay, and W. A. Wallis, New York: McGraw-Hill.
SAS Institute. 1985. *SAS User's Guide*. Cary, NC: SAS Institute.
Searle, S. R. 1971. *Linear Models*. New York: Wiley.
Snedecor, G. W. and Cochran, W. G. 1967. *Statistical Methods*. Ames, IA: Iowa State University Press.
Thompson, W. A. Jr. 1962. The problem of negative estimates of variance components. *Annals of Mathematical Statistics* **33** (March): 273–89.
Tukey, J. W. 1949. One degree of freedom for nonadditivity. *Biometrics* **5** (September): 232–42.
Tukey, J. W. 1977. *Exploratory Data Analysis*. Reading, MA: Addison-Wesley.
van der Waerden, B. L. 1953. Ein neuer test fur das problem der zwei stichproben. *Mathematische Annalen* **126**: 93–107.

van der Waerden, B. L. and Nievergelt, E. 1956. *Tafeln zum Vergleich zweier Stichprobem mittels X test und Zeichentest*. Berlin: Springer-Verlag.

Welch, B. L. 1947. The generalization of "Student's" problem when several different population variances are involved. *Biometrika* **34** (January): 28–35.

Westenberg, J. 1952. A tabulation of the median test with comments and corrections to previous papers. *Proceedings Koninklijke Nederlandse Akademie van Wetenschappen* **55**: 10–14.

Wilcoxon, F. 1945. Individual comparisons by ranking methods. *Biometrics Bulletin* **1**: 80–3.

Yates, F. 1934. The analysis of multiple classifications with unequal numbers in the different classes. *Journal of the American Statistical Association* **29** (March): 51–66.

CHAPTER 8

Analysis of Cross-Classified Data

"Old statisticians never die ... they just get broken down by age and sex."

S. Smith

0. Introduction

Suppose we have a population wherein associated with each member of the population is characteristic from the set of r_1 characteristics $c_{11}, \ldots, c_{1r_1}$ and a characteristic from the set of r_2 characteristics $c_{21}, \ldots, c_{2r_2}$. For example, if the population is that of all adults (age 18 or over) in the United States, then characteristic 1 might be sex, so that $r_1 = 2$, $c_{11} =$ male, and $c_{12} =$ female, and characteristic 2 might be age, so that $r_2 = 100$, and $c_{21} = 18$, $c_{22} = 19, \ldots, c_{2,100} = 117$. We can associate with each member of the population the pair (i_i, i_2), where i_1 denotes the index of characteristic 1 and i_2 denotes the index of characteristic 2 of that member.

When we select a random sample of n members of the population and record their characteristics, our random variables are $\mathbf{I}_1 = (\mathbf{i}_1, \mathbf{i}_2)$, $\mathbf{I}_2 = (\mathbf{i}_{12}, \mathbf{i}_{22}), \ldots, \mathbf{I}_n = (\mathbf{i}_{1n}, \mathbf{i}_{2n})$. We can thus count the number of observations in our sample for which $\mathbf{I}$ takes on the value (i_1, i_2), where $i_1 = 1, \ldots, r_1$, $i_2 = 1, \ldots, r_2$, and denote this number by $\mathbf{n}_{i_1 i_2}$. Clearly, $\sum_{i_1=1}^{r_1} \sum_{i_2=1}^{r_2} \mathbf{n}_{i_1 i_2} = n$.

We can further posit that the random variables measured from our sample are the $\mathbf{n}_{i_1 i_2}$'s, and that these random variables are generated from a multinomial distribution where $\theta_{i_1 i_2} = \Pr\{\mathbf{I}_l = (i_1, i_2)\}$ for each $l = 1, \ldots, n$ where $\theta_{i_1 i_2} \geq 0$ for all i_1, i_2 and $\sum_{i_1=1}^{r_1} \sum_{i_2=1}^{r_2} \theta_{i_1 i_2} = 1$. Sometimes the observed values of the $\mathbf{n}_{i_1 i_2}$'s are recorded in a "cross-classification" or "cross-tabulation", a table such as the following:

$$
\begin{array}{cc|cccc|c}
 & & \multicolumn{4}{c}{\overbrace{\hphantom{n_{11}\quad n_{12}\quad \cdots \quad n_{1r_2}}}^{i_2}} & \\
 & & 1 & 2 & \cdots & r_2 & \\
\hline
\multirow{4}{*}{i_1} & 1 & n_{11} & n_{12} & \cdots & n_{1r_2} & n_{1\cdot} \\
 & 2 & n_{21} & n_{22} & \cdots & n_{2r_2} & n_{2\cdot} \\
 & \vdots & \vdots & \vdots & & \vdots & \\
 & r_1 & n_{r_11} & n_{r_12} & \cdots & n_{r_1r_2} & n_{r_1\cdot} \\
\hline
 & & n_{\cdot 1} & n_{\cdot 2} & \cdots & n_{\cdot r_2} & n
\end{array}
$$

where

$$n_{\cdot i_2} = \sum_{i_1=1}^{r_1} n_{i_1 i_2} \quad \text{and} \quad n_{i_1 \cdot} = \sum_{i_2=1}^{r_2} n_{i_1 i_2}.$$

The analysis of cross-classified categorical data is a study of models which structure this multinomial distribution (and its generalization when we record m characteristics, and not merely the $m = 2$ characteristics as exemplified above) in a parsimonious parametric fashion. In the above example, we have a multinomial distribution with $r_1 r_2$ parameters ($r_1 r_2 - 1$ of them being linearly independent). If we had postulated that having characteristic 1 is statistically independent of having characteristic 2, then we would instead postulate two separate sets of multinomial parameters, $\{\psi_{1i_1}, i_1 = 1, \ldots, r_1\}$ and $\{\psi_{2i_2}, i_2 = 1, \ldots, r_2\}$, so that

$$\theta_{i_1 i_2} = \psi_{1i_1}\psi_{2i_2}$$

for all i_1, i_2. Thus we have only $r_1 + r_2$ parameters (and $r_1 + r_2 - 2$ linearly independent parameters) governing the distribution of the $\mathbf{n}_{i_1 i_2}$'s.

In this chapter we will study procedures for estimating parameters, given various assumptions about parsimonious models of the parametric structure of the multinomial distribution of the $\mathbf{n}_{i_1 i_2 \ldots i_m}$, and testing the hypothesis that the data are consistent with the postulated model. In Section 1 we study the classic case which forms the basis for our introductory example, namely $m = 2$ and $\theta_{i_1 i_2} = \psi_{1i_1}\psi_{2i_2}$. In Section 2 we introduce the notion of "interaction", i.e., the introduction of additional parameters beyond the ψ's, then model the θ's, illustrating both interpretation and estimation of these parameters in the case $m = 2$. This is followed in Section 3 by an analysis of the case $m = 3$, wherein the analysis of Sections 1 and 2 are generalized and the groundwork is set for the analysis of the case of general m. Section 4 goes back to the case $m = 2$ and, instead of including more than the $r_1 + r_2$ ψ parameters, we postulate some order conditions on the ψ's, such as $\psi_{11} \le \psi_{12} \le \cdots \le \psi_{1r_1}$ and $\psi_{21} \le \psi_{22} \cdots \le \psi_{2r_2}$. Section 5 considers the latent class model, one in which our population is a mixture of subpopulations, each with its own set of multinomial parameters with its own internal parametric structure (though typically the structure is the independence structure described above).

This chapter is a compressed overview of material treated in greater depth in books by Bishop, Fienberg, and Holland [1975] and Haberman [1978, 1979], as well as, in an introductory fashion, in Fienberg [1977]. Many of Goodman's seminal research papers on topics discussed in this chapter appear in the two collections by Goodman [1978] and [1984].

1. Independence in the $r_1 \times r_2$ Table

(a) General

Prescription

The independence hypothesis, as described above,[1] is expressed as $\theta_{ij} = \psi_{1i}\psi_{2j}, i = 1, \ldots, r_1, j = 1, \ldots, r_2$. One approach to testing the hypothesis is to:

(1) estimate the marginal parameters ψ_{1i} and ψ_{2j} by $\hat{\boldsymbol{\psi}}_{1i} = \mathbf{n}_{i\cdot}/n$ and $\hat{\boldsymbol{\psi}}_{2j} = \mathbf{n}_{\cdot j}/n$;
(2) assume that the $\mathbf{n}_{ij}$ have a multinomial distribution with $\mathscr{E}\mathbf{n}_{ij} = n\theta_{ij}$, and
(3) calculate the Pearson [1904] chi-square "goodness-of-fit" statistic for the table, namely

$$\begin{aligned}
\mathbf{x} &= \sum_{i=1}^{r_1}\sum_{j=1}^{r_2} (\mathbf{n}_{ij} - n\hat{\boldsymbol{\theta}}_{ij})^2/n\hat{\boldsymbol{\theta}}_{ij} \\
&= \sum_{i=1}^{r_1}\sum_{j=1}^{r_2} (\mathbf{n}_{ij} - n\hat{\boldsymbol{\psi}}_{i\cdot}\hat{\boldsymbol{\psi}}_{\cdot j})^2/n\hat{\boldsymbol{\psi}}_{i\cdot}\hat{\boldsymbol{\psi}}_{i\cdot j} \\
&= \sum_{i=1}^{r_1}\sum_{j=1}^{r_2} (\mathbf{n}_{ij} - \mathbf{n}_{i\cdot}\mathbf{n}_{\cdot j}/n)^2/(\mathbf{n}_{i\cdot}\mathbf{n}_{\cdot j}/n) \\
&= n\left(\sum_{i=1}^{r_1}\sum_{j=1}^{r_2} \frac{\mathbf{n}_{ij}^2}{\mathbf{n}_{i\cdot}\mathbf{n}_{\cdot j}} - 1\right).
\end{aligned}$$

It has been shown (see Bishop, Fienberg, and Holland [1975], pp. 472–73) that $\mathbf{x}$ has a chi-square distribution with $(r_1 - 1)(r_2 - 1)$ degrees of freedom for large samples.

As an example, consider the following table, taken from Srole *et al.* [1962] and studied by Haberman [1974] and Goodman [1979]:

[1] In this section we, for ease of notation, replace i_1 with the single subscript i and i_2 with the single subscript j. The sub-subscripting will prove useful, though, in subsequent sections of this chapter.

Mental health status	Socioeconomic status A	B	C	D	E	F	
Well	64	57	57	72	36	21	307
Mild symptoms	94	94	105	141	97	71	602
Moderate symptoms	58	54	65	77	54	54	362
Impaired	46	40	60	94	78	71	389
	262	245	287	384	265	217	1660

Here

$$x = 1660\left(\frac{64^2}{307 \times 262} + \frac{57^2}{307 \times 245} + \cdots + \frac{71^2}{389 \times 217} - 1\right)$$

$$= 45.99$$

is to be compared with the upper $100\alpha\%$ point of a chi-squared distribution with $(r_1 - 1)(r_2 - 1) = 3 \times 6 = 18$ degrees of freedom.

Theoretical Background

An alternative approach is to calculate the likelihood ratio and use its asymptotic distribution as the basis for a test. The likelihood function of the data is the multinomial

$$\frac{n!}{\prod_{i=1}^{r_1}\prod_{j=1}^{r_2} n_{ij}!}\prod_{i=1}^{r_1}\prod_{j=1}^{r_2}\theta_{ij}^{n_{ij}}.$$

Under the general hypothesis, $\hat{\boldsymbol{\theta}}_{ij} = \mathbf{n}_{ij}/n$. Under the null hypothesis, the likelihood function can be rewritten as

$$\frac{n!}{\prod_{i=1}^{r_1}\prod_{j=1}^{r_2} n_{ij}!}\prod_{i=1}^{r_1}\prod_{j=1}^{r_2}(\psi_{i.}\psi_{.j})^{n_{ij}} = \frac{n!}{\prod_{i=1}^{r_1}\prod_{j=1}^{r_2} n_{ij}!}\prod_{i=1}^{r_1}\psi_{i.}^{n_{i.}}\prod_{j=1}^{r_2}\psi_{.j}^{n_{.j}},$$

so that $\hat{\boldsymbol{\psi}}_{i.} = \mathbf{n}_{i.}/n$ and $\hat{\boldsymbol{\psi}}_{.j} = \mathbf{n}_{.j}/n$. Thus the logarithm of the likelihood ratio is

$$\mathbf{u} = \sum_{i=1}^{r_1}\mathbf{n}_{i.}\log(\mathbf{n}_{i.}/n) + \sum_{j=1}^{r_2}\mathbf{n}_{.j}\log(\mathbf{n}_{.j}/n) - \sum_{i=1}^{r_1}\sum_{j=1}^{r_2}\mathbf{n}_{ij}\log(\mathbf{n}_{ij}/n)$$

$$= \sum_{i=1}^{r_1}\mathbf{n}_{i.}\log\mathbf{n}_{i.} + \sum_{j=1}^{r_2}\mathbf{n}_{.j}\log\mathbf{n}_{.j} - \sum_{i=1}^{r_1}\sum_{j=1}^{r_2}\mathbf{n}_{ij}\log\mathbf{n}_{ij} - n\log n.$$

Under the null hypothesis of independence, the asymptotic distribution of $-2\mathbf{u}$ is the chi-square distribution with $(r_1 r_2 - 1) - ([r_1 - 1] - [r_2 - 1]) = (r_1 - 1)(r_2 - 1)$ degrees of freedom, as is the goodness-of-fit statistic given above. In our above example, u is -23.71, so that $-2u = 47.42$.

(b) 2 × 2 Tables

Prescription

A special case which has had an independent history of analysis procedures is that of the 2 × 2 table. First, algebraically the chi-square goodness-of-fit test statistic reduces to

$$\mathbf{X} = \frac{n(\mathbf{n}_{11}\mathbf{n}_{22} - \mathbf{n}_{21}\mathbf{n}_{12})^2}{\mathbf{n}_{1}.\mathbf{n}_{2}.\mathbf{n}._{1}\mathbf{n}._{2}}$$

and has asymptotically a $\chi^2(1)$ distribution.

An alternative approach to the analysis of the 2 × 2 table, and indeed the "correct" way of testing for independence in such tables, is the Fisher–Irwin exact test (Fisher [1934] and Irwin [1935]). The test involves the enumeration of all 2 × 2 tables with row and column sums equal to $n_1.$, $n_2.$, $n._1$, and $n._2$ and counting the fraction of such tables which are more inconsistent with the hypothesis of independence than the particular table observed. If that fraction is smaller than the level of significance α, the independence hypothesis is rejected. For example, suppose our 2 × 2 table were

	Reaction	No reaction	
Drug A	3	7	10
Drug B	12	6	18
	15	13	28

and our null hypothesis is that whether a member of the population reacts to a drug is independent of whether the drug is drug A or drug B. Keeping the marginals fixed, we see that the tables even more inconsistent with this null hypothesis are

2	8
13	5

1	9
14	4

0	10
15	3

.

It can be shown that the probability of a given 2 × 2 table being observed, conditional on the marginals being fixed, is

$$\frac{n_1.!\, n_2.!\, n._1!\, n._2!}{n!\, n_{11}!\, n_{12}!\, n_{21}!\, n_{22}!}.$$

Thus we can calculate the probability of the observed and more extreme tables from this formula.

In the above example the probability of the observed table is given by

$$\frac{10!\ 18!\ 15!\ 13!}{28!\ 3!\ 7!\ 12!\ 6!} = 0.0595,$$

and that of the three more extreme tables are 0.0103, 0.0008, and 0.00002, respectively. Thus if the observed table were

2	8
13	5

the probability of observing a table, at least as inconsistent with the null hypothesis as that one, would be $0.0103 + 0.0008 + 0.00002 = 0.0111$, so that the null hypothesis would be rejected for $\alpha = 0.05$. Since the probability associated with the table of the example is itself greater than 0.05, the null hypothesis would be accepted, and the nominal probability of that table or one more extreme is 0.0706.

An aspect of this test which has elicited some research activity is the impact of the discreteness of the hypergeometric distribution on the actual significance level of the test. If one fixes α, the nominal significance level, then α is an upper bound on the significance level actually achieved if one rejects the null hypothesis if $\mathbf{n}_{11}$ is less than some critical value x, where x is chosen so that

$$\Pr\{\mathbf{n}_{11} \geq x\} \leq \alpha \leq \Pr\{\mathbf{n}_{11} \leq x + 1\}.$$

One approach which achieves an actual significance level equal to α is the use of an "auxiliary randomization device" when $\mathbf{n}_{11} = x + 1$. This entails tossing a coin with probability

$$\phi = \frac{\alpha - \Pr\{\mathbf{n}_{11} \leq x\}}{\Pr\{\mathbf{n}_{11} = x + 1\}}$$

of heads and rejecting the null hypothesis if heads occurs. Tocher [1950] has shown that with this enhancement this procedure is the uniformly most powerful unbiased test of the independence hypothesis for 2×2 tables. (Variants of the Fisher–Irwin procedure which do not use such a randomization device but which achieve a fixed significance level α are the procedures of Boschloo [1970] and Garside [1971, 1972]. To use these, though, one needs special tables given in their respective papers. Moreover, their procedures do not have the optimality property that the Fisher–Irwin procedure augmented by randomization has.)

Another approach is to find a continuous approximation to the hypergeometric distribution. One such candidate is the chi-square statistic for testing for independence in the 2×2 table. Yates [1934] has proposed a modification of this statistic, a "correction for continuity", to make this statistic a better approximation to the hypergeometric probability. This modified statistic is

$$\mathbf{u}' = \frac{n(\mathbf{n}_{11}\mathbf{n}_{22} - \mathbf{n}_{12}\mathbf{n}_{21} \pm n/2)^2}{n_1.n_2.n_{.1}n_{.2}},$$

where the + sign is used if $n_1.n_{.1} > nn_{11}$ and the − sign is used otherwise.

The chi-square statistic $\mathbf{u}$ associated with the observed table is $\mathbf{u} = 3.475$, corresponding to an associated probability level of 0.0623; $\mathbf{u}'$ is equal to 2.157, corresponding to an associated probability level of 0.1419. Note that here the probability level associated with $\mathbf{u}$ (0.0623) was closer to the true probability (0.0706) than was the probability level associated with $\mathbf{u}'$. This casual observation was noted more systematically in papers by Plackett [1964] and Grizzle [1967]. A general discussion on whether or not to use the Yates continuity correction is provided in both Conover [1974] and the various discussion papers that follow it. The general conclusion is that if one wants to approximate the exact hypergeometric probability then use the correction, but if one wants a $\chi^2(1)$ distributed test statistic, one may as well use $\mathbf{u}$ and eschew the continuity correction.

Tables of the hypergeometric distribution for $n = 1(1)50(10)100$ are given in Lieberman and Owen [1961]. Tables to facilitate the Fisher–Irwin exact test are given in Finney *et al.* [1963] for $n_1. \le n_2. = 1(1)40$. Fisher and Yates [1948] present a table (Table VIII) to be used in testing significance of a 2×2 table. Their table is based on the $\mathbf{u}'$ statistic and gives, for $\alpha = 0.025$ and 0.005, exact critical values of $\mathbf{u}'$ as a function of two variables, the smallest expected value of any cell (m) and the ratio of the smallest expected value to the smallest marginal total (p). That is,

$$m = \min_{i,j}\left(\frac{n_i.n_{.j}}{n}\right),$$

$$p = \frac{m}{\min(n_1., n_2., n_{.1}, n_{.2})}.$$

Tabulated values are $m = 1(1)6, 8, 12, 24, 48, 96$, $p = 0, 0.25, 0.5$.

Theoretical Background

To develop the exact test more fully, we first reparametrize by defining $\delta = \theta_{12}\theta_{21}/\theta_{11}\theta_{22}$. Then

$$\theta_{11} = \theta_1.\theta_{.1} + \left(\frac{1-\delta}{\delta}\right)\theta_{12}\theta_{21},$$

$$\theta_{12} = \theta_1.\theta_{.2} - \left(\frac{1-\delta}{\delta}\right)\theta_{12}\theta_{21},$$

$$\theta_{21} = \theta_2.\theta_{.1} - \left(\frac{1-\delta}{\delta}\right)\theta_{12}\theta_{21},$$

$$\theta_{22} = \theta_2.\theta_{.2} + \left(\frac{1-\delta}{\delta}\right)\theta_{12}\theta_{21},$$

where $\theta_{i\cdot} = \theta_{i1} + \theta_{i2}$ and $\theta_{\cdot j} = \theta_{1j} + \theta_{2j}$. The independence hypothesis then can be seen as equivalent to the parametric hypothesis $\delta = 1$.

Now let us consider the distribution of the elements of the table, given the marginal totals. It is given by

$$\Pr\{\mathbf{n}_{11} = n_{11}, \mathbf{n}_{21} = n_{21} | n_{1\cdot}, n_{2\cdot}\} = \binom{n_{1\cdot}}{n_{11}}\left(\frac{\theta_{11}}{\theta_{1\cdot}}\right)^{n_{11}}\left(\frac{\theta_{12}}{\theta_{1\cdot}}\right)^{n_{12}}\binom{n_{2\cdot}}{n_{21}}\left(\frac{\theta_{21}}{\theta_{2\cdot}}\right)^{n_{21}}\left(\frac{\theta_{22}}{\theta_{2\cdot}}\right)^{n_{22}},$$

so that

$$\Pr\{\mathbf{n}_{11} = n_{11} | n_{\cdot 1}\} = \frac{\binom{n_{1\cdot}}{n_{11}}\binom{n_{2\cdot}}{n_{\cdot 1} - n_{11}}\delta^{n_{11}}}{\sum_{x=0}^{n_{\cdot 1}}\binom{n_{1\cdot}}{x}\binom{n_{2\cdot}}{n_{\cdot 1} - x}\delta^{x}}$$

(see Lehmann [1959], pp. 140–46). Under the null hypothesis of independence, $\delta = 1$, and so the distribution of $\mathbf{n}_{11}$ is the hypergeometric distribution

$$\Pr\{\mathbf{n}_{11} = n_{11} | n_{\cdot 1}\} = \frac{\binom{n_{1\cdot}}{n_{11}}\binom{n_{2\cdot}}{n_{21}}}{\binom{n}{n_{\cdot 1}}} = \frac{n_{1\cdot}!\, n_{2\cdot}!\, n_{\cdot 1}!\, n_{\cdot 2}!}{n!\, n_{11}!\, n_{12}!\, n_{21}!\, n_{22}!}.$$

2. Log-Linear Models in the $r_1 \times r_2$ Table

The independence hypothesis $\theta_{ij} = \psi_{i\cdot}\psi_{\cdot j}$ can be expressed as $\log \theta_{ij} = \log \psi_{i\cdot} + \log \psi_{\cdot j}$. If we let $\eta_{ij} = \log \theta_{ij}$, we can, in analogy with the analysis of variance decomposition, express η_{ij} as

$$\eta_{ij} = \bar{\eta}_{\cdot\cdot} + (\bar{\eta}_{i\cdot} - \bar{\eta}_{\cdot\cdot}) + (\bar{\eta}_{\cdot j} - \bar{\eta}_{\cdot\cdot}) + (\bar{\eta}_{ij} - \bar{\eta}_{i\cdot} - \bar{\eta}_{\cdot j} + \bar{\eta}_{\cdot\cdot}),$$

where

$$\bar{\eta}_{i\cdot} = \sum_{j=1}^{r_2} \eta_{ij}/r_2, \qquad \bar{\eta}_{\cdot j} = \sum_{i=1}^{r_1} \eta_{ij}/r_1, \qquad \bar{\eta}_{\cdot\cdot} = \sum_{i=1}^{r_1}\sum_{j=1}^{r_2} \eta_{ij}/r_1 r_2.$$

The term

$$\eta_{ij} - \bar{\eta}_{i\cdot} - \bar{\eta}_{\cdot j} + \bar{\eta}_{\cdot\cdot}$$

is called the "interaction", and the terms $\bar{\eta}_{i\cdot} - \bar{\eta}_{\cdot\cdot}$ and $\bar{\eta}_{\cdot j} - \bar{\eta}_{\cdot\cdot}$ are called "main effects" in the analysis of variance jargon. The independence hypothesis, in this re-expression, is equivalent to the statement

$$\eta_{ij} - \bar{\eta}_{i\cdot} - \bar{\eta}_{\cdot j} + \bar{\eta}_{\cdot\cdot} = 0, \quad \text{all } i, j,$$

i.e., that all interactions are equal to zero.

This analogy leads us to the general multinomial reparametrization which forms the cornerstone of the so-called "log-linear model". Let $\xi_{..} = \bar{\eta}_{..}$, $\xi_{i.} = \bar{\eta}_{i.} - \bar{\eta}_{..}$, $\xi_{.j} = \bar{\eta}_{.j} - \bar{\eta}_{..}$, and $\xi_{ij} = \eta_{ij} - \bar{\eta}_{i.} - \bar{\eta}_{.j} + \bar{\eta}_{..}$, with the constraints that

$$\sum_{i=1}^{r_1} \xi_{i.} = \sum_{j=1}^{r_2} \xi_{.j} = \sum_{i=1}^{r_1} \xi_{ij} = \sum_{j=1}^{r_2} \xi_{ij} = 0.$$

Thus there are $r_1 - 1$ independent $\xi_{i.}$'s, $r_2 - 1$ independent $\xi_{.j}$'s, $r_1 r_2 - r_1 - r_2 + 1$ independent ξ_{ij}'s, and so, including $\xi_{..}$, a set of $(r_1 - 1) + (r_2 - 1) + (r_1 r_2 - r_1 - r_2 + 1) + 1 = r_1 r_2$ parameters, a set equivalent in number to the number of θ_{ij}'s. With this reparametrization, we can entertain more specific hypotheses than the general independence hypothesis. For example, we can postulate independence holding only in a subtable of the $r_1 \times r_2$ table.

Let us consider once again our earlier example. Here $r_1 = 3$, $r_2 = 4$, and the independence hypothesis can be rewritten as $\xi_{ij} = 0$ and, in addition, we might want to test the hypotheses (a) $\xi_{i.} = 0$, $\xi_{ij} = 0$ and (b) $\xi_{.j} = 0$, $\xi_{ij} = 0$, as well as the hypothesis that the expected value of each cell of the table is identical, i.e., (c) $\xi_{i.} = 0$, $\xi_{.j} = 0$, $\xi_{ij} = 0$. The hypothesis that the $\xi_{i.} = 0$ as well as that the $\xi_{ij} = 0$ implies that $\mathscr{E}\mathbf{n}_{ij} = n_{i.}/r_2$, so that the chi-square test of this hypothesis is given by

$$\mathbf{x} = r_2 \sum_{i=1}^{r_1} \left(\frac{\sum_{j=1}^{r_2} \mathbf{n}_{ij}^2}{n_{i.}} \right) - n.$$

For our example, $x = 100.27$ and is to be compared with the $100\alpha\%$ point of a $\chi^2(r_1 - 1) = \chi^2(2)$ variable.

The hypothesis that $\xi_{.j} = 0$ as well as that the $\xi_{ij} = 0$ implies that $\mathscr{E}\mathbf{n}_{ij} = n_{.j}/r_1$, so that for this hypothesis the test statistic is

$$\mathbf{x} = r_1 \sum_{j=1}^{r_2} \left(\frac{\sum_{i=1}^{r_1} \mathbf{n}_{ij}^2}{n_{.j}} \right) - n$$

and has a $\chi^2(r_2 - 1)$ distribution. In our example, $\mathbf{x} = 160.189$.

Finally, the hypothesis that $\xi_{i.} = 0$, $\xi_{.j} = 0$, and $\xi_{ij} = 0$ implies that $\mathscr{E}\mathbf{n}_{ij} = n/r_1 r_2$. This hypothesis is tested by the statistic

$$\mathbf{x} = \frac{r_1 r_2}{n} \sum_{i=1}^{r_1} \sum_{j=1}^{r_2} \mathbf{n}_{ij}^2 - n$$

(in our example, $\mathbf{x} = 220.76$), and has a $\chi^2(1)$ distribution.

For each of these hypotheses the expected value of each of the $\mathbf{n}_{ij}$'s was expressible in a simple closed form, thereby making the expression of the χ^2 goodness-of-fit statistic

$$\mathbf{x} = \sum_{i=1}^{r_1} \sum_{j=1}^{r_2} (\mathbf{n}_{ij} - \mathscr{E}\mathbf{n}_{ij})^2/\mathscr{E}\mathbf{n}_{ij}$$

merely a matter of algebraic substitution. Such, however, is not the case for the generalizations of this hypothesis for higher dimensional tables, as will be seen in the next section. It is there that the essential elements of log-linear modeling will become more evident.

3. The Three-Dimensional Table

In the case of $m = 3$, we observe $\mathbf{n}_{i_1 i_2 i_3}$, for $i_1 = 1, \ldots, r_1$, $i_2 = 1, \ldots, r_2$, and $i_3 = 1, \ldots, r_3$, with

$$n\theta_{i_1 i_2 i_3} = \mathscr{E}\mathbf{n}_{i_1 i_2 i_3} \quad \text{and} \quad n = \sum_{i_1=1}^{r_1} \sum_{i_2=1}^{r_2} \sum_{i_3=1}^{r_3} \mathbf{n}_{i_1 i_2 i_3}.$$

As in the case of $m = 2$, we can write

$$\log \mathscr{E}\mathbf{n}_{i_1 i_2 i_3} = \xi_{\cdots} + \xi_{i_1 \cdot\cdot} + \xi_{\cdot i_2 \cdot} + \xi_{\cdot\cdot i_3} + \xi_{i_1 i_2 \cdot} + \xi_{i_1 \cdot i_3} + \xi_{\cdot i_2 i_3} + \xi_{i_1 i_2 i_3}.$$

(Notation for this model has not been standardized. Bishop [1967] originated the notation used in Bishop, Fienberg, and Holland [1975], who express the model as

$$\log \theta_{ijk} = u + u_{1(i)} + u_{2(j)} + u_{3(k)} + u_{12(ij)} + u_{13(ik)} + u_{23(jk)} + u_{123(ijk)},$$

and Goodman [1970] and Haberman [1978] express the model as

$$\log \theta_{ijk} = \lambda + \lambda_i^A + \lambda_j^B + \lambda_k^C + \lambda_{ij}^{AB} + \lambda_{jk}^{AC} + \lambda_{jk}^{BC} + \lambda_{ijk}^{ABC}.$$

The advantages of our notation are that it is more consistent with the analysis of variance notation from which the model was adapted, and that the position of the data and the subscripts define the interaction parameter, obviating the need for the mnemonic superscripts and subscripts).

The term "interaction" is used by different authors to refer to different constructs. One way to interpret the parameters of the log-linear model is as follows. If we break the $r_1 \times r_2 \times r_3$ table up into r_3 $r_1 \times r_2$ tables, stratifying on the third category, then each of the $r_1 \times r_2$ tables itself can be expressed in a log-linear model, namely

$$\log \mathscr{E}\mathbf{n}_{i_1 i_2 i_3} = \xi_{\cdot\cdot(i_3)} + \xi_{i_1 \cdot (i_3)} + \xi_{\cdot i_2 (i_3)} + \xi_{i_1 i_2 (i_3)},$$

where the i_3 subscript is parenthesized to distinguish the subscripted parameter from that for the full log-linear model. Then the various interactions in the three-way table are defined as:

$$\xi_{\cdots} = \frac{1}{r_3} \sum_{i_3=1}^{r_3} \xi_{\cdot\cdot(i_3)},$$

$$\xi_{i_1 \cdot\cdot} = \frac{1}{r_3} \sum_{i_3=1}^{r_3} \xi_{i_1 \cdot (i_3)},$$

$$\xi_{\cdot i_2 \cdot} = \frac{1}{r_3} \sum_{i_3=1}^{r_3} \xi_{\cdot i_2(i_3)},$$

$$\xi_{i_1 i_2 \cdot} = \frac{1}{r_3} \sum_{i_3=1}^{r_3} \xi_{i_1 i_2(i_3)}.$$

Thus the two-way interaction $\xi_{i_1 i_2 \cdot}$ in a three-way table is the average of the two-way interactions. If *each* two-way table satisfies the independence hypothesis, then the two-way interaction in the three-way table is equal to zero. But this condition is not necessary for zero two-way interaction in a three-way table.

We use as our example the classic $2 \times 2 \times 2$ table of Bartlett [1935], where

$$n_{111} = 156,$$

$$n_{211} = 84,$$

$$n_{121} = 84,$$

$$n_{221} = 156,$$

$$n_{112} = 107,$$

$$n_{212} = 133,$$

$$n_{122} = 31,$$

$$n_{222} = 209.$$

Appendix II illustrates the use of the program of Appendix I to fit four log-linear models to these data. The fitted models, in the notation of this section, respectively, are

I: $\xi_{i_1 i_2 i_3} = 0;$
II: $\xi_{\cdot i_2 i_3} = 0, \xi_{i_1 i_2 i_3} = 0;$
III: $\xi_{i_1 \cdot i_3} = 0, \xi_{i_1 i_2 i_3} = 0;$
IV: $\xi_{i_1 i_2 \cdot} = 0, \xi_{i_1 i_2 i_3} = 0.$

The program, by contrast, asks for the interactions to be fit, which in these examples are

I: $\xi_{i_1 i_2 \cdot}, \xi_{i_1 \cdot i_3}, \xi_{\cdot i_2 i_3};$
II: $\xi_{i_1 i_2 \cdot}, \xi_{i_1 \cdot i_3};$
III: $\xi_{i_1 i_2 \cdot}, \xi_{\cdot i_2 i_3};$
IV: $\xi_{i_1 \cdot i_3}, \xi_{\cdot i_2 i_3}.$

As can be seen from the output of the program, the estimated expected values of each of the $\mathbf{n}_{i_1 i_2 i_3}$'s is as summarized in the following table:

$i_1 i_2 i_3$	$n_{i_1 i_2 i_3}$	Estimated Expected Value I	II	III	IV
1 1 1	156	161.06	166.98	131.5	120
2 1 1	84	78.94	89.48	108.5	120
1 2 1	84	78.91	73.02	57.5	120
2 2 1	156	161.09	150.52	182.5	120
1 1 2	107	101.91	96.02	131.5	69
2 1 2	133	138.09	127.52	108.5	171
1 2 2	31	36.12	41.98	57.5	69
2 2 2	209	203.88	214.48	182.5	171
Degrees of freedom		1	2	2	2
Chi-square statistic		2.27	7.42	52.32	101.94

To reconstruct estimates of the ξ's from the $\hat{\theta}_{i_1 i_2 i_3}$'s, we note that

$$\sum_{i_3=1}^{r_3} \log \theta_{i_1 i_2 i_3} = r_3(\xi_{\cdots} + \xi_{i_1 \cdot\cdot} + \xi_{\cdot i_2 \cdot} + \xi_{\cdot\cdot i_3}),$$

$$\sum_{i_2=1}^{r_2} \sum_{i_3=1}^{r_3} \log \theta_{i_1 i_2 i_3} = r_2 r_3(\xi_{\cdots} + \xi_{i_1 \cdot\cdot}),$$

$$\sum_{i_1=1}^{r_1} \sum_{i_2=1}^{r_2} \sum_{i_3=1}^{r_3} \log \theta_{i_1 i_2 i_3} = r_1 r_2 r_3 \xi_{\cdots}.$$

We can therefore work backward from these equations to successively estimate $\xi_{\cdots}$, then the first-order terms (exemplified by $\xi_{i_1 \cdot\cdot}$), next the second-order terms (exemplified by $\xi_{i_1 i_2 \cdot}$), and finally $\xi_{i_1 i_2 i_3}$. Applying this process to the previously derived estimates of the $\theta_{i_1 i_2 i_3}$'s, produces the following estimates of the ξ's:

	I	II	III	IV
$\hat{\xi}_{\cdots}$	−2.19729	−2.18756	−2.16086	−2.12925
$\hat{\xi}_{1\cdot\cdot}$	−0.25438	−0.25178	−0.24068	−0.22689
$\hat{\xi}_{\cdot 1 \cdot}$	0.08098	0.07681	0.07681	0
$\hat{\xi}_{\cdot\cdot 1}$	0.05555	0.04981	0	0.04980
$\hat{\xi}_{11\cdot}$	0.35670	0.33682	0.33681	0
$\hat{\xi}_{1\cdot 1}$	0.25424	0.22691	0	0.22689
$\hat{\xi}_{\cdot 11}$	−0.08093	0	0	0
$\hat{\xi}_{111}$	0	0	0	0

(That $\hat{\xi}_{\cdot 1 \cdot} = 0$ and $\hat{\xi}_{\cdot 11} = 0$ in case IV and that $\hat{\xi}_{\cdot\cdot 1} = 0$ and $\hat{\xi}_{\cdot 11} = 0$ in case III are a coincidental consequence of the nature of the Bartlett data. The other $\hat{\xi}$'s that are equal to 0 in the four cases are a result of the constraints defining the respective cases.)

By contrast, if there were no constraints on the interactions, but if the independence hypothesis holds, then the estimates of the ξ's are given by the following:

$$\hat{\xi}_{\cdots} = -2.20433,$$

$$\hat{\xi}_{1\cdot\cdot} = -0.26573,$$

$$\hat{\xi}_{\cdot 1\cdot} = 0.98357,$$

$$\hat{\xi}_{\cdot\cdot 1} = 0.77731,$$

$$\hat{\xi}_{11\cdot} = 0.36611,$$

$$\hat{\xi}_{1\cdot 1} = 0.26573,$$

$$\hat{\xi}_{\cdot 11} = -0.09836,$$

$$\hat{\xi}_{111} = -0.05659.$$

(These estimators are developed in exactly the same way as those for case I–IV above, but using $\hat{\theta}_{ijk} = \hat{\theta}_{i\cdot\cdot}\hat{\theta}_{\cdot j\cdot}\hat{\theta}_{\cdot\cdot k}$, where the marginal θ's are estimated by the marginal observed fractions, e.g., $\hat{\theta}_{i\cdot\cdot} = n_{i\cdot\cdot}/n$.)

We might be interested in testing whether a particular interaction is at all present, within the context of some log-linear model (i.e., a model which may in fact specify that some other interactions are equal to 0). One can estimate the variances of the interaction estimator, using formulas given in Section 4.4.2 of Bishop, Fienberg, and Holland [1975]. But, as they point out in Section 4.4.3, comparison of a standardized interaction estimator (i.e., one divided by its estimated standard deviation) with zero is not a proper test of the hypothesis of zero interaction. Some suggested procedures for accomplishing such a test are given in Section 4.4.3 of Bishop, Fienberg, and Holland [1975]. But the recommended approach (see Goodman [1969]) is to fit two models, one including and one excluding the interaction suspected to be zero, and compare the associated chi-square statistics for the two models.

Theoretical Background

We have described "interaction" by analogy with the analysis of variance. First, one attempts to model $\log \theta_{i_1 i_2 i_3}$ as a sum of "main effects", i.e., parameters involving i_1, i_2, and i_3 separately. Then, if that model fails, one adds additional parameters involving pairs of subscripts (in analogy to first-order interaction in anova). An alternative interpretation of "interaction", which leads to exactly the same ultimate model for $\log \theta_{i_1 i_2 i_3}$, is the following. One might wish to consider the hypothesis of "multiple independence" that (i_1, i_2) is independent of i_3, i.e., that

$$\mathscr{E}\mathbf{n}_{i_1 i_2 i_3} = n\theta_{i_1 i_2 i_3} = n\theta_{i_1 i_2 \cdot}\theta_{\cdot\cdot i_3}.$$

Now this hypothesis implies that

$$\theta_{i_1 \cdot i_3} = \theta_{i_1\cdot\cdot}\theta_{\cdot\cdot i_3},$$

$$\theta_{\cdot i_2 i_3} = \theta_{\cdot i_2 \cdot}\theta_{\cdot\cdot i_3},$$

i.e., that i_3 is pairwise independent of i_1 and i_2. But the converse is not true. The most general set of conditions which, when supplemented by the pairwise independence condition, implies the hypothesis of multiple independence is that $\theta_{i_1 i_2 i_3}$ is expressible as

$$\theta_{i_1 i_2 i_3} = \pi_{i_1 i_2 \cdot} \pi_{i_1 \cdot i_3} \pi_{\cdot i_2 i_3}$$

(see Roy and Kastenbaum [1956]). This implies, when $r_1 = r_2 = r_3 = 2$, that

$$\frac{\theta_{111}\theta_{221}}{\theta_{211}\theta_{121}} = \frac{\theta_{112}\theta_{222}}{\theta_{212}\theta_{122}},$$

the Bartlett [1935] condition of "no interaction". In general, there will be a set of $(r_1 - 1)(r_2 - 1)(r_3 - 1)$ "no interaction" conditions, namely

$$\frac{\theta_{r_1 r_2 r_3}\theta_{i_1 i_2 r_3}}{\theta_{i_1 r_2 r_3}\theta_{r_1 i_2 r_3}} = \frac{\theta_{r_1 r_2 i_3}\theta_{i_1 i_2 i_3}}{\theta_{i_1 r_2 i_3}\theta_{r_1 i_2 i_3}}$$

for $i_1 = 1, 2, \ldots, r_1 - 1$, $i_2 = 1, 2, \ldots, r_2 - 1$, $i_3 = 1, 2, \ldots, r_3 - 1$.

The hypothesis of multiple independence is called the "no interaction" hypothesis by Roy and Kastenbaum. Rather than defining "interaction" parameters, they focus on estimation of the θ's subject to the above constraints. An algorithm for finding the estimates is given in Kastenbaum and Lamphiear [1959]. One should note that this algorithm does not lead directly to the appropriate test for the hypothesis of "no three-factor interaction". Such a test, based on a partitioning of chi-square into components (one of which tests for this interaction), was introduced in Goodman [1969].

The term "interaction", and the associated terminology "partitioning of chi-square", has unfortunately taken on an alternative meaning in this context, due mainly to the work of Lancaster [1949, 1951]. This confusion of terminology and procedures has led to the query published in Snedecor [1958] and the explications of Lewis [1962], Plackett [1962], and Darroch [1962].

The essence of Lancaster's approach is to define random variables

$$\mathbf{x}_{ijk} = \frac{\mathbf{n}_{ijk} - n\theta_{i\cdot\cdot}\theta_{\cdot j\cdot}\theta_{\cdot\cdot k}}{\sqrt{n\theta_{i\cdot\cdot}\theta_{\cdot j\cdot}\theta_{\cdot\cdot k}}},$$

and note that the usual chi-square statistic for testing independence is merely the squared length of the vector of $\mathbf{x}_{ijk}$'s. Thus if one can find an orthogonal transformation of this vector, such that all of the components of the transformed vector are mutually independent, then one can partition the overall chi-square statistic into the sum of independent chi-square statistics, each with one degree of freedom. The components of the transformed vector are then dubbed as a set of main effects and interactions associated with the cross-classification. Goodman [1964], though, points out that the statistic associated with Lancaster's "interaction" is not necessarily asymptotically distributed as chi-square, nor is it even asymptotically valid as a test of the hypothesis of no three-factor interaction as defined earlier.

4. Analysis of Cross-Classifications with Ordered Categories

The procedures described in the last three sections view the cross-classification categories as nominal, rather than ordinal, variables. That is, one can permute the order of presentation of the categories in any dimension without affecting the sense of the table and/or the analysis. If, however, there is an order to the category elements in any dimension, such information is totally neglected by the methods presented heretofore. In this section we consider approaches for modeling and analyzing such cross-classifications. For simplicity of exposition, we consider the $m = 2$ case.

One classical approach to the analysis of the ordered cross-classification (see Kendall and Stuart [1973]) assumes that the ordering of the categories is reflected in a set of scores for each category. If scores are assigned to each of the categories, then the problem of testing for independence of the two categories is one of testing for independence of grouped bivariate data. The standard approach for such a test is to calculate the sample correlation coefficient from these data, and rely on whatever distributional theory is at hand for sample correlation coefficients based on the bivariate distribution of scores.

One common choice of scores is the set of "normal scores". To develop these scores, first put the r_1 $\mathbf{n}_{i\cdot}$'s in rank order. Then the row with the lth largest $\mathbf{n}_{i\cdot}$ ($l = 1, \ldots, r_1$) is assigned score $e(l, r_1)$, where $e(l, r_1)$ is the expected value of the lth order statistic from a random sample of r_1 $N(0, 1)$ variables. Similarly, put the r_2 $\mathbf{n}_{\cdot j}$'s in rank order and assign $e(k, r_2)$ to the column with the kth largest $\mathbf{n}_{\cdot j}$ ($k = 1, \ldots, r_2$). Tables of the expected value of $N(0, 1)$ order statistics are given in Fisher and Yates [1948], Pearson and Hartley [1972], and Harter [1969].

The reason for this choice is that the sample correlation coefficient $\mathbf{r}$ based on these scores has as limiting distribution that of the sample correlation coefficient $\mathbf{r}$ from a sample from a normal distribution. Thus, for example, $\tanh^{-1} \mathbf{r}$ is asymptotically distributed as $N(\tanh^{-1} \rho, 1/r_1 r_2)$, where ρ is the true correlation between the categories.

For the example of Section 1, if (as done by Haberman [1974]) row scores are set as $(3, 1, -1, -3)$ and column scores are set as $(5, 3, 1, -1, -3, -5)$, then the sample correlation coefficient is $\mathbf{r} = 0.1496$. Since $\tanh^{-1} \mathbf{r} = 0.1507$ and $1/r_1 r_2 = 0.042$, we see that this table departs significantly from independence. If row scores and column scores are set as normal scores, namely $(1.03, 0.30, -0.30, -1.03)$ and $(1.27, 0.64, 0.20, -0.20, -0.64, -1.27)$, respectively, then $\mathbf{r} = 0.1504$. Since $\tanh^{-1} \mathbf{r} = 0.1515$, this calculation also indicates that the row and column categories are associated.

Suppose the row and column categories in an $r_1 \times r_2$ table are ordered but no specific scores are associated with the categories. One can then assign "worst case" scores $x_1, \ldots, x_{r_1}$ to the rows and $y_1, \ldots, y_{r_2}$ to the columns and

Table 1. Percentiles of the Maximum Eigenvalue for a Wishart Distributed Matrix.

ν \ α	0·05	0·01	0·05	0·01	0·05	0·01	0·05	0·01	0·05	0·01	0·05	0·01	0·05	0·01	0·05	0·01	0·05	0·01
	$p = 2$		$p = 3$		$p = 4$		$p = 5$		$p = 6$		$p = 7$		$p = 8$		$p = 9$		$p = 10$	
2	8·594	12·16	10·74	14·57	12·68	16·73	14·49	18·73	16·21	20·64	17·88	22·47	19·49	24·23	21·06	25·95	22·61	27·63
3	10·74	14·57	13·11	17·18	15·24	19·50	17·21	21·65	19·09	23·69	20·88	25·64	22·62	27·52	24·31	29·34	25·96	31·12
4	12·68	16·73	15·24	19·50	17·52	21·96	19·63	24·24	21·62	26·38	23·53	28·43	25·37	30·41	27·15	32·32	28·90	34·18
5	14·49	18·73	17·21	21·65	19·63	24·24	21·85	26·62	23·95	28·86	25·96	31·00	27·88	33·05	29·75	35·04	31·57	36·97
6	16·21	20·64	19·09	23·69	21·62	26·38	23·95	28·86	26·14	31·19	28·23	33·40	30·24	35·53	32·18	37·59	34·08	39·59
7	17·88	22·47	20·88	25·64	23·53	28·43	25·96	31·00	28·23	33·40	30·40	35·69	32·48	37·89	34·50	40·01	36·45	42·07
8	19·49	24·23	22·62	27·52	25·37	30·41	27·88	33·05	30·24	35·53	32·48	37·89	34·63	40·15	36·70	42·33	38·72	44·45
9	21·06	25·95	24·31	29·34	27·15	32·32	29·75	35·04	32·18	37·59	34·50	40·01	36·70	42·33	38·84	44·57	40·91	46·74
10	22·61	27·63	25·96	31·12	28·90	34·18	31·57	36·97	34·08	39·59	36·45	42·07	38·72	44·45	40·91	46·74	43·04	48·95
11	24·12	29·28	27·58	32·86	30·60	36·00	33·35	38·86	35·93	41·54	38·36	44·08	40·69	46·51	42·93	48·85	45·10	51·11
12	25·61	30·89	29·17	34·56	32·27	37·78	35·09	40·71	37·73	43·45	40·22	46·04	42·60	48·52	44·90	50·91	47·12	53·22
15	29·96	35·59	33·80	39·52	37·13	42·94	40·15	46·05	42·96	48·96	45·61	51·70	48·15	54·32	50·58	56·85	52·94	59·28
20	36·94	43·08	41·18	47·37	44·84	51·10	48·14	54·49	51·21	57·63	54·10	60·60	56·86	63·43	59·50	66·14	62·05	68·77
25	43·67	50·27	48·27	54·89	52·22	58·88	55·78	62·51	59·07	65·87	62·17	69·03	65·12	72·04	67·94	74·93	70·67	77·72
30	50·24	57·24	55·15	62·15	59·37	66·40	63·16	70·23	66·66	73·79	69·95	77·13	73·07	80·31	76·05	83·35	78·93	86·29
40	63·02	70·75	68·50	76·18	73·18	80·86	77·37	85·08	81·24	88·98	84·86	92·64	88·29	96·11	91·57	99·43	94·72	102·6
50	75·46	83·84	81·44	89·73	86·53	94·79	91·08	99·34	95·27	103·6	99·18	107·5	102·9	111·2	106·4	114·8	109·8	118·2
60	87·66	96·72	94·09	102·9	99·55	108·3	104·4	113·2	108·9	117·7	113·1	121·9	117·0	125·8	120·8	129·6	124·4	133·3
70	99·70	109·2	106·5	115·9	112·3	121·6	117·5	126·8	122·3	131·5	126·7	135·9	130·9	140·1	134·8	144·1	138·7	147·9
80	111·6	121·6	118·8	128·6	124·9	134·7	130·4	140·1	135·4	145·0	140·0	149·7	144·4	154·0	148·6	158·2	152·6	162·2
90	123·4	133·8	130·9	141·2	137·4	147·5	143·1	153·2	148·3	158·3	153·2	163·2	157·8	167·8	162·1	172·1	166·3	176·3
100	135·0	145·9	143·0	153·6	149·7	160·2	155·6	166·1	161·1	171·5	166·1	176·5	170·9	181·3	175·4	185·8	179·8	190·1
120	158·1	169·9	166·7	178·2	173·9	185·2	180·4	191·5	186·2	197·3	191·6	202·7	196·8	207·8	201·6	212·7	206·3	217·3
140	181·0	193·5	190·1	202·3	197·8	209·8	204·7	216·5	211·0	222·7	216·7	228·4	222·2	233·8	227·3	239·0	232·3	243·9
160	2·037	216·9	213·3	226·2	221·5	234·1	228·8	241·2	235·4	247·7	241·4	253·7	247·2	259·4	252·6	264·8	257·8	270·0
180	226·2	240·0	236·3	249·8	244·9	258·1	252·6	265·6	259·5	272·4	265·8	278·7	271·9	284·6	277·6	290·3	283·0	295·7
200	248·6	263·0	259·2	273·3	268·2	282·0	276·2	289·7	283·4	296·8	290·0	303·4	296·4	309·6	302·4	315·5	308·0	321·2

If $20 \leq \nu \leq 200$, approximate percentage points. $C(\alpha)$, for $\alpha = 0.10, 0.025, 0.005$, may be found as follows:
$C(0.10) = 1.50C(0.05) - 0.50C(0.01)$, $C(0.025) = 0.55C(0.05) + 0.45C(0.01)$,
$C(0.005) = -0.39C(0.05) + 1.39C(0.01)$.

SOURCE: Pearson and Hartley [1972]. Reprinted from *Biometrika Tables for Statisticians*, Vol. 2, with the permission of the Biometrika Trustees.

test for independence using one of the standard tests basd on row and column category scores given above. The procedures for developing these "worst case" scores is to view the $r_1 \times r_2$ table of observed proportions $\mathbf{p}_{ij} = \mathbf{n}_{ij}/n$ as a matrix $\mathbf{P}$ and to find the vectors X and Y such that $X'\mathbf{P}Y$ is maximum subject to the constraints that:

(1) the average row and column scores are both equal to 0, i.e.,

$$\sum_{i=1}^{r_1} \mathbf{p}_{i\cdot}x_i = 0, \qquad \sum_{j=1}^{r_2} \mathbf{p}_{\cdot j}y_j = 0,$$

and;

(2) the observed variances of the row and column scores are both equal to 1, i.e.,

$$\sum_{i=1}^{r_1} \mathbf{p}_{i\cdot}x_i^2 = 1, \qquad \sum_{j=1}^{r_2} \mathbf{p}_{\cdot j}y_j^2 = 1.$$

Kendall and Stuart ([1973], p. 591) show that the maximum of $X'\mathbf{P}Y$ is achieved via the following set of computations:

(1) Let E_s be the s-vector, all of whose elements are 1's.
(2) Let $\mathbf{G} = \mathbf{P}E_{r_2}E_{r_1}\mathbf{P}/n$, i.e., the maximum likelihood estimate of the matrix of θ_{ij} under the assumption of independence.
(3) Let $\mathbf{H}$ be the matrix whose (i, j)th element is

$$\mathbf{h}_{ij} = \mathbf{p}_{ij}/\sqrt{\mathbf{g}_{ij}},$$

and $\mathbf{F}$ be the matrix whose (i, j)th element is

$$\mathbf{f}_{ij} = \sqrt{\mathbf{g}_{ij}}$$

(4) Let $\mathbf{D} = \mathbf{H} - \mathbf{F}$ and $\mathbf{B} = \mathbf{DD}'$.

Then the maximum value of $X'\mathbf{P}Y$ is given by $\mathbf{r}$, the largest eigenvalue of $\mathbf{B}$. Haberman [1981] has shown that $n\mathbf{r}$ is asymptotically distributed as the distribution of the maximum eigenvalue of a Wishart $(I, r_1 - 1, r_2 - 1)$ distributed matrix. Percentiles of this distribution are given in Table 1.

For the example cited in Section 1 above, we find that

$P =$	0.03855	0.03434	0.03434	0.04337	0.02169	0.01265
	0.05663	0.05663	0.06325	0.08494	0.05843	0.04277
	0.03494	0.03253	0.03916	0.04639	0.03253	0.03253
	0.02771	0.02410	0.03614	0.05663	0.04699	0.04277
$G =$	0.02919	0.02730	0.03197	0.04278	0.02952	0.02418
	0.05724	0.05352	0.06270	0.08389	0.05789	0.04741
	0.03442	0.03219	0.03770	0.05045	0.03481	0.02851
	0.03699	0.03459	0.04051	0.05421	0.03741	0.03063
	0.22566	0.20784	0.19203	0.20970	0.12621	0.08136

$$H = \begin{cases} 0.23669 & 0.24476 & 0.25261 & 0.29326 & 0.24286 & 0.19644 \\ 0.18833 & 0.18132 & 0.20166 & 0.20652 & 0.17435 & 0.19267 \\ 0.14409 & 0.12957 & 0.17957 & 0.24321 & 0.24294 & 0.24437 \end{cases}$$

$$F = \begin{cases} 0.17085 & 0.16521 & 0.17881 & 0.20684 & 0.17182 & 0.15549 \\ 0.23924 & 0.23135 & 0.25040 & 0.28964 & 0.24061 & 0.21773 \\ 0.18552 & 0.17940 & 0.19417 & 0.22460 & 0.18658 & 0.16884 \\ 0.19232 & 0.18597 & 0.20128 & 0.23283 & 0.19341 & 0.17502 \end{cases}$$

and so

$$B = \begin{bmatrix} 0.01258 & 0.00195 & -0.00093 & -0.01270 \\ & 0.00066 & -0.00057 & -0.00201 \\ & & 0.00111 & 0.00045 \\ & & & 0.01335 \end{bmatrix}.$$

The largest eigenvalue of B is $r = 0.02602$. Now $n\mathbf{r} = 43.1932$, which, when compared with 17.21, the 95% point of the distribution of the maximum eigenvalue of a Wishart $(I, 3, 5)$ distributed matrix, indicates strongly that the row and column categories are associated.

An alternative approach to such tables, due to Goodman [1979], models the odds-ratios for 2×2 subtables formed from adjacent rows and adjacent columns, i.e.,

$$\eta_{ij} = \frac{\theta_{ij}\theta_{i+1,j+1}}{\theta_{i,j+1}\theta_{i+1,j}},$$

in a number of ways. The independence model posits that $\eta_{ij} = 1$ for all i, j. Goodman presents five alternative models of association:

$$\begin{array}{rl} \text{uniform association} & \eta_{ij} = \eta, \\ \text{row-effect association} & \eta_{ij} = \eta_{i\cdot}, \\ \text{column-effect association} & \eta_{ij} = \eta_{\cdot j}, \\ \text{model I} & \eta_{ij} = \eta_{i\cdot}\eta_{\cdot j}, \\ \text{model II} & \log \eta_{ij} = \phi_{i\cdot}\phi_{\cdot j} \end{array}$$

for $i = 1, \ldots, r_1 - 1$, $j = 1, \ldots, r_2 - 1$.

Consider in particular model I. Translating this model into a model of the θ_{ij} produces the characterization

$$\theta_{ij} = \alpha_i \beta_j \gamma_i^j \delta_j^i.$$

One then estimates these model parameters using iterative proportional fitting or some other algorithm which produces a best asymptotically normal (BAN) set of estimates $\hat{\theta}_{ij}$ of the θ_{ij}. Since $\mathbf{g}_{ij} = \mathbf{p}_{i\cdot}\mathbf{p}_{\cdot j}$ is the maximum likelihood estimate of θ_{ij} under independence, the statistic

$$\mathbf{x} = n \sum_{i=1}^{r_1} \sum_{j=1}^{r_2} (\hat{\boldsymbol{\theta}}_{ij} - \mathbf{g}_{ij})^2/\mathbf{g}_{ij}$$

has a chi-square distribution with an appropriate number of degrees of freedom, namely $(r_1 - 1)(r_2 - 1)$ minus the number of additional constraints imposed by the model type.

Using the procedure of Goodman [1979] on the data of our example, we find that the expected frequencies are

63.78	59.18	56.82	68.27	36.38	22.57
94.42	91.93	106.35	143.50	94.10	71.69
57.82	51.60	62.84	83.18	58.64	47.93
45.98	42.29	60.99	89.04	75.87	74.82

with an associated value of x for this model of 3.0569, clearly not significant. For additional approaches to modeling cross-classifications with ordered categories, see Goodman [1984].

5. Latent Class Model

It may be the case that the $r_1 \times r_2$ table does not exhibit independence but that it is the sum of l $r_1 \times r_2$ tables, each of which satisfies the independence hypothesis. The model describing the relationship can be expressed as

$$\theta_{ij} = \sum_{k=1}^{l} v_k \theta_{i\cdot}^{(k)} \theta_{\cdot j}^{(k)},$$

where v_k is the probability that a member of our sample comes from the subpopulation (or "class") corresponding to subtable k, $\theta_{i\cdot}^{(k)}$ is the probability of a member of our sample being in category i on item 1 given that he is from class k, and $\theta_{\cdot j}^{(k)}$ is the probability that he is in category j on item 2 given that he is from class k. The original model had $r_1 r_2 - 1$ parameters θ_{ij}; the new model has $l - 1$ v_k's, $l(r_1 - 1)$ $\theta_{i\cdot}^{(k)}$'s, and $l(r_2 - 1)$ $\theta_{\cdot j}^{(k)}$'s, for a total of $l(r_1 + r_2 - 1) - 1$ parameters. Clearly, only if $l \le r_1 r_2/(r_1 + r_2 - 1)$ can this model be identified, i.e., its parameters be uniquely determined from the θ_{ij}.

The Bartlett $2 \times 2 \times 2$ table data may be considered as an example for construction of a latent class model. But in this example $p_{\cdot\cdot 1} = 480/960 = 0.5$ and $p_{\cdot 11} = 480/960 = 0.5$. The two equations

$$0.5 = p_{\cdot\cdot 1} = v_1 \hat{\theta}_{\cdot\cdot 1}^{(1)} + (1 - v_1)\hat{\theta}_{\cdot\cdot 1}^{(2)},$$

$$0.5 = p_{\cdot 11} = v_1 \hat{\theta}_{\cdot 1 \cdot}^{(1)} \hat{\theta}_{\cdot\cdot 1}^{(1)} + (1 - v_1)\hat{\theta}_{\cdot 1 \cdot}^{(2)} \hat{\theta}_{\cdot\cdot 1}^{(2)},$$

cannot hold for $0 < \hat{\theta}_{\cdot 1 \cdot}^{(1)} < 1$ and $0 < \hat{\theta}_{\cdot 1 \cdot}^{(2)} < 1$, since each term in the second equation will be less than the comparable term in the first equation. Further analysis of the Bartlett data show additional inconsistencies with the latent class model. Thus, even if in general for given $r_1, r_2, \ldots, r_m$ and l a latent class model may exist, the model may still not exist for a particular set of θ's.

To illustrate a latent class model for a 2 × 2 × 2 situation, consider the following data:

i_1	i_2	i_3	$n_{i_1 i_2 i_3}$	$\hat{\theta}_{i_1 i_2 i_3}$
1	1	1	2	0.050
1	2	1	4	0.100
2	1	1	3	0.075
2	2	1	4	0.100
1	1	2	4	0.100
1	2	2	7	0.175
2	1	2	7	0.175
2	2	2	9	0.225
			40	1.000

These data can be reconstructed from two latent classes with the parameters given in the following table:

k	1	2
ν_k	0.23	0.77
$\theta^{(k)}_{1..}$	0.82	0.30
$\theta^{(k)}_{.1.}$	0.23	0.45
$\theta^{(k)}_{..1}$	0.42	0.29

For example,

$$\hat{\theta}_{122} = 0.23 \times 0.82 \times 0.77 \times 0.58 + 0.77 \times 0.30 \times 0.55 \times 0.71$$
$$= 0.084 + 0.090.$$

Suppose, though, that the true θ's were given by the following table, just slightly different from the observed $\hat{\theta}$'s. (This could happen, for example, if our observations were exactly equal to 40 $\theta_{i_1 i_2 i_3}$, except that they were rounded to the nearest integer.)

i_1	i_2	i_3	$\theta_{i_1 i_2 i_3}$
1	1	1	10/192 = 0.0521
1	2	1	17/192 = 0.0885
2	1	1	14/192 = 0.0729
2	2	1	19/192 = 0.0990
1	1	2	22/192 = 0.1146
1	2	2	35/192 = 0.1823
2	1	2	34/192 = 0.1771
2	2	2	41/192 = 0.2135

A latent class analysis of the θ's would yield:

k	1	2
v_k	0.75	0.25
$\theta_{1..}^{(k)}$	0.50	0.25
$\theta_{.1.}^{(k)}$	0.33	0.67
$\theta_{..1}^{(k)}$	0.33	0.25

For example,

$$\begin{aligned}\theta_{122} &= \tfrac{35}{192} \\ &= \tfrac{3}{4} \times \tfrac{1}{2} \times \tfrac{2}{3} \times \tfrac{2}{3} + \tfrac{1}{4} \times \tfrac{1}{4} \times \tfrac{1}{3} \times \tfrac{3}{4} \\ &= \tfrac{1}{6} + \tfrac{1}{64}.\end{aligned}$$

Thus small variations in the $\hat{\theta}$'s can produce wide variations in the latent class parameter estimates. Madansky [1959] (see also Madansky [1978]) has found in an example that the information-matrix-based asymptotic variance of the latent class estimators will be less than 1 only if $n > 1200$. Care should therefore be taken in constructing latent class models for multiway contingency tables based on small sample sizes.

Theoretical Background

The latent class model for the $r_1 \times r_2$ table is really a statement about the rank of the matrix of θ_{ij}'s. The model can be written matricially as

$$\Theta = \sum_{k=1}^{l} v_k \Theta_{1k} \Theta'_{2k},$$

where Θ is the $r_1 \times r_2$ matrix of θ_{ij}'s, Θ_{1k} is the r_1 vector of $\theta_{i.}^{(k)}$'s, and Θ_{2k} is the r_2 vector of $\theta_{.j}^{(k)}$'s. Since each summand is a matrix of rank 1, the latent class model implies that the rank of Θ is at most l.

Now let N be the $l \times l$ diagonal matrix of v_k's, let Θ_1 be the $r_1 \times l$ matrix whose columns are the Θ_{1k}'s, and let Θ_2 be the $r_2 \times l$ matrix whose columns are the Θ_{2k}'s. Then $\Theta = \Theta_1 N \Theta'_2$. But let G be an orthogonal matrix and let $N^{1/2}$ denote the diagonal matrix whose diagonal elements are the square roots of the v_k's. We see that Θ is also expressible as

$$\Theta = \Theta_1 N^{1/2} G G' N^{1/2} \Theta'_2.$$

Now let C be a diagonal matrix whose entries are the column sums of $N^{1/2}G$. Define

$$\begin{aligned}\Theta_1^* &= \Theta_1 N^{1/2} G C^{-1}, \\ \Theta_2^* &= \Theta_2 N^{1/2} G C^{-1}, \\ N^* &= C^2.\end{aligned}$$

Then Θ is expressible as $\Theta = \Theta_1^* N^* \Theta_2^{*\prime}$ as well, so that the latent class model for the $r_1 \times r_2$ table is unidentifiable. We see from this that we cannot merely rely on the parameter counting argument given above to insure us of the identifiability of the model.

The general case is that of an $r_1 \times r_2 \times \cdots \times r_m$ table with latent class model

$$\theta_{i_1 \ldots i_m} = \sum_{k=1}^{l} v_k \theta_{i_1}^{(k)} \ldots \theta_{i_2}^{(k)} \ldots \theta_{i_m}^{(k)} \ldots .$$

The unrestricted maximum likelihood estimator of $\theta_{i_1 i_2 \ldots i_m}$ is

$$\mathbf{p}_{i_1 i_2 \ldots i_m} = \mathbf{n}_{i_1 i_2 \ldots i_m}/n.$$

Let

$$\theta_{i_1 \ldots i_m}^{(k)} = v_k \theta_{i_1 \ldots}^{(k)} \theta_{\cdot i_2 \ldots}^{(k)} \cdots \theta_{\ldots i_m}^{(k)},$$

and

$$\theta_{i_1 \ldots i_m | k} = \theta_{i_1 \ldots i_m}^{(k)} / \theta_{i_1 i_2 \ldots i_m}.$$

Goodman [1974] points out that the maximum likelihood estimators of the v_k and the various $\theta^{(k)}$'s satisfy the $l + 1$ equations

$$\hat{\mathbf{v}}_k = \sum_{i_1=1}^{r_1} \cdots \sum_{i_m=1}^{r_m} \mathbf{p}_{i_1 \ldots i_m} \hat{\boldsymbol{\theta}}_{i_1 \ldots i_m | k},$$

$$\hat{\boldsymbol{\theta}}_{\ldots i_r \ldots}^{(k)} = \sum_{\substack{\text{all } i_j \\ \text{except } i_r}} \sum \mathbf{p}_{i_1 \ldots i_m} \hat{\boldsymbol{\theta}}_{i_1 \ldots i_m | k} / \hat{\mathbf{v}}_k, \qquad r = 1, \ldots, l.$$

Goodman further suggests an iterative estimation procedure, beginning with initial values of the $\theta^{(k)}$'s and the $\hat{\mathbf{v}}_k$ to produce values of $\hat{\theta}_{i_1 \ldots i_m | k}$, and then re-evaluating the $\hat{\theta}^{(k)}$'s and $\hat{v}_k$ via these equations. A computer program for this procedure is given in Clogg [1977].

There are a number of proposed "starter sets" for this procedure. Madansky [1960], generalizing an approach due to Anderson [1954], describes a procedure involving eigenvalues and eigenvectors of nonsymmetric matrices. Gilula [1983] has derived another method which involves the singular value decomposition of a nonsymmetric matrix. Lazarsfeld and Henry [1968] and McHugh [1956, 1958] describe a "scoring" method for iteratively solving the likelihood equations, and Haberman [1979] provides a program using a variant of this method. Green [1951] has developed estimates based on a factor analysis of a matrix derived from the contingency table.

Appendix I

```
      REAL MARG(147),U(147),TABLE(343),FIT(343),MAXDEV,DEV(10)
      INTEGER DIM(7),DIM1(7),CONFIG(7,120),LOCMAR(120)
      DIMENSION KT(7)
      DATA KT/7*1/
      DATA MAXIT/10/,MAXDEV/.25/
      DATA CONFIG/840*0/
      WRITE(5,800)
800   FORMAT(' NUMBER OF VARIABLES (AT LEAST 2, NO MORE THAN 7)?')
3000  READ (5,*) NVAR
      IF(NVAR.LT.2.OR.NVAR.GT.7) GO TO 3000
      DO 80 I=1,NVAR
      WRITE(5,801)I
801   FORMAT(' NUMBER OF CATEGORIES FOR VARIABLE ',I2,'?')
      READ (5,*)DIM(I)
80    CONTINUE
      DO 83 I=1,NVAR
      IJ=NVAR+1-I
83    DIM1(I)=DIM(IJ)
      NVAR1=NVAR-1
      NTAB=1
      DO 40 I=1,NVAR
40    NTAB=NTAB*DIM(I)
      IF(NTAB.GT.343)WRITE(5,777)NTAB
777   FORMAT(' YOUR TABLE HAS ',I5,' CELLS, EXCEEDING OUR MAXIMUM OF 343
     +.')
      IF(NTAB.GT.343) GO TO 2000
      NMAR=0
      DO 30 I=1,NVAR1
      I1=I+1
      DO 30 J=I1,NVAR
30    NMAR=NMAR+DIM(I)*DIM(J)
      IF(NMAR.GT.147)WRITE(5,778)NMAR
778   FORMAT(' YOU HAVE ',I5,' MARGINAL CONSTRAINTS, EXCEEDING OUR 147 L
     +IMIT.')
      IF(NMAR.GT.147) GO TO 2000
      NU=NMAR
      WRITE(5,920)
920   FORMAT(' NOW ENTER THE DATA ONE CELL AT A TIME ')
      WRITE(5,921)
921   FORMAT(' IN ACCORDANCE WITH THE CELL SUBSCRIPT CUE')
      K=1
      MARKER=NVAR
1111  CONTINUE
      GO TO (2000,8222,8223,8224,8225,8226,8227),NVAR
8222  WRITE(5,9222)KT(2),KT(1)
      GO TO 4000
8223  WRITE(5,9223)KT(3),KT(2),KT(1)
      GO TO 4000
8224  WRITE(5,9224)KT(4),KT(3),KT(2),KT(1)
      GO TO 4000
8225  WRITE(5,9225)KT(5),KT(4),KT(3),KT(2),KT(1)
      GO TO 4000
8226  WRITE(5,9226)KT(6),KT(5),KT(4),KT(3),KT(2),KT(1)
      GO TO 4000
8227  WRITE(5,9227)KT(7),KT(6),KT(5),KT(4),KT(3),KT(2),KT(1)
9222  FORMAT(1H I2,',',I2)
9223  FORMAT(1H 2(I2,','),I2)
9224  FORMAT(1H 3(I2,','),I2)
9225  FORMAT(1H 4(I2,','),I2)
9226  FORMAT(1H 5(I2,','),I2)
```

```
9227  FORMAT(1H 6(I2,','),I2)
4000  READ(5,*)IC
      TABLE(K)=IC
      K=K+1
      KT(MARKER)=KT(MARKER)+1
      IF(KT(MARKER).LE.DIM1(MARKER)) GO TO 1111
1113  MARKER=MARKER-1
      IF(MARKER.EQ.0) GO TO 60
      KT(MARKER)=KT(MARKER)+1
      IF(KT(MARKER).LE.DIM1(MARKER)) GO TO 1112
      GO TO 1113
1112  MARK1=MARKER+1
      DO 1114 I=MARK1,NVAR
1114  KT(I)=1
      MARKER=NVAR
      GO TO 1111
60    CONTINUE
1000  CONTINUE
8802  WRITE(5,802)
802   FORMAT(' NUMBER OF INTERACTIONS TO BE FIT?')
      READ(5,*)NCON
      IF (NCON.EQ.0) GO TO 881
      MCON=2**NVAR-1-NVAR
      IF(NCON.GT.MCON)WRITE(5,779)MCON
779   FORMAT(' YOU CAN FIT A MAXIMUM OF ',I3,' INTERACTIONS FOR THIS TAB
     +LE.')
      IF(NCON.GT.MCON) GO TO 8802
      WRITE (5,600)
      WRITE(5,900)
900   FORMAT(' WHEN ENTERING INTERACTION SUBSCRIPTS, ENTER THEM')
      GO TO (2000,4902,4903,4904,4905,4906,4907),NVAR
4902  WRITE(5,1902)
      GO TO 2004
1902  FORMAT(' AS XX,XX  ')
4903  WRITE(5,1903)
      GO TO 2004
1903  FORMAT(' AS XX,XX,XX  ')
4904  WRITE(5,1904)
      GO TO 2004
1904  FORMAT(' AS XX,XX,XX,XX  ')
4905  WRITE(5,1905)
      GO TO 2004
1905  FORMAT(' AS XX,XX,XX,XX,XX  ')
4906  WRITE(5,1906)
      GO TO 2004
1906  FORMAT(' AS XX,XX,XX,XX,XX,XX  ')
4907  WRITE(5,1907)
1907  FORMAT(' AS XX,XX,XX,XX,XX,XX,XX  ')
2004  WRITE(5,902)
902   FORMAT(' THE USE OF 00 FOR THE EXTRANEOUS SUBSCRIPTS IS ADVISED.')
      WRITE(5,903)
903   FORMAT(' FOR EXAMPLE, THE (1,2) INTERACTION ')
      GO TO (2000,3902,3903,3904,3905,3906,3907),NVAR
3902  WRITE(5,2902)
      GO TO 2001
2902  FORMAT(' IS ENTERED AS 01,02  ')
3903  WRITE(5,2903)
      GO TO 2001
2903  FORMAT(' IS ENTERED AS 01,02,00  ')
3904  WRITE(5,2904)
      GO TO 2001
2904  FORMAT(' IS ENTERED AS 01,02,00,00  ')
3905  WRITE(5,2905)
      GO TO 2001
```

```
2905  FORMAT(' IS ENTERED AS 01,02,00,00,00  ')
3906  WRITE(5,2906)
      GO TO 2001
2906  FORMAT(' IS ENTERED AS 01,02,00,00,00,00  ')
3907  WRITE(5,2907)
2907  FORMAT(' IS ENTERED AS 01,02,00,00,00,00,00  ')
2001  IF(NVAR.EQ.2) GO TO 2002
      WRITE(5,2003)
2003  FORMAT(' OR IS ENTERED AS 01,02  ')
2002  WRITE(5,600)
      DO 81 I=1,NCON
      WRITE(5,803)I
803   FORMAT(' SUBSCRIPTS OF INTERACTION ',I2,' TO BE FIT:')
      READ(5,804)(CONFIG(J,I),J=1,NVAR)
804   FORMAT(7(I2,X))
81    CONTINUE
881   DO 82 I=1,NTAB
82    FIT(I)=1.
      DO 890 I=1,NTAB
      IF(TABLE(I).EQ.0) GO TO 907
890   CONTINUE
      GO TO 901
907   WRITE(5,9021)
9021  FORMAT(' DO YOU WISH TO FORCE ALL 0 OBSERVED CELLS TO HAVE 0 FIT?'
     +)
      WRITE(5,86)
      READ(5,87)IYN
      IF(IYN.EQ.1) GO TO 905
      DO 906 I=1,NTAB
      IF(TABLE(I).EQ.0)FIT(I)=0
906   CONTINUE
      GO TO 901
905   DO 88 I=1,NTAB
      IF(TABLE(I).NE.0) GO TO 88
      WRITE (5,84)I
84    FORMAT(' THE ',I3,'-TH ELEMENT OF THE TABLE IS 0. ')
      WRITE(5,85)
85    FORMAT(' DO YOU WANT TO FORCE THE FIT OF THAT CELL TO BE 0?  ')
      WRITE (5,86)
86    FORMAT(' ANSWER 0 FOR YES, 1 FOR NO.')
      READ (5,87)IYN
87    FORMAT(I1)
      IF(IYN.EQ.0)FIT(I)=0.
88    CONTINUE
901   CONTINUE
      IFAULT=0
      CALL LOGLIN(NVAR,DIM,NCON,CONFIG,NTAB,TABLE,FIT,LOCMAR,NMAR,MARG,
     +  NU,U,MAXDEV,MAXIT,DEV,NLAST,IFAULT)
      WRITE(5,501)IFAULT
501   FORMAT(' ERROR CONDITION:',I2)
      K=0
600   FORMAT(1H )
      WRITE(5,600)
      MARKER=NVAR
      DO 2009 I=1,7
2009  KT(I)=1
2010  CONTINUE
      K=K+1
      GO TO (2000,5102,5103,5104,5105,5106,5107),NVAR
5102  WRITE(5,6102)KT(1),KT(2),TABLE(K),FIT(K)
6102  FORMAT(1H 2I1,1X,F5.0,1X,F8.2)
      GO TO 6108
5103  WRITE(5,6103)KT(1),KT(2),KT(3),TABLE(K),FIT(K)
6103  FORMAT(1H 3I1,1X,F5.0,1X,F8.2)
      GO TO 6108
```

```
5104  WRITE(5,6104)KT(1),KT(2),KT(3),KT(4),TABLE(K),FIT(K)
6104  FORMAT(1H 4I1,1X,F5.0,1X,F8.2)
      GO TO 6108
5105  WRITE(5,6105)(KT(J),J=1,5),TABLE(K),FIT(K)
6105  FORMAT(1H 5I1,1X,F5.0,1X,F8.2)
      GO TO 6108
5106  WRITE(5,6106)(KT(J),J=1,NVAR),TABLE(K),FIT(K)
6106  FORMAT(1H 6I1,1X,F5.0,1X,F8.2)
      GO TO 6108
5107  WRITE(5,6107)(KT(J),J=1,NVAR),TABLE(K),FIT(K)
6107  FORMAT(1H 7I1,1X,F5.0,1X,F8.2)
6108  CONTINUE
      KT(MARKER)=KT(MARKER)+1
      IF(KT(MARKER).LE.DIM(MARKER)) GO TO 2010
2013  MARKER=MARKER -1
      IF(MARKER.EQ.0)GO TO 31
      KT(MARKER)=KT(MARKER)+1
      IF(KT(MARKER).LE.DIM(MARKER)) GO TO 2012
      GO TO 2013
2012  MARK1=MARKER+1
      DO 2011 I=MARK1,NVAR
2011  KT(I)=1
      MARKER=NVAR
      GO TO 2010
31    CONTINUE
      NZ=0
      CHI=0.
      DO 10 I=1,NTAB
      IF(TABLE(I).EQ.0..AND.FIT(I).EQ.0.) NZ=NZ+1
      IF(FIT(I).EQ.0.) GO TO 10
      CHI=CHI+(FIT(I)-TABLE(I))**2/FIT(I)
10    CONTINUE
      IDF=NTAB-1-NVAR-NCON-NZ
      WRITE(5,502)CHI,IDF
502   FORMAT(1HO'CHI-SQUARE = ',F8.2,2X,I3,' DEGREES OF FREEDOM')
      WRITE(5,3001)
3001  FORMAT(' DO YOU WISH TO TRY TO FIT ANOTHER SET OF INTERACTIONS?')
      WRITE(5,3002)
3002  FORMAT(' ENTER 0 IF NO, 1 IF YES')
      READ(5,*)ITRY
      IF(ITRY.EQ.1) GO TO 1000
2000  CONTINUE
      END
      INCLUDE 'LOGLIN.FTN'
      INCLUDE 'ADJUST.FTN'
      INCLUDE 'COLLAP.FTN'
```

STRUCTURE

SUBROUTINE LOGLIN (NVAR, DIM, NCON, CONFIG, NTAB, TABLE, FIT, LOCMAR, NMAR, MARG, NU, U, MAXDEV, MAXIT, DEV, NLAST, IFAULT)

Formal parameters

NVAR	Integer	input:	the number of variables d in the table.
DIM	Integer array (*NVAR*)	input:	the number of categories r_j in each variable of the table.
NCON	Integer	input:	the number s of marginal totals to be fit.
CONFIG	Integer array (*NVAR, NCON*)	input:	the sets C_k, $k = 1, \ldots, s$, indicating marginal totals to be fit.
NTAB	Integer	input:	the number of elements in the table.

TABLE	Real array (*NTAB*)	input:	the table to be fit.
FIT	Real array (*NTAB*)	input and output:	the fitted table.
LOCMAR	Integer array (*NCON*)	output:	pointers to the tables in *MARG*.
NMAR	Integer	input:	the dimension of *MARG*.
MARG	Real array (*NMAR*)	output:	the marginal tables to be fit.
NU	Integer	input:	the dimension of *U*.
U	Real array (*NU*)	output:	a work area used to store fitted marginal tables.
MAXDEV	Real	input:	the maximum permissible difference between an observed and fitted marginal total.
MAXIT	Integer	input:	the maximum permissible number of iterations.
DEV	Real array (*MAXIT*)	output:	*DEV*(*I*) is the maximum observed difference encountered in iteration cycle *I* between an observed and fitted marginal total.
NLAST	Integer	output:	the number of the last iteration.
IFAULT	Integer	output:	an error indicator. See *failure indications* below.

Failure indications

If *IFAULT* = 0, then no error was detected.
If *IFAULT* = 1, then *CONFIG* contains an error.
If *IFAULT* = 2, then *MARG* or *U* is too small.
If *IFAULT* = 3, then the algorithm did not converge to the desired accuracy within *MAXIT* iterations.
If *IFAULT* = 4, then *DIM*, *TABLE*, *FIT* or *NVAR* is erroneously specified.

SOURCE: Haberman [1972]. Reprinted from *Applied Statistics*, Vol. 21, with the permission of the Royal Statistical Society.

```
      SUBROUTINE LOGLIN(NVAR,DIM,NCON,CONFIG,NTAB,TABLE,FIT,LOCMAR,
     *  NMAR,MARG,NU,U,MAXDEV,MAXIT,DEV,NLAST,IFAULT)
C
C        ALGORITHM AS 51  APPL.STATIST. (1972) VOL.21, NO.2
C
C        PERFORMS AN ITERATIVE PROPORTIONAL FIT OF THE MARGINAL TOTALS
C        OF A CONTINGENCY TABLE.
C
C        THIS VERSION PERMITS UP TO SEVEN VARIABLES. IF THIS LIMIT IS
C        TOO SMALL, CHANGE THE VALUE OF MAXVAR, AND OF THE DIMENSION
C        IN THE DECLARATION OF CHECK AND ICON - SEE ALSO THE CHANGES
C        NEEDED IN AS 51.1 AND AS 51.2
C
      REAL MARG(NMAR),U(NU),TABLE(NTAB),FIT(NTAB),MAXDEV,DEV(MAXIT)
      INTEGER DIM(NVAR),CHECK(7),CONFIG(NVAR,NCON),LOCMAR(NCON),
     *  POINT,SIZE,ICON(7)
C
      DATA MAXVAR/7/
C
C        CHECK VALIDITY OF NVAR, THE NUMBER OF VARIABLES,
C        AND OF MAXIT, THE MAXIMUM NUMBER OF ITERATIONS
C
      IF(NVAR.GT.0.AND.NVAR.LE.MAXVAR.AND.MAXIT.GT.0)GOTO10
```

```
    5 IFAULT=4
      RETURN
C
C         LOOK AT TABLE AND FIT CONSTANTS
C
   10 SIZE=1
      DO 30 J=1,NVAR
      IF(DIM(J).LE.0)GOTO5
   20 SIZE=SIZE*DIM(J)
   30 CONTINUE
      IF(SIZE.LE.NTAB)GOTO40
   35 IFAULT=2
      RETURN
   40 X=0.0
      Y=0.0
      DO 60 I=1,SIZE
      IF(TABLE(I).LT.0.0.OR.FIT(I).LT.0.0)GOTO5
      X=X+TABLE(I)
      Y=Y+FIT(I)
   60 CONTINUE
C
C          MAKE A PRELIMINARY ADJUSTMENT TO OBTAIN THE FIT TO AN
C          EMPTY CONFIGURATION LIST
C
      IF(Y.EQ.0.0)GOTO5
      X=X/Y
      DO 80 I=1,SIZE
   80 FIT(I)=X*FIT(I)
C
C         ALLOCATE MARGINAL TABLES
C
      POINT=1
      DO 150 I=1,NCON
C
C         A ZERO BEGINNING A CONFIGURATION INDICATES THAT THE LIST IS
C         COMPLETED
C
      IF(CONFIG(1,I).EQ.0)GOTO160
C
C         GET MARGINAL TABLE SIZE. WHILE DOING THIS TASK, SEE IF THE
C         CONFIGURATION LIST CONTAINS DUPLICATIONS OR ELEMENTS OUT OF
C         RANGE
C
      SIZE=1
      DO 90 J=1,NVAR
   90 CHECK(J)=0
      DO 120 J=1,NVAR
      K=CONFIG(J,I)
C
C         A ZERO INDICATES THE END OF THE STRING
C
      IF(K.EQ.0)GOTO130
C
C         SEE IF ELEMENT VALID
C
      IF(K.GE.0.AND.K.LE.NVAR)GOTO100
   95 IFAULT=1
      RETURN
C
C         CHECK FOR DUPLICATION
C
  100 IF(CHECK(K).EQ.1)GOTO95
  110 CHECK(K)=1
C
C         GET SIZE
C
      SIZE=SIZE*DIM(K)
  120 CONTINUE
C
C         SINCE U IS USED TO STORE FITTED MARGINALS, SIZE MUST NOT
C         EXCEED NU
C
  130 IF(SIZE.GT.NU)GOTO35
```

```
C
C        LOCMAR POINTS TO MARGINAL TABLES TO BE PLACED IN MARG
C
  140 LOCMAR(I)=POINT
      POINT=POINT+SIZE
  150 CONTINUE
C
C        GET N, NUMBER OF VALID CONFIGURATIONS
C
      I=NCON+1
  160 N=I-1
C
C        SEE IF MARG CAN HOLD ALL MARGINAL TABLES
C
      IF(POINT.GT.NMAR+1)GOTO35
C
C        OBTAIN MARGINAL TABLES
C
  170 DO 190 I=1,N
      DO 180 J=1,NVAR
  180 ICON(J)=CONFIG(J,I)
      CALL COLLAP(NVAR,TABLE,MARG,1,LOCMAR(I),NTAB,NMAR,DIM,ICON,.TRUE.)
  190 CONTINUE
C
C        PERFORM ITERATIONS
C
      DO 220 K=1,MAXIT
C
C        XMAX IS MAXIMUM DEVIATION OBSERVED BETWEEN FITTED AND TRUE
C        MARGINAL DURING A CYCLE
C
      XMAX=0.0
      DO 210 I=1,N
      DO 200 J=1,NVAR
  200 ICON(J)=CONFIG(J,I)
      CALL COLLAP(NVAR,FIT,U,1,1,NTAB,NU,DIM,ICON,.TRUE.)
      CALL ADJUST(NVAR,FIT,U,MARG,1,1,LOCMAR(I),NTAB,NU,NMAR,DIM,
     *  ICON,XMAX)
  210 CONTINUE
C
C        TEST CONVERGENCE
C
      DEV(K)=XMAX
      IF(XMAX.LT.MAXDEV)GOTO240
  220 CONTINUE
      IF(MAXIT.GT.1)GOTO230
      NLAST=1
      RETURN
C
C        NO CONVERGENCE
C
  230 IFAULT=3
      NLAST=MAXIT
      RETURN
C
C        NORMAL TERMINATION
C
  240 NLAST=K
      RETURN
      END
      SUBROUTINE COLLAP(NVAR,X,Y,LOCX,LOCY,NX,NY,DIM,CONFIG,OPTION)
C
C        ALGORITHM AS 51.1 APPL.STATIST. (1972), VOL.21, NO.2
C
C        IF OPTION IS TRUE, COMPUTES A MARGINAL TABLE FROM A COMPLETE
C        TABLE. IF OPTION IS FALSE, REMOVES AN EFFECT FROM A TABLE.
C        ALL PARAMETERS ARE ASSUMED VALID - THERE IS NO TEST FOR ERRORS
C        IN INPUT.
C        IF THE VALUE OF NVAR IS TO BE GREATER THAN 7, THE DIMENSIONS IN
C        THE DECLARATIONS OF SIZE AND COORD MUST BE INCREASED TO THE
C        VALUES OF NVAR+1 AND NVAR, RESPECTIVELY
C
      INTEGER SIZE(8),DIM(NVAR),CONFIG(NVAR),COORD(7)
      LOGICAL OPTION
```

```
C
C         THE LARGER TABLE IS X AND THE SMALLER ONE IS Y
C
      REAL X(NX),Y(NY)
C
C         INITIALISE ARRAYS
C
      SIZE(1)=1
      DO 10 K=1,NVAR
      L=CONFIG(K)
      IF(L.EQ.0)GOTO20
      SIZE(K+1)=SIZE(K)*DIM(L)
   10 CONTINUE
C
C         FIND NUMBER OF VARIABLES IN CONFIGURATION
C
      K=NVAR+1
   20 N=K-1
C
C         IF MARGINAL TABLE DESIRED, INITIALISE Y. FIRST CELL OF
C         MARGINAL TABLE IS AT Y(LOCY) AND TABLE HAS SIZE(K) ELEMENTS
C
      IF(.NOT.OPTION)GOTO40
      LOCU=LOCY+SIZE(K)-1
      DO 30 J=LOCY,LOCU
   30 Y(J)=0.0
C
C         INITIALISE COORDINATES
C
   40 DO 50 K=1,NVAR
   50 COORD(K)=0
C
C         FIND LOCATIONS IN TABLES
C
      I=LOCX
   60 J=LOCY
      DO 70 K=1,N
      L=CONFIG(K)
      J=J+COORD(L)*SIZE(K)
   70 CONTINUE
      IF(OPTION)Y(J)=Y(J)+X(I)
      IF(.NOT.OPTION)X(I)=X(I)-Y(J)
C
C         UPDATE COORDINATES
C
      I=I+1
      DO 80 K=1,NVAR
      COORD(K)=COORD(K)+1
      IF(COORD(K).LT.DIM(K))GOTO60
      COORD(K)=0
   80 CONTINUE
      RETURN
      END
      SUBROUTINE ADJUST(NVAR,X,Y,Z,LOCX,LOCY,LOCZ,NX,NY,NZ,DIM,CONFIG,D)
C
C         ALGORITHM AS 51.2 APPL.STATIST. (1972), VOL.21, NO.2
C
C         MAKES PROPORTIONAL ADJUSTMENT CORRESPONDING TO CONFIG. ALL
C         PARAMETERS ARE ASSUMED VALID - NO TEST IS MADE FOR ERRORS.
C         IF THE VALUE OF NVAR IS TO BE GREATER THAN 7,THE DIMENSIONS IN
C         THE DECLARATIONS OF SIZE AND COORD MUST BE INCREASED TO THE
C         VALUES OF NVAR+1 AND NVAR, RESPECTIVELY
C
      INTEGER SIZE(8),DIM(NVAR),CONFIG(NVAR),COORD(7)
      REAL X(NX),Y(NY),Z(NZ)
C
C         SET SIZE ARRAY
C
      SIZE(1)=1
      DO 10 K=1,NVAR
      L=CONFIG(K)
      IF(L.EQ.0)GOTO20
      SIZE(K+1)=SIZE(K)*DIM(L)
   10 CONTINUE
```

```
C
C        FIND NUMBER OF VARIABLES IN CONFIGURATION
C
      K=NVAR+1
   20 N=K-1
C
C        TEST SIZE OF DEVIATION
C
      L=SIZE(K)
      J=LOCY
      K=LOCZ
      DO 30 I=1,L
      E=ABS(Z(K)-Y(J))
      IF(E.GT.D)D=E
      J=J+1
      K=K+1
   30 CONTINUE
C
C        INITIALIZE COORDINATES
C
      DO 40 K=1,NVAR
   40 COORD(K)=0
      I=LOCX
C
C        PERFORM ADJUSTMENT
C
   50 J=0
      DO 60 K=1,N
      L=CONFIG(K)
      J=J+COORD(L)*SIZE(K)
   60 CONTINUE
      K=J+LOCZ
      J=J+LOCY
C
C        NOTE THAT Y(J) SHOULD BE NON-NEGATIVE
C
      IF(Y(J).LE.0.0)X(I)=0.0
      IF(Y(J).GT.0.0)X(I)=X(I)*Z(K)/Y(J)
C
C        UPDATE COORDINATES
C
      I=I+1
      DO 70 K=1,NVAR
      COORD(K)=COORD(K)+1
      IF(COORD(K).LT.DIM(K))GOTO50
      COORD(K)=0
   70 CONTINUE
      RETURN
      END
```

Appendix II

```
NUMBER OF VARIABLES (AT LEAST 2, NO MORE THAN 7)?
3
NUMBER OF CATEGORIES FOR VARIABLE  1?
2
NUMBER OF CATEGORIES FOR VARIABLE  2?
2
NUMBER OF CATEGORIES FOR VARIABLE  3?
2
NOW ENTER THE DATA ONE CELL AT A TIME
IN ACCORDANCE WITH THE CELL SUBSCRIPT CUE
 1, 1, 1
156
 2, 1, 1
84
```

```
 1, 2, 1
84
 2, 2, 1
156
 1, 1, 2
107
 2, 1, 2
133
 1, 2, 2
31
 2, 2, 2
209
NUMBER OF INTERACTIONS TO BE FIT?
3

WHEN ENTERING INTERACTION SUBSCRIPTS, ENTER THEM
AS XX,XX,XX
THE USE OF 00 FOR THE EXTRANEOUS SUBSCRIPTS IS ADVISED.
FOR EXAMPLE, THE (1,2) INTERACTION
IS ENTERED AS 01,02,00
OR IS ENTERED AS 01,02

SUBSCRIPTS OF INTERACTION  1 TO BE FIT:
01,02,00
SUBSCRIPTS OF INTERACTION  2 TO BE FIT:
01,03,00
SUBSCRIPTS OF INTERACTION  3 TO BE FIT:
02,03,00
ERROR CONDITION: 0

111  156.   161.06
112   84.    78.94
121   84.    78.91
122  156.   161.09
211  107.   101.91
212  133.   138.09
221   31.    36.12
222  209.   203.88
CHI-SQUARE =      2.27    1 DEGREES OF FREEDOM
DO YOU WISH TO TRY TO FIT ANOTHER SET OF INTERACTIONS?
ENTER 0 IF NO, 1 IF YES
1
NUMBER OF INTERACTIONS TO BE FIT?
2

WHEN ENTERING INTERACTION SUBSCRIPTS, ENTER THEM
AS XX,XX,XX
THE USE OF 00 FOR THE EXTRANEOUS SUBSCRIPTS IS ADVISED.
FOR EXAMPLE, THE (1,2) INTERACTION
IS ENTERED AS 01,02,00
OR IS ENTERED AS 01,02

SUBSCRIPTS OF INTERACTION  1 TO BE FIT:
01,02
SUBSCRIPTS OF INTERACTION  2 TO BE FIT:
01,03
ERROR CONDITION: 0

111  156.   166.98
112   84.    89.48
121   84.    73.02
122  156.   150.52
211  107.    96.02
```

```
212  133.    127.52
221   31.     41.98
222  209.    214.48
CHI-SQUARE =      7.42     2 DEGREES OF FREEDOM
DO YOU WISH TO TRY TO FIT ANOTHER SET OF INTERACTIONS?
ENTER 0 IF NO, 1 IF YES
1
NUMBER OF INTERACTIONS TO BE FIT?
2

WHEN ENTERING INTERACTION SUBSCRIPTS, ENTER THEM
AS XX,XX,XX
THE USE OF 00 FOR THE EXTRANEOUS SUBSCRIPTS IS ADVISED.
FOR EXAMPLE, THE (1,2) INTERACTION
IS ENTERED AS 01,02,00
OR IS ENTERED AS 01,02

SUBSCRIPTS OF INTERACTION  1 TO BE FIT:
01,02
SUBSCRIPTS OF INTERACTION  2 TO BE FIT:
02,03
ERROR CONDITION: 0

111  156.    131.50
112   84.    108.50
121   84.     57.50
122  156.    182.50
211  107.    131.50
212  133.    108.50
221   31.     57.50
222  209.    182.50
CHI-SQUARE =     52.32     2 DEGREES OF FREEDOM
DO YOU WISH TO TRY TO FIT ANOTHER SET OF INTERACTIONS?
ENTER 0 IF NO, 1 IF YES
1
NUMBER OF INTERACTIONS TO BE FIT?
2

WHEN ENTERING INTERACTION SUBSCRIPTS, ENTER THEM
AS XX,XX,XX
THE USE OF 00 FOR THE EXTRANEOUS SUBSCRIPTS IS ADVISED.
FOR EXAMPLE, THE (1,2) INTERACTION
IS ENTERED AS 01,02,00
OR IS ENTERED AS 01,02

SUBSCRIPTS OF INTERACTION  1 TO BE FIT:
01,03
SUBSCRIPTS OF INTERACTION  2 TO BE FIT:
02,03
ERROR CONDITION: 0

111  156.    120.00
112   84.    120.00
121   84.    120.00
122  156.    120.00
211  107.     69.00
212  133.    171.00
221   31.     69.00
222  209.    171.00
CHI-SQUARE =    101.94     2 DEGREES OF FREEDOM
DO YOU WISH TO TRY TO FIT ANOTHER SET OF INTERACTIONS?
ENTER 0 IF NO, 1 IF YES
0
```

References

Anderson, T. W. 1954. On estimation of parameters in latent structure analysis. *Psychometrika* **19** (March): 1–10.

Bartlett, M. S. 1935. Contingency table interactions. *Journal of the Royal Statistical Society, Supplement* **2** (No. 2): 248–52.

Bishop, Y. M. M. 1967. Multidimensional contingency tables: cell estimates. Ph.D. dissertation, Department of Statistics, Harvard University. Ann Arbor: University Microfilms.

Bishop, Y. M., Fienberg, S. E., and Holland, P. W. 1975. *Discrete Multivariate Analysis: Theory and Practice.* Cambridge, MA: MIT Press.

Boschloo, R. D. 1970. Raised conditional level of significance for the 2 × 2 table when testing for the equality of two probabilities. *Statistica Neerlandica* **21** (No. 1): 1–35.

Clogg, C. C. 1977. Unrestricted and restricted maximum likelihood latent structure analysis: a manual for users. Working paper No. 1977–09, Population Issues Research Office, The Pennsylvania State University.

Conover, W. J. 1974. Some reasons for not using the Yates continuity correction on 2 × 2 contingency tables. *Journal of the American Statistical Association* **69** (June): 374–82.

Darroch, J. N. 1962. Interactions in multi-factor contingency tables. *Journal of the Royal Statistical Society, Series B* **24** (No. 1): 251–63.

Fienberg, S. E. 1977. *The Analysis of Cross-Classified Categorical Data.* Cambridge, MA: MIT Press.

Finney, D. J., Latscha, A., Bennett, B. M., and Hsu, P. 1963. *Tables for Testing Significance in a 2 × 2 Contingency Table.* Cambridge: Cambridge University Press.

Fisher, R. A. 1934. *Statistical Methods for Research Workers*, 5th edn. Edinburgh: Oliver and Boyd.

Fisher, R. A. and Yates, F. 1948. *Statistical Tables for Biological, Agricultural, and Medical Research*, 6th edn. New York: Hafner.

Garside, G. R. 1971. An accurate correction for the χ^2-test in the homogeneity case of 2 × 2 contingency tables. *New Journal of Statistics and Operational Research* **7** (Part 1): 1–26.

Garside, G. R. 1972. Further tables of an accurate correction for the χ^2-test in the homogeneity case of the 2 × 2 contingency table. *New Journal of Statistics and Operational Research* **8** (Part 1): 6–25.

Gilula, Z. 1983. Latent conditional independence in two-way contingency tables: a diagnostic approach. *British Journal of Mathematical and Statistical Psychology* **36** (May): 114–22.

Goodman, L. A. 1964. Simple methods of analyzing three-factor interaction in contingency tables. *Journal of the American Statistical Association* **59** (June): 319–52.

Goodman, L. A. 1969. On partitioning χ^2 and detecting partial association in three-way contingency tables. *Journal of the Royal Statistical Society, Series B* **31** (No. 3): 486–98.

Goodman, L. A. 1970. The multivariate analysis of qualitative data: Interactions among multiple classifications. *Journal of the American Statistical Association* **65** (March): 226–56.

Goodman, L. A. 1974. The analysis of systems of qualitative variables when some of the variables are unobservable. Part I: A modified latent structure approach. *American Journal of Sociology* **79** (March): 1179–1259.

Goodman, L. A. 1978. *Analyzing Qualitative Categorical Data: Log Linear Models and Latent Structure Analysis.* Cambridge, MA: Abt Books.

Goodman, L. A. 1979. Simple models for the analysis of association in cross-classifica-

tions having ordered categories. *Journal of the American Statistical Association* **34** (September): 537–52.
Goodman, L. A. 1984. *The Analysis of Cross-Classified Data Having Ordered Categories.* Cambridge, MA: Harvard University Press.
Green, B. F. Jr. 1951. A general solution for the latent class model of latent structure analysis. *Psychometrika* **16** (June): 151–66.
Grizzle, J. 1967. Continuity correction in the χ^2-test for 2×2 tables. *The American Statistician* **21** (October): 28–32.
Haberman, S. J. 1972. Log-linear fit for contingency tables. *Applied Statistics* **21** (No. 2): 218–25.
Haberman, S. J. 1974. Log-linear models for frequency tables with ordered classifications. *Biometrics* **30** (December): 589–600.
Haberman, S. J. 1978. *Analysis of Qualitative Data*, Vol. I: *Introductory Topics.* New York: Academic Press.
Haberman, S. J. 1979. *Analysis of Qualitative Data*, Vol. II. *Introductory Topics.* New York: Academic Press.
Haberman, S. J. 1981. Tests for independence in two-way contingency tables based on canonical correlation and on linear-by-linear interaction. *Annals of Statistics* **9** (November): 1178–86.
Harter, H. L. 1969. *Order Statistics and Their Use in Testing and Estimation*, Vol. 2. Washington, DC: U.S. Government Printing Office.
Irwin, J. O. 1935. Tests of significance for differences between percentages based on small numbers. *Metron* **12** (No. 2): 83–94.
Kastenbaum, M. A. and Lamphiear, D. E. 1959. Calculation of chi-square to test the no three-factor interaction hypothesis. *Biometrics* **15** (March): 107–15.
Kendall, M. G. and Stuart, A. 1973. *The Advanced Theory of Statistics*, Vol. 2, 3rd edn. New York: Hafner.
Lancaster, H. O. 1949. The derivation and partition of χ^2 in certain discrete distributions *Biometrika* **36** (June): 117–29.
Lancaster, H. O. 1951. Complex contingency tables treated by the partitioning of χ^2. *Journal of the Royal Statistical Society, Series B* **13** (No. 2): 242–49.
Lazarsfeld, P. F. and Henry, N. W. 1968. *Latent Structure Analysis.* New York: Houghton Mifflin.
Lehmann, E. L. 1959. *Testing Statistical Hypotheses.* New York: Wiley.
Lewis, B. N. 1962. On the analysis of interaction in multi-dimensional contingency tables. *Journal of the Royal Statistical Society, Series A* **125** (Part 1): 88–117.
Lieberman, G. J. and D. B. Owen. 1961. *Tables of the Hypergeometric Probability Distribution.* Stanford, CA: Stanford University Press.
McHugh, R. B. 1956. Efficient estimation and local identification in latent class analysis. *Psychometrika* **21** (December): 331–47.
McHugh, R. B. 1958. Note on 'Efficient estimation and local identification in latent class analysis.' *Psychometrika* **23** (September): 273–74.
Madansky, A. 1959. Partitioning methods in latent class analysis. Paper P-1644. Santa Monica, CA: RAND Corporation.
Madansky, A. 1960. Determinantal methods in latent class analysis. *Psychometrika* **25** (June): 183–98.
Madansky, A. 1978. Latent structure. In *International Encyclopedia of Statistics*, Vol. I, ed. W. H. Kruskal and J. M. Tanur. New York: Free Press, pp. 499–505.
Pearson, K. 1904. *On the Theory of Contingency and Its Relation to Association and Normal Correlation.* London: Drapers' Co. Memoirs, Biometric Series No. 1.
Pearson, E. S. and Hartley, H. O. 1972. *Biometrika Tables for Statisticians*, Vol. 2. Cambridge: Cambridge University Press.
Plackett, R. L. 1962. A note on interactions in contingency tables. *Journal of the Royal Statistical Society, Series B* **24** (No. 1): 162–66.

Plackett, R. L. 1964. The continuity correction in 2 × 2 tables. *Biometrika* **51** (December): 327–37.

Roy, S. N. and Kastenbaum, M. A. 1956. On the hypothesis of no "interaction" in a multi-way contingency table. *Annals of Mathematical Statistics* **27** (September): 749–57.

Snedecor, G. W. 1958. Chi-squares of Bartlett, Mood, and Lancaster in a 2 × 2 contingency table. *Biometrics* **14** (December): 560–62.

Srole, L., Langner, T. S., Michael, S. T., Opler, M. K., and Rennie, T. A. C. 1962. *Mental Health in the Metropolis: The Midtown Manhatten Study*. New York: McGraw-Hill.

Tocher, K. D. 1950. Extension of Neyman–Pearson theory of tests to discontinuous variates. *Biometrika* **37** (June): 130–44.

Yates, F. 1934. Contingency tables involving small numbers and the χ^2 test. *Journal of the Royal Statistical Society, Supplement* **1** (No. 2): 217–35.

Index of Reference Tables

Index